MEMBRANE PHYSIOLOGY

Richard A. Nystrom

Department of Biological Sciences
University of Delaware

Prentice-Hall, Inc.
Englewood Cliffs, New Jersey

ISBN: 0-13-574087-8

Library of Congress Catalog Card Number: 72-6153

Printed in the United States of America

PRENTICE-HALL INTERNATIONAL, INC., *London*
PRENTICE-HALL OF AUSTRALIA, PTY. LTD., *Sydney*
PRENTICE-HALL OF CANADA, LTD., *Toronto*
PRENTICE-HALL OF INDIA PRIVATE LIMITED, *New Delhi*
PRENTICE-HALL OF JAPAN, INC., *Tokyo*

To Karin
  Karl
  Kris
  Kari

ACKNOWLEDGMENTS

The author wishes to acknowledge the kind permission granted to reproduce figures and tables included in this work. The figures have all been redrawn and are taken from the following sources:

Figs. 1, 2, 3. Courtesy of P. D. Lunger, Department of Biological Sciences, University of Delaware, Newark, Delaware.

Figs. 4, 5, 7, 10 from J. D. Robertson, 1964. Unit membranes: a review with recent new studies of experimental alterations and a new subunit structure in synaptic membranes, pp. 1–81 in M. Locke (ed.), Cellular Membranes in Development. New York: Academic Press.

Figs. 6, 8, 11 from J. D. Robertson, 1960. The molecular structure and contact relationships of cell membranes. Prog. Biophys. *10*: 343–418.

Fig. 9 from J. B. Finean, 1962. Nature and stability of the plasma membrane, Symposium on the Plasma Membrane, A. P. Fishman (ed.), 1151–1162. By permission of the American Heart Association, Inc.

Fig. 12 from J. D. Robertson, 1964. Unit membranes: a review with recent new studies of experimental alterations and a new subunit structure in synaptic membranes, pp. 1–81 in M. Locke (ed.), Cellular Membranes in Development. New York: Academic Press. After J. B. Finean, 1956. Recent ideas on the structure of myelin, pp. 127–31 in Biochemical Problems of Lipids (Proc. 2nd Internat. Conf. Univ. Ghent). London: Butterworths.

Fig. 13(a) after J. B. Finean, 1953. Phospholipid-cholesterol complex in the structure of myelin. Experientia 9: 17–19. Basel, Switzerland: Verlag Birkhäuser. Fig. 13(b) after J. B. Finean, 1957. The molecular organization of nerve myelin. Acta Neurol. Psychiat. Belgium 5: 462–71.

Fig. 14(a) from A. A. Benson, 1966. On the orientation of lipids in chloroplast and cell membranes. J. Am. Oil Chem. Soc. 43: 265–70. Fig. 14(b) (top) from D. E. Green and J. F. Perdue, 1966. Membranes as expressions of repeating units. Proc. Natl. Acad. Sci. US 55: 1295–1302. Fig. 14(b) (bottom) from M. T. Tourtellotte, R. N. McElhaney, and R. L. Rader, 1968. Current concepts of membrane structure. Bull. Inst. Cell Biol. 9: 1–14 (No. 5). Fig. 14(c) from J. A. Lucy, 1964. Globular lipid micelles and cell membranes. J. Theoret. Biol. 7: 360–73. Fig. 14(d) from J. Lenard and S. J. Singer, 1966. Protein conformation in cell membrane preparations as studied by optical rotatory dispersion and circular dichroism. Proc. Natl. Acad. Sci. US 56: 1828–35. Fig. 14(e) from V. Luzzati, F. Reiss-Husson, E. Rivas and T. Galik-Krzywicki, 1966. Structure and polymorphism in lipid-water systems and their possible biological implications. Ann. N. Y. Acad. Sci. 137: 409–413.

Fig. 15 from J. L. Kavanau, 1965. Structure and Function in Biological Membranes, Vol. 1. San Francisco: Holden-Day, Inc. 322 pp.

Fig. 16 from F. O. Schmitt and P. F. Davison, 1965. Role of protein in neural function: an essay. Neurosci. Res. Prog. Bull. 3: 55–76.

Figs. 17, 19, 20 from A. C. Giese, 1968. Cell Physiology, 3rd ed. Philadelphia: W. B. Saunders. 671 pp.

Figs. 18, 21 from H. Davson and J. F. Danielli, 1952. Permeability of Natural Membranes, 2nd ed. New York: Cambridge University Press. 365 pp.

Fig. 22 from W. D. Stein, 1967. The Movement of Molecules Across Cell Membranes. New York: Academic Press. 369 pp.

Fig. 23 from T. Hoshiko, 1961. Electrogenesis in frog skin, pp. 31–47 in A. M. Shanes (ed.), Biophysics of Physiological and Pharmacological Actions. Publication Number 69. Copyright 1961 by the American Association for the Advancement of Science, Washington, D.C.

Fig. 24 from Dr. D. W. Martin, Department of Zoology, University of Arkansas, Fayetteville, Arkansas.

Figs. 25, 26, 27 from V. Koefoed-Johnsen and H. H. Üssing, 1958. The nature of the frog skin potential. Acta Physiol. Scand. 42: 298–308.

Fig. 28 from P. F. Curran, F. C. Herrera, and W. J. Flanigan, 1963. The effect of Ca and antidiuretic hormone on Na transport across frog skin. II. Sites and mechanisms of action. J. gen. Physiol. 46: 1011–27.

Fig. 29 from D. C. Tosteson and J. F. Hoffman, 1960. Regulation of cell volume by active cation transport in high and low potassium sheep red cells. J. gen. Physiol. 44: 169–94.

Figs. 30, 31 from J. F. Hoffman, 1962. The active transport of sodium by ghosts of human red blood cells. J. gen. Physiol. *45*: 832–59.

Fig. 32 from I. M. Glynn, 1957. The ionic permeability of the red cell membrane. Prog. Biophys. *8*: 242–307.

Fig. 33 from L. E. Hokin and M. R. Hokin, 1960. Studies on the carrier function of phosphatidic acid in sodium transport. J. gen. Physiol. *44*: 61–85.

Fig. 34 from J. F. Hoffman, 1961. Molecular mechanisms of active cation transport, pp. 3–17 in A. M. Shanes (ed.), Biophysics of Physiological and Pharmacological Actions. Publication Number 69. Copyright 1961 by the American Association for the Advancement of Science, Washington, D.C.

Fig. 35 from D. Bodian, 1967. Neurons, circuits, and neuroglia, pp. 6–24 in G. C. Quarton, T. Melnechuk, and F. O. Schmitt (eds.), The Neurosciences — A Study Program. New York: The Rockefeller University Press; (left diagram) from H. Grundfest, 1957. Electrical inexcitability of synapses and some consequences in the central nervous system. Physiol. Rev. *37*: 337–61; (right figures) from D. Bodian, 1962. The generalized vertebrate neuron. Science *137*: 323–26 (3 Aug.). Copyright 1962 by the American Association for the Advancement of Science, Washington, D.C.

Figs. 36, 37, 38 from Dr. R. A. Meiss, Department of Biological Sciences, University of Delaware, Newark, Delaware.

Fig. 39 from Dr. D. F. Wilson, Department of Biological Sciences, University of Delaware, Newark, Delaware.

Fig. 40 from A. Bortoff, 1961. Slow potential variations of small intestine. Am. J. Physiol. 201: 203–208.

Fig. 41 from J. W. Woodbury, 1966. The cell membrane: ionic and potential gradients and active transport, pp. 1–25 in T. C. Ruch and H. D. Patton (eds.), Physiology and Biophysics. Philadelphia: W. B. Saunders.

Fig. 42 from J. D. Woodbury, 1966. The cell membrane: ionic and potential gradients and active transport, pp. 1–25 in T. C. Ruch and H. D. Patton (eds.), Physiology and Biophysics. Philadelphia: W. B. Saunders. Compiled from: J. C. Eccles, 1957. The Physiology of Nerve Cells. Baltimore: The Johns Hopkins Press, 270 pp; A. L. Hodgkin and P. Horowicz, 1959. Movements of Na and K in single muscle fibres. J. Physiol. *145*: 405–32; and R. D. Keynes and R. C. Swann, 1959. The effect of external sodium concentration on the sodium fluxes in frog skeletal muscle. J. Physiol. *147*: 591–625.

Fig. 43. (Upper) from E. J. Conway, 1957. Nature and significance of concentration relations of potassium and sodium ions in skeletal muscle. Physiol. Rev. *37*: 84–132. (Lower) from A. L. Hodgkin and P. Horowicz, 1959. The influence of potassium and chloride ions on the membrane potential of single muscle fibres. J. Physiol. *148*: 127–60.

Figs. 44, 45 from T. L. Jahn, 1962. A theory of electronic conduction through membranes and of active transport of ions based on redox transmembrane potentials. J. Theoret. Biol. *2*: 129–38.

Fig. 46 from G. N. Ling, 1960. The interpretation of selective ionic permeability and cellular potentials in terms of the fixed-charge-induction hypothesis. J. gen. Physiol. suppl. *43*: 149–74.

Fig. 47 from A. C. Brown, 1966. Passive and active transport, pp. 820–42 in T. C. Ruch and H. D. Patton (eds.), Physiology and Biophysics. Philadelphia: W. B. Saunders.

Fig. 48 from T. H. Bullock and G. A. Horridge, Copyright © 1965. Structure and Function in the Nervous Systems of Invertebrates, Vol. 1. San Francisco: W. H. Freeman and Company. 798 pp.

Fig. 49(a) from J. W. Woodbury, 1966. Action potential: properties of excitable membranes, pp. 26–58 in T. C. Ruch and H. D. Patton (eds.) Physiology and Biophysics. Philadelphia: W. B. Saunders. 1242 pp. Fig. 49(b) from A. L. Hodgkin and A. F. Huxley, 1952. Currents carried by sodium and potassium ions through the membrane of the giant axon of *Loligo*. J. Physiol. *116*: 449–72. Fig. 49(c) from A. L. Hodgkin, 1958. Ionic movements and electrical activity in giant nerve fibres. Proc. Roy. Soc. London B *148*: 1–37.

Fig. 50 from G. A. Horridge, Copyright © 1968. Interneurons — Their Origin, Action, Specificity, Growth, and Plasticity. San Francisco: W. H. Freeman and Company. 436 pp. (a), (b), and (c) from E. J. Furshpan and D. D. Potter, 1959. Transmission at the giant motor synapses of the crayfish. J. Physiol. *145*: 289–325. (d) from T. H. Bullock and G. A. Horridge, Copyright © 1965. Structure and Function in the Nervous Systems of Invertebrates, Vol. 1. San Francisco: W. H. Freeman and Company. 798 pp.

Fig. 51(a) from D. R. Curtis and J. C. Eccles, 1959. The time course of excitatory and inhibitory synaptic action. J. Physiol. *145*: 529–46. Fig. 51(b) from J. C. Eccles, 1957. The Physiology of Nerve Cells. Baltimore: The Johns Hopkins Press. 270 pp. Fig. 51(c) from J. C. Eccles, R. M. Eccles, and A. Lundberg, 1957. Synaptic actions on motoneurons in relation to the two components of the group I muscle afferent volley. J. Physiol. *136*: 527–46. Fig. 51(d) from J. C. Coombs, J. C. Eccles, and P. Fatt, 1955. Excitatory synaptic action in motoneurons. J. Physiol. *130*: 374–95.

Fig. 52 from J. C. Eccles, 1959. Neurophysiology — Introduction, pp. 59–74 in J. Field, H. W. Magoun, and V. E. Hall (eds.), Handbook of Physiology, Section I. Neurophysiology, Vol. 1. Bethesda, Maryland: American Physiological Society.

Fig. 53(a) from D. R. Curtis and J. C. Eccles, 1959. The time course of excitatory and inhibitory synaptic actions. J. Physiol. *145*: 529–46. Fig. 53(b,d) from J. S. Coombs, J. C. Eccles, and P. Fatt, 1955. The specific ionic conductances and the ionic movements across the motoneuronal membrane that produce the inhibitory post-synaptic potential. J. Physiol. *130*: 326–73. Fig. 53(c) from J. S. Coombs, J. C. Eccles, and P. Fatt, 1955. The inhibitory suppression of reflex discharges from motoneurons. J. Physiol. *130*: 396–413.

Fig. 54 (a–c, g–h) from R. D. Keynes and H. Martins-Ferreira, 1953. Membrane potentials in the electroplates of the electric eel. J. Physiol. *119*: 315–51. Fig. 54(d–f) from M. V. L. Bennett, M. Wurzel, and H. Grundfest, 1961. Properties of electroplaques of *Torpedo nobiliana*. J. gen. Physiol. *44*: 757–801.

Figs. 55, 56 from H. Grundfest, 1967. Synaptic and ephaptic transmission, pp. 353–72 in G. C. Quarton, T. Melnechuk, and F. O. Schmitt (eds.), The Neurosciences — A Study Program. New York: The Rockefeller University Press.

Fig. 57 from D. Nachmansohn and I. B. Wilson, 1951. The enzymatic hydrolysis and synthesis of acetylcholine. Adv. Enzymol. *12*: 259–339.

Fig. 58 from I. Tasaki and I. Singer, 1966. Membrane macromolecules and nerve excitability: a physico-chemical interpretation of excitation in squid giant axon. Ann. N.Y. Acad. Sci. *137*: 792–806.

Fig. 59(a–d) from P. Mueller and D. O. Rudin, 1963. Induced excitability in reconstituted cell membrane structure. J. Theoret. Biol. *4*: 268–80. Fig. 59(e) from P. Mueller and D. O. Rudin, 1968. Action potentials induced in bimolecular lipid membranes. Nature *217*: 713–19.

Fig. 60(a) from A. Katchalsky, 1967. Membrane thermodynamics, pp. 326–43 in G. C. Quarton, T. Melnechuk, and F. O. Schmitt (eds.), The Neurosciences — A Study Program. New York: The Rockefeller University Press. Fig. 60(b) from V. E. Shashoua, 1967. Electrically active poly-electrolyte membranes. Nature *215*: 846–47. Fig. 60(c–d) from Dr. V. E. Shashoua, Department of Biology, Massachusetts Institute of Technology, Cambridge, Massachusetts.

Table 1 from E. D. Korn, 1966. Structure of biological membranes. Science *153* (September, 1966), 1497. Copyright 1966 by the American Association for the Advancement of Science, Washington, D.C.

Table 2 from H. McLennan, 1957. The diffusion of potassium, sodium, sucrose, and inulin in the extracellular spaces of mammalian tissues. Biochim. Biophys. Acta *24*: 1–8.

Table 3 after R. M. Dowben, 1969. General Physiology: A Molecular Approach. New York: Harper & Row. By permission of Harper & Row.

Tables 4 and 5 from H. Davson and J. F. Danielli, 1952. Permeability of Natural Membranes, 2nd ed. New York: Cambridge University Press. 365 pp.

Table 6 from J. W. Woodbury, 1966. The cell membrane: ionic and potential gradients and active transport, pp. 1–25 in T. C. Ruch and H. D. Patton (eds.), Physiology and Biophysics. Philadelphia: W. B. Saunders.

CONTENTS

PREFACE

This book is an integrated and relatively compact presentation of cytological, physiological, and molecular data on plasma membranes of animal cells. For a long time the study of the structure and function of biological membranes has been one of the more active endeavors of biological investigators. So many disciplines are involved — biophysics, biochemistry, morphology, pharmacology, physiology, physical chemistry, molecular biology — and the literature is so extensive that it is difficult for workers of diverse training and experience to fully appreciate the level of understanding and the advances within neighboring fields. A young biologist contemplating investigations on membranes and an older worker trained in another field both require a conceptual insight into the nature of membrane problems and a scientific intuition into their solution. The recent scientific conferences and symposiums dealing with general aspects of membranology, a multitude of research reviews and books, and at least one new journal devoted solely to membrane biology have served to unify the field. However, much of this material characteristically contains articles that are restricted in scope and written for scientifically sophisticated readers. None is sufficiently broad to embrace all aspects of the work on biological membranes and at the same time sufficiently integrated to provide a background necessary to relate one finding to another. None is exceptionally noted for its readability. This book may help as an introduction to that literature in several ways.

By discussing the current state of our understanding of the nature of membrane structure, permeation, and excitability of both living and artificial systems, the book covers a broad range of knowledge, but it is clearly not comprehensive. Within the areas covered, only work that I consider important or interesting is included. Investigations on cytoplasmic membranes and on plant and bacterial membranes are barely mentioned; pinocytosis and phagocytosis are only acknowledged; and the role of the cell surface in cellular contact and immunological relationships is ignored completely — to the dismay of virologists, immunologists, pathologists, tissue culturists, and cancer researchers, among others. This book is restricted but not superficial. The intent is not only to present a readable account of the major ideas of the various approaches to membrane study and their broader implications but also to include the complications, limitations, alternative explanations, and important minor points to these investigations. This book is concise and elementary but not simple. The intent is to inform the reader so that he will be able to more effectively glean the sophisticated books, reviews, and research papers on membranes and evaluate their interpretations in his own mind. Sufficient knowledge is extant to permit an uncomplicated summary of only the widely held conclusions and to tell a pretty story, but a beginner would not be well served and a seasoned veteran would not be well satisfied by such an approach. Difficulties with arguments are readily presented, but experimental and language technicalities and ponderous derivations are reduced. Experimental procedures are discussed where they are important to the proper interpretation of the data that they yield. The important quantitative analyses of membrane data are hinted at, but full and, hence, proper treatment is to be found in the publications cited.

The many references cited are intended more as a guide through the literature than as a historical account of authenticity; so a large number refer to reviews and interpretative work. The examples selected for a more detailed analysis represent my own favorites rather than an attempt to tell all or to give the most recent. Experienced investigators will undoubtedly be unhappy about the treatment given some topics as well as the exclusion of others. The success of my approach will be determined by the usefulness of this small volume to you.

Historically, this manuscript is derived in part from a series of discussions held many years ago with members of the Engineering Physics Department of the E. I. du Pont de Nemours Company, Wilmington, Delaware. The time to finish it became more readily available as a result of the freedom from academic duties beyond normal teaching and research provided by a sabbatical leave from the University of Delaware and a Special Fellowship from the National Institutes of Neurological Diseases and Stroke to work on other things at the Massachusetts Institute of Technology and the Marine Biological Laboratory at Woods Hole.

While the responsibility for this manuscript is clearly mine, I am most grateful and thankful for the encouragement, advice, assistance, and especially the criticism of A. Bortoff, Z. de Schauensee, P. B. Dunham, A. Gould, C. M. Lent, R. A. Levy, P. D. Lunger, D. W. Martin, R. A. Meiss, I. Nadelhaft, V. Shashoua, L. Smucker, W. M. Trippeer, J. T. Tupper, D. F. Wilson, R. A. Yates, and the anonymous referees found by J. R. Riina of Prentice-Hall, Inc. I am also pleased to acknowledge K. D. Roeder, C. L. Prosser, R. R. Ronkin, and F. O. Schmitt, all of whom helped to influence my scientific career.

RICHARD A. NYSTROM

Newark, Delaware

INTRODUCTION

chapter one

Cellular activities are characterized by their occurrence at or near membranes located throughout the cell (Figs. 1–3). For example, much of cellular metabolism is associated with enzymes aligned on cristae of mitochondria; protein synthesis is associated with ribosomes attached to endoplasmic reticulum; lipid transport, with Golgi apparatus; photosynthesis, with grana within chloroplasts; light reception, with retinal rods. The interest here, however, is concerned with the surface of the cell—the cellular membrane or, more properly, the plasma membrane—the barrier or interface between the living inside and the dead outside of the cell.

A variety of physical measurements indicates that the cell surface is about 75Å thick although the published figures vary greatly. This thin structure is made of protein, some carbohydrate, and lipid, mostly sterol and phospholipid. For more than 30 years the molecular arrangement has been described as a double layer of lipid between two protein layers. Studies using X-ray diffraction, polarized light, electron microscopy, and permeability data have in general confirmed this view although newer techniques and approaches have exposed globular substructures and have confused the satisfaction with this description. Pores in membrane structures have been proposed with regularity, but, as the supporting experiments are all indirect ones and as investigators become more familiar with dynamic concepts, the need for the existence of a pore as a fixed morphological structure to understand membrane permeation is becoming less important.

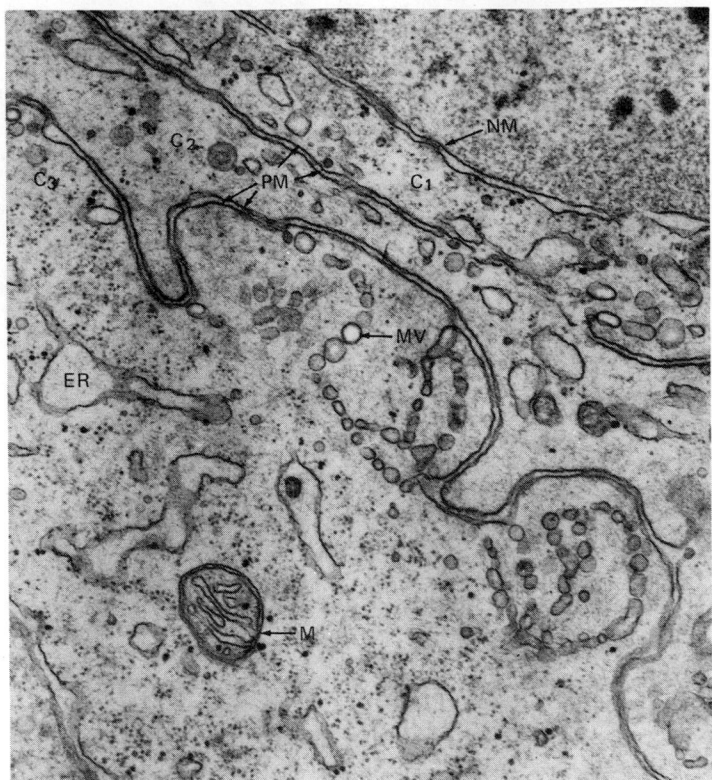

Figure 1. *Examples of membranous structures in cells:* three adjacent frog kidney cells showing at least five types of membrane. C_1, C_2, and C_3 are portions of three cells separated by plasma membranes (PM) and a small amount of extracellular space. The double nuclear membrane (NM) appears at the upper right. A mitochondrion (M) and its cristae are seen at the lower left. Smooth endoplasmic reticulum (ER) and ribosomal particles appear in the cytoplasm of C_3 especially. The rows of membranous vesicles (MV) near the plasma membrane of C_3 are considered by some workers to be plasma membrane that is either forming or degenerating ($\times$ 40,000).

Life exists in a liquid phase. Cells contain and surround themselves with liquid media; water is the solvent. Materials continuously exchange between the inside and outside of cells, dead or alive. Obviously, the cell surface is not an absolute barrier. In living cells, the plasma membrane regulates this exchange with care by processes best described by nonequilibrium thermodynamics; only a dead cell is in complete equilibrium with the solutes in its environment. Larger molecules permeate the membrane more slowly than do smaller ones. Molecules soluble in lipid cross the cell surface barrier faster than others. The transfer of polar molecules is impeded. But water

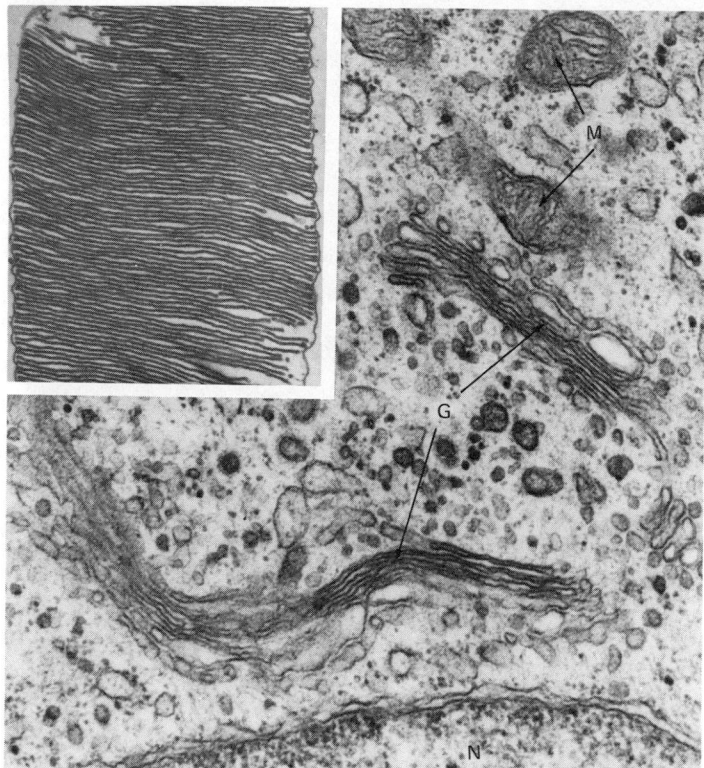

Figure 2. *Examples of membranous structures in cells:* rod membrane in turtle retina and Golgi membrane in frog kidney cells. Portions of three mitochondria (M) can be seen at the upper right and a small part of nucleus (N) and the double nuclear membrane at the bottom as well as Golgi apparatus (G). The endoplasmic reticulum in these cells is highly vesicular ($\times$ 40,000). A portion of the outer segment of a retinal rod from turtle eye (insert) illustrates the densely packed membranes characteristic of this light receptor ($\times$ 14,500).

molecules are among the fastest to permeate cells. Often the transport of molecules across the plasma membrane cannot be explained solely on the basis of concentration and charge gradients. Metabolic energy appears necessary for some transport processes and a cell will accumulate large concentrations of particular molecules while excluding others. Indeed the pronounced separation of charged molecules across the cell surface is the basis for normal functioning of excitable cells in nerve and muscle. Energy-requiring enzymes, carrier molecules, gates, and special membrane properties become involved in the explanations for the permeation by many substances even in cases where the steady state distribution appears to be adequately explained by the rules of simple diffusion.

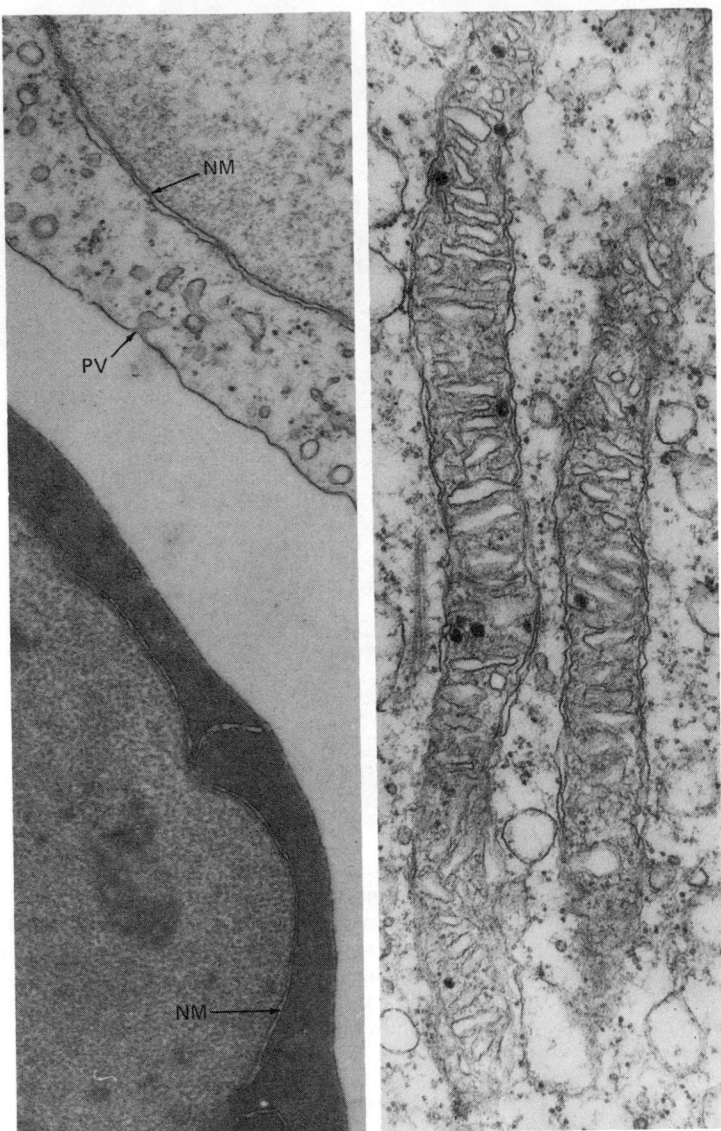

Figure 3. *Examples of membranous structures in cells:* erythrocytes and mitochondria in frog kidney. The left figure shows a nucleated frog red blood cell
at the bottom and an endothelial cell with its nucleus at the top. The double
nuclear membrane (NM) is clearly discerned in both cells, but the magnification
of this electron micrograph ($\times$ 27,000) is not sufficient to show the trilaminar
structure of the plasma membrane (PM). Note the pinocytotic vesicle (PV)
opening into the lumen of the blood vessel. The right figure illustrates two large
mitochondria filled with cristae and matrix ($\times$ 41,000).

One of the most distinctive properties of the plasma membrane is that of irritability, that is, the capacity to react to changes in the environment. This property is especially prominent in nerve and muscle cells. The excitability of these cells is determined in part by the ability of the membrane to segregate molecules, especially charged ones. The ionic composition inside cells is different from that outside. Sodium is the major cation in the cells' environment, usually tissue fluids; potassium is the major cation inside most cells. An electrical potential difference, commonly thought to result from this unequal distribution of ions, exists across the membrane of resting cells. Upon excitation by any appropriate means, the properties of the membrane change reversibly, often in a sudden and explosive manner, to produce changes in this potential difference. Cells or even parts of a single cell differ in their responses: the propagated impulse of nerve axons, the transducer action of receptor cells, the transmission function at synaptic and neuromuscular junctions. Some cells excite others, some inhibit, and others have more subtle influences. The origin and significance of the electrical phenomena generated by membranes have been vigorously studied but important questions remain.

In addition to the plasma membrane, there are a number of extraneous membranes and coats around cells. These coats are frequently but not always characterized by a high porosity; only large molecules do not pass freely. Some striking examples are the jelly layers, vitelline membranes, and fertilization membranes of some eggs; cellular cement or hyaline plasma layers between cells; the highly sculptured pellicles of protozoans; and plant cell walls. The important functional structure of cells that separates inside from outside, however, is the plasma membrane.

A discussion of the plasma membrane should include these questions: What is it? What does it do? How does it do it? The excitement in this field, which has led to extensive and rigorous study for nearly a century, has produced many excellent ideas, ingenious experiments, and useful models. The voluminous literature that has resulted from this activity, however, is often contradictory, confusing, and incomplete. The basic questions remain.

STRUCTURE OF THE
PLASMA MEMBRANE

chapter two

In 1961 Eric Ponder [106] wrote that it is not far wrong to describe him as being "not convinced about the structure, or even the necessary existence, of the cell membrane as it is generally described" and, in asking himself "whether he believes that a cell membrane, lipid, sievelike, or mosaic in structure, perhaps with enzymes incorporated in it, is *solely* responsible for the entrance and egress of substances in the case of the typical cell, he would have to reply that he does not know, and that, on the basis of the existing evidence, he cannot know." He continues, "it is certainly true that many of the conclusions about the cell membrane and its permeability are based on preexisting ideas, on unallowable simplifications, as well as on a disregard of both physical chemistry and of the results of experiments on the cells themselves." As recently as 1966, Loewenstein [79] introduced a conference on biological membranes with the observation that no one has been able to adequately define a biological membrane or even give its space limits.

Most students of cellular surfaces are convinced, however, that such things as membranes exist, that plasma membranes from a number of cells have been isolated, and that their properties have, at least in part, been described. Warnings such as those above are meant to emphasize that (1) it is extremely difficult to distinguish clearly between cell wall, surface ultrastructure, and plasma membrane; (2) it is often possible to describe membrane phenomena as properties of an interface; and (3) it is not possible

6

at present to relate with each other the different experiments performed by different investigators on different material using different methods. It is easy to envisage the membrane as a discrete package of static molecules arranged in a highly organized and uniform structure. It is more probable that membranes are not uniform but are labile and dynamic structures somewhat disordered and undergoing continuous change in the living state. Variations in the composition of membranes may not alter function or even structure significantly; there may be many ways to build membranes that behave alike. On the other hand, subtle differences, presently unknown, can result in important changes in the functioning of different regions in a single cell, such as a neuron. It is unlikely that membranes are completely distinct from the cellular environment or the cytoplasm and, because of this, membranes cannot be expected to behave in a fully physiological fashion when isolated from their neighboring molecular environment or when this environment is altered in some. manner. Yet much can be inferred about membrane structure and function by doing just that.

As long as it is recognized that interpretations of data must be made with caution, it is at least convenient to assume that plasma membranes of cells exist as real structures. This is clearly the view held by the majority of workers in this field.

2.1 Chemical Composition

There are three levels of molecular organization to be considered in describing the architecture of the plasma membrane: (1) thin layers containing a single class of compounds (e.g., lipid, protein, carbohydrate) parallel to the cell surface; (2) specific molecular types (e.g., cholesterol, phosphatidylethanolamine, lecithin, ATPase) that are organized or distributed in particular patterns within each layer; and (3) specific reactive groups (e.g., carboxyl, amino, receptor, antigenic) located at the cell surface and throughout the membrane layers. The earlier work on membrane structure, as well as much of the recent work (e.g., electron microscopy, X-ray diffraction), has dealt mainly with the first level of organization. Only in the past few years has work on isolated membrane preparations been providing useful information in quantity on the other two levels of molecular organization that probably determine permeability properties, specific enzymatic activities, immunochemical specificities, excitation, impulse propagation, and other important membrane phenomena [15, 70, 71, 116].

Chemical analysis of membranes has been accomplished for a number of organelles, cells, and organisms, and the major constituents are lipid, protein, and polysaccharide. One of the earliest and still popular sources of material for all kinds of membrane studies has been mammalian erythrocytes. They are copiously available in pure cell type at the ends of the investigator's fingertips and can be treated in a variety of simple ways to lose their hemoglobin and to leave only membrane and not much more. Moreover, red cells have no internal membrane structures to contaminate these plasma membrane preparations. Neville [100] opened the way to other membrane preparations by devising a method to isolate plasma membranes from rat liver cells. Neville's techniques have been modified or extended to isolate plasma membrane fractions from cells of liver, intestinal brush border, kidney, muscle, amoebae, Ehrlich ascites carcinomas, mouse fibroblast tissue, HeLa cultures, *Mycoplasm*, and nerve, among many others.

Most of the methods employed to isolate membrane fragments generally treat the cells in some fashion to harden the surface membrane and to block sulfhydryl groups [168]. The membranes are then pulled away from the cytoplasm by, for example, swelling the cells in hypotonic media, homogenizing them, and separating the membrane fragments from other cellular components by physical means, e.g., separation in sucrose density gradients. How these treatments may alter details in the molecular organization of membranes is not clear. Obvious difficulties have not appeared, but few of the analyses can be accomplished on untreated membranes. It is unlikely, however, that these isolation procedures significantly alter the analysis of the individual molecular species that compose membranes.

Membranes are among the most stable of cellular structures. Maddy [88] found no detectable turnover of the mechanical component of membranes, which is consistent with the low turnover of both lipid and protein reported for various plasma membranes [107, 167].

Generally 50% or more of the dry weight of plasma membrane is protein [123]. The low protein content of nerve myelin makes it an unusual membrane preparation. This protein can serve a structural role to confer mechanical stability to membranes or it can possess catalytic functions. A protein molecule itself need not have a specific enzymatic function since it can interact with lipid to provide a matrix for the organization of catalytic factors.

The original suggestion that membranes contained protein was derived from the low surface tensions of biological membranes contrasted to the high surface tensions of pure lipid films on water. Since it is now known that phospholipids also possess low surface tensions in water, other evidence must be presented to demonstrate the value of protein to the mechanical stability of membranes. Proteolytic enzymes, which interfere minimally with the enzymatic and permeability properties of plasma membranes, deform the cellular surface [176] or reduce the force necessary to deform the cell [88]. Divalent cations may also serve to stabilize membranes, however [123].

Evidence has accumulated over the years to show that membranes are asymmetric or that the outer surface is or behaves differently from the inner surface. The inner membrane of mitochondria possesses an isotropic organization of the electron transport chain [93]; the sodium pump in plasma membranes is outwardly directed in most instances [177]. Robertson's electron micrographs of myelin membranes show structural asymmetry [116] (see Fig. 11, p. 24) and the action of many pharmacological and immunological agents on cells can be demonstrated only by application to one or the other of the two membrane surfaces (e.g., see Chapter 4). Sialic acid is found bound to proteins on the surfaces of many cell types [23] and is located on the outer surface of the membrane as indicated by the action of neuraminidase on red cells [36] and electron micrographs of liver cells [7]. Sialic acid has also been proposed to conver mechanical strength to membranes [175] as it does to mucoproteins [58]. No evidence is currently available to distinguish a difference in the role of protein or lipid in generating and maintaining the anisotropic membrane conditions.

The protein components of human erythrocyte membranes outweigh the lipid components and can be separated into several different fractions. Rosenberg and Guidotti [123] separated eight fractions containing at least twelve different proteins that were present in significant amount. Molecular weights ranged from 10,000 to 150,000 daltons. The fractions differed in their proportions of nonpolar and acidic amino acids. Four of the eight fractions contained large amounts of sialic acid, but Rosenberg and Guidotti did not find a major protein component in red cell membrane that has the characteristics of the structural protein isolated from mitochondrial membranes [77]. Proteins extracted from plasma membranes of rat kidney cells fall into two major fractions [49]. One fraction contained at least five different proteins of molecular weights greater than 1 million daltons. The Na-K-dependent, ouabain-sensitive, ATPase activity that is unique to plasma membranes was found in this fraction. The other fractions contained at least eight different proteins of molecular weight near 45,000 daltons. Some of these smaller proteins were insoluble in butanol and may be structural components. Also, after incorporation into the membranes of living cells, radioactive leucine becomes unevenly distributed among the different protein fractions, suggestive of a heterogeneity of membrane protein constituents.

Roughly 20–30% of the dry weight of plasma membranes is lipid [161], about 50% of the total solids of mammalian brain is lipid, and isolated myelin fractions contain an even higher proportion of lipid [71]. Clearly, lipids are a major component of cells and are specifically concentrated in membranes. Since other membrane components, such as proteins, appear in a variety of cellular structures, lipid can be considered as the characteristic component of membranes [85, 124]. The presence of lipid in membranes was originally postulated to explain the permeability behavior of cells, and much of the behavior of membranes can be mimicked by purely lipid model systems (see Chapter 5). While lipids are an essential component of membranes, the large proportion of protein present suggests that protein is important in the manner in which the lipid performs its roles. Maddy [88] has speculated upon this question. Lipid sheets can provide nonaqueous barriers to compartmentalize an essentially aqueous cell or organism, yet the tertiary structure of proteins can allow hydrophobic zones to exist and so allow the same sort of compartmentalization. Hydrophobically bonded protein-lipid sheets are relatively stable structures. The polypeptide backbone of proteins limits the possibilities for different molecular conformations and the separation of two macromolecules requires the simultaneous rupture of many bonds. Purely protein membranes may be too stable to permit the delicately poised transitions that are associated with metabolically active membranes. The potential of phospholipid-water systems to serve in modulating systems is illustrated by the complexity of the phase diagrams obtained by Luzzati [85]. Indeed, lipids characteristically display a variety of structures under conditions (water content, temperature, ionic strength, etc.) that are not far from those prevailing in a living cell.

As much as 95 % of the total lipid in plasma membranes can be accounted for using current extraction procedures [174]. Table 1 illustrates some of the variations in lipid composition among membranes of different organisms. The lipid composition of a single membrane type is known to vary with species, as is the case in mammalian erythrocytes [158]; however, patterns do appear. The lipid composition of the surface membrane of a cell differs from other membrane systems in that cell [1, 11, 51]. For example, glycolipid is present in most subcellular particulates (mitochondria, microsomes, lysosomes, etc.) but is absent from most plasma membranes [109]. Cardiolipin is a major mitochondrial lipid but is not found in plasma membranes [174]. The lipid composition of animal plasma membranes shows characteristic features such as greater (two to three times greater) cholesterol and neutral lipid content than other membranes from the same source [174]. Lipid composition of isolated surface membrane preparations obtained from rat liver [31], erythrocytes [162], and cultured mouse fibroblasts [174] shows a greater sphingomyelin content and lower lecithin and phosphatidylethanolamine content than preparations of internal cytomembranes. Fatty acids found in liver plasma membranes have longer chains and are more saturated than comparable molecules found in the remainder of the cell [109]. Similarly, the hydrocarbons of plasma membranes have longer and less-branched chains than cytoplasmic hydrocarbons. Individual membrane lipids may contain specific fatty acid patterns. Van Golde, Tomasi, and van Deenen [162] found 20 different molecular species of fatty acids in lecithin (phosphatidylcholine) and a total of between 150 and 200 different lipid molecules in human red cell membranes. Weinstein [174] identified ten highly saturated fatty acids (linolenic, palmitic, oleic, stearic, lignoceric, palmitoleic, myriolic,

TABLE 1

LIPID COMPOSITION OF ANIMAL AND BACTERIAL MEMBRANES
VALUES GIVEN ARE APPROXIMATE PERCENTAGES OF TOTAL LIPID

Lipid	Source *Myelin*	*Erythrocyte*	*Mitochondria*	*Microsome*	*Microsome*	*Azotobacter agilis*	*Escherichia coli*	*Agrobacterium tumefaciens*	*Bacillus megaterium*
Cholesterol	25	25	5	6	*	0	0	0	0
Phosphatidylethanolamine	14	20	28	17	18	100	100	90	45
Phosphatidylserine	7	11	0	0	9	0	0	0	0
Phosphatidylcholine	11	23	48	64	48	0	0	10	0
Phosphatidylinositol	0	2	8	11	6	0	0	0	0
Phosphatidylglycerol	0	0	1	2	0	0	0	0	45
Cardiolipin	0	0	11	0	2	0	0	0	0
Lysyl phosphatidylglycerol	0	0	0	0	0	0	0	0	10
Sphingomyelin	6	18	0	0	9	0	0	0	0
Cerebroside	21	0	0	0	0	0	0	0	0
Cerebroside sulfate	4	0	0	0	0	0	0	0	0
Ceramide	1	0	0	0	0	0	0	0	0
Unknown or other	12	2	0	0	0	0	0	0	0

* Not analyzed.

behenic, linoleic, and arachidonic, listed in order of abundance in L cells of mouse fibroblasts) and sizable fractions of unsaturated fatty acids.

Lipid parameters may also prove useful in the classification of surface membrane preparations. Plasma membranes of mouse fibroblasts and liver cells have basically similar lipid composition. About 40% of the lipids in these membranes are neutral lipids, about half of which is cholesterol. The majority of the lipid is phospholipid: mostly lecithin, sphingomyelin, and phosphatidylethanolamine [109, 174]. The plasma membranes of microvilli in intestinal epithelia, erythrocytes, myelin, and HeLa cells have high ratios of cholesterol to phospholipid and are unusual in possessing glycolipid. Differences also appear in the composition of the neutral fat and phospholipid. On a molar basis, the lipid in erythrocyte membrane is about half cholesterol and half phospholipid—primarily lecithin, phosphatidylserine, and phosphatidylethanolamine [158]. Differences in membrane properties and functions have been blamed on the variation in lipid composition although a few workers emphasize the role of protein. While the magnitude of roles remains to be determined, there are studies on lipid-protein interactions that indicate that the organization of membrane structure is dependent on the lipid composition of the membrane [91]. In contrast to proteins, however, membrane lipids cannot be safely considered as genetically determined species-representative components: Dietary changes cause significant alterations in chain length and unsaturation in the membrane lipids of erythrocytes [37, 99] and mitochondria [113].

Nerve myelin has also been extensively studied. Similar to other membranes, myelin has long been known to contain protein, lipid, and some polysaccharide. The principal lipids of nerve myelin include cerebroside, cholesterol, sphingomyelin, and the phosphatidyl derivatives of choline, ethanolamine, and serine [101]. Some of these lipids are more loosely bonded in the membrane than others [122]. Finean [44] summarized the chemical data on myelin in an emperical formula: $(cholesterol_2$-$phospholipid_2$-$cerebroside)_n$-$protein_p$-$polysaccharide_q$-rH_2O.

The small amount of mucopolysaccharide present in red cell membranes possesses a number of strong negatively charged groups that appear to determine some of the immunological behavior [97], the electrical potential difference across the erythrocyte surface [135], and cellular adhesion phenomena [98]. Galactose, mannose, glucose, and acid polysaccharides as well as amino acids but no sialic acid occur in the surface membranes of *Amoeba proteus* [102, 112].

In reviewing morphological, chemical, enzymatic, and antigenic characteristics of membranes, Benedetti and Emmelot [8] list ten enzymes that are known to be located at cellular surfaces. These include the Na-K-dependent, ouabain-sensitive ATPase characteristic of the membrane sodium pump (see Chapter 3), membrane-bound acetylcholinesterase, and alkaline phosphatase.

Van Deenen [157] emphasizes the "bewildering number" of both lipid and protein constituents in plasma membranes. The enormous possibilities for molecular associations between lipid and lipid, lipid and protein, and protein and protein make formidable indeed the problem of constructing a detailed picture of the molecular organization of a biological membrane and its variations that can explain unambiguously the diverse specialized transport processes, catalytic abilities, immunological relationships, cellular contact phenomena, excitable phenomena, and physical properties that are observed in living cells. Some of the physical techniques of X-ray diffraction, optical rotatory dispersion, circular dichroism, infrared spectroscopy, and nuclear magnetic resonance have produced information on the physical state of membrane molecules and functional groups and have allowed study of relatively unaltered membrane preparations. But these techniques have not yet produced significant evidence on lipid and protein interactions that will allow problems of molecular organization to be solved. How far we have come in understanding the molecular organization of plasma membranes is the major theme for the remainder of this chapter.

2.2 Classical Concept of Membrane Structure

From a review of early biological history, Homer Smith [141] concluded that a clear concept of a plasma membrane, and the name itself, originated with Carl Nägeli in 1855. Nägeli, working with algae, fungi, mosses, and unicellular plants, noted that the cell surface is impermeable to intracellular or extracellular pigments and is responsible for the cell's osmotic properties. He derived the name *plasma membrane* because he found the cellular surface to be more dense, viscous, and otherwise different from cytoplasm. The concept was used by the better known Pfeffer and de Vries in their experiments on osmotic pressure during the late 1800's. Although Bernstein described the surface of a cell as a thin structure permeable to some ions and molecules in 1902, the concept of a plasma membrane did not achieve real status as a physical structure until the microdissection experiments were performed in the early 1930's by Robert Chambers, who actually pulled off the plasma membrane from a starfish egg with a microdissection needle.

The work on the molecular structure of the plasma membrane began with the experiments of Overton in 1895. Because certain lipoid substances penetrated cells with relative ease, Overton said that lipids are important in the nature of the molecular organization of cellular membranes. His conclusion has subsequently been supported and extended by numerous studies on the permeability of cell membranes to a variety of compounds. In addition to lipid solubility, both molecular size and certain kinetic energy requirements have to be met before diffusion of molecules through membranes can occur.

The work of Langmuir and others suggested a way in which lipids might be arranged in cellular membranes. Lipids, when compressed into a monolayer on a water surface, arrange themselves perpendicularly to the surface with their polar groups directed toward the water and their nonpolar groups toward the air. Cell membranes contain lipids but are boundaries between two aqueous phases. Gorter and Grendel [57] extracted the lipid from red blood cell ghosts with large quantities of acetone, evaporated the residue, dissolved it in benzene, and spread it as a monolayer on the surface of a Langmuir trough. Sliding a glass barrier along the surface of the water that fills the trough reduces the area of the lipid film until all the exposed area is covered by the lipid film and can thus be measured. Gorter and Grendel found that the extracted lipid occupied twice the calculated surface area of red blood cells. Hence, they suggested that the cell membrane consists of a bimolecular lipid leaflet with the hydrophilic groups of the lipid molecules located at the surfaces and the hydrophobic carbon chains making up the interior of the leafet. Even though later studies [4] have shown that Gorter's and Grendel's extraction procedure did not remove all the lipid within the membrane and that their method for estimating the area of erythrocyte surface was also too small, the concept of a bimolecular leaflet of lipid has not changed.

If the cellular surface is a lipid barrier in an aqueous environment, then the surface tension at the interface should be comparable to that of pure lipid in water, which is high. When this hypothesis was originally tested on plasma membrane preparations, however, the situation appeared more complicated. Cole [20] obtained values for cell surface tension much less than 0.2 dyne/cm by measuring the force required to compress intact marine eggs. Harvey and Shapiro [65] calculated surface tension values of 0.2 dyne/cm by using a centrifuge-microscope technique and intracellular oil droplets. Most pure lipoid substances give surface tension values in the range of 10 dyne/cm or more. To explain the low surface tension measurements, Danielli and Harvey [25] said that lipid-water interfaces did not occur but that these low values were due to the presence of some surface active agent, probably protein, and the lipid polar surfaces were covered by at least one layer of protein. Since then, however, it has been found that, in some membrane systems at least [60, 104], lipids with highly polar groups may assume the emulsifier role that Danielli attributed to proteins. Phospholipid components of membranes can themselves provide the low values of interfacial tension and mechanical stability at membranes surfaces [68]. For example, artificial membranes formed by a bilayer of phospholipid molecules alone without any proteins present possess a surface tension near 1 dyne/cm [153]. Proteins are generally known as unusually fine emulsifiers, however: They lower interfacial tension, increase the tightness of packing of molecules at interfaces, and provide mechanical stability to interfaces by conferring a high structural viscosity upon them [20]. Moreover, proteins added to phospholipid films will lower their already low surface tension even more [88]. In short, proteins are still considered important constituents of biological membranes despite the weakness of the original Danielli argument.

The concept of the *paucimolecular* membrane was first proposed in 1935 by Danielli [24, 25, 64] and was fully described in the classical monograph of Davson and Danielli first published in 1943 [27, 28, 48]. The general model of the structure of the plasma membrane [Fig. 5(a), p. 16] is that of one or more bimolecular leaflets of lipid and a single layer of protein covering each exposed polar surface of the lipid. They used four main lines of evidence to support their theory. (1) The evidence from permeability studies is extensive [28]. (2) Polarization microscopy of red blood cell membranes demonstrated bimolecular leaflets and carbon chains radially oriented [127, 132].

(3) By comparing the light reflection from red cell ghosts with that of films of built-up monolayers of barium stearate, Waugh and Schmitt [170] concluded that red cell membranes have a thickness of about 200 Å, of which only 50–100 Å is made up of lipid. The remaining material was thought to be a nonlipid component and perhaps an adherent cytoplasmic-stromal constituent. (4) Cole and Cole [21, 22] suggested a membrane thickness of 50–100 Å by comparing measurements of electrical capacity of lipid films of known thickness with that of cellular membranes, which is about 1 $\mu F/cm^2$. In fact, Dean, Curtis, and Cole [30] found a capacity of 1 $\mu F/cm^2$ for a model film made from a tanned monolayer of egg albumen and a mixture of amyl acetate and lecithin.

2.3 The Unit Membrane
Theory of Robertson

J. David Robertson [116, 118, 120, 121] has championed a "unit membrane" hypothesis for the structure of biological membranes. The development of

Figure 4. *Electron micrograph of a plasma membrane:* this is a classic picture obtained by Robertson of a portion of a human red blood cell that, when fixed with permanganate and sectioned, shows the typical railroad track appearance of the membrane structure that bounds the cell. Pictures like this have provided strong support for the unit membrane hypothesis.

high-resolution electron microscopy has permitted the fine structure of cells, including the plasma membrane, to be visualized (Fig. 4). Competent interpretation of electron microscopic and X-ray diffraction data led Robertson [115], in 1959, to conclude that a trilaminar structure, similar to that proposed by Davson and Danielli, can be seen in a wide variety of separate cell types in different kinds of animals of diverse phyla if permanganate fixation is properly used. (Other fixatives used in electron microscopy since

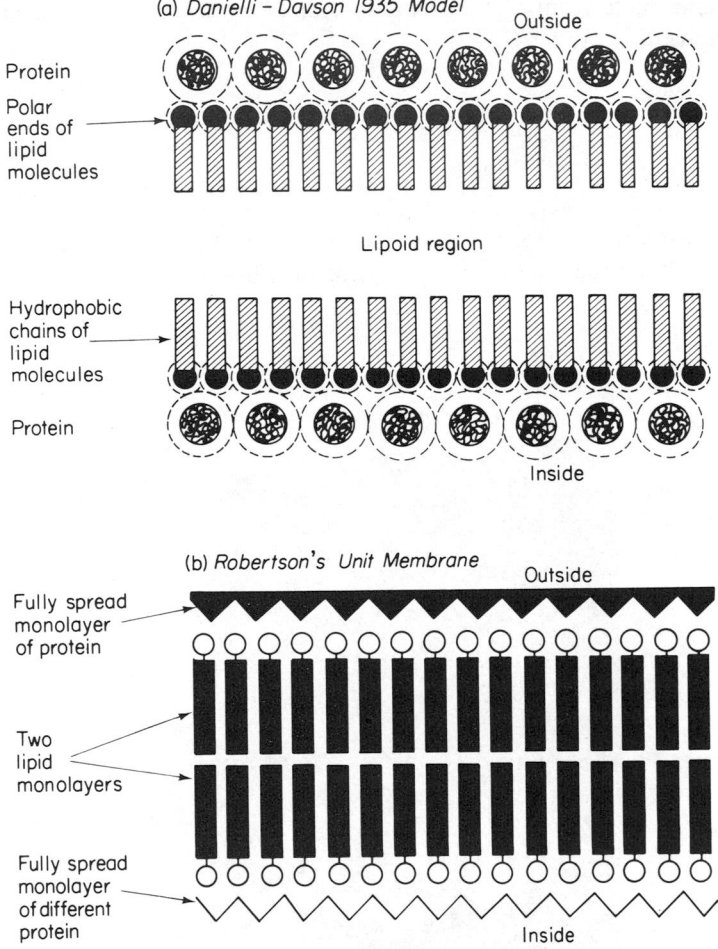

Figure 5. *Diagram of bilayer membrane models:* Robertson's model corresponds closely to that originally presented by Danielli in 1935. The major differences are that Danielli and Davson could not specify the number of monomolecular layers of lipid, the form of the nonlipid components, and the chemical differences between the outside and the inside protein layers.

1959 also show clearly the railroad track appearance of membranes, especially if the material is cut into thin sections.) Robertson's early work was with nerve myelin, a peculiar system of membranes tightly wrapped around axons, but he has defended his views with later work on more varied sources of membranes.

Basically, Robertson has extended the Davson-Danielli paucimolecular model (Fig. 5) for membrane structure by specifying (1) the number of lipid monomolecular layers in the central zone—two; (2) the form of the nonlipid components in the outer and inner layers—fully spread monolayers of macromolecular material, primarily protein (the X-ray data do not allow enough room for globular protein in compact myelin); and (3) the asymmetry of membrane—the outside protein layer is chemically different from the inside layer. The polar region of the lipid interacts electrostatically with the protein layer, a portion of which is in an extended conformation. The interactions within the nonpolar region of the lipid involve London-van der Waals forces and other hydrophobic reactions.

While Robertson's unit membrane hypothesis is consistent with the classical Davson-Danielli model and is a reasonable description of myelin structure, his emphasis on the unitary, composite nature of membrane is progressively more difficult to defend as data accumulate from the many laboratories. Still, Robertson's contributions are real: His astute experimentation and vigorous writing have led the onslaught on the problem of membrane structure. The Davson-Danielli-Robertson theory, despite the criticism that will be discussed later, is supported by quantities of evidence: from polarization microscopy [129], electron microscopy [33], X-ray diffraction studies [45], permeability and electrical conductivity measurements [89], calorimetric data [111], and work on artificial systems [155].

2.4 Supporting Studies—
Early Work on Myelin

Evidence provided by physical studies on myelinated nerve fibers also supports the view that plasma membranes are constructed by a series of tangentially oriented protein molecules and radially oriented lipid molecules. The white myelin sheaths that cover most vertebrate axons are thick, especially when compared to the thin plasma membrane, and contain lipid (predominantly cerebrosides, cholesterol, and phospholipid), protein, and some polysaccharide. Beginning more than a century ago, nerve myelin has been repeatedly examined in polarized light by a number of workers. Myelin shows a radial positive birefringence but the sign reverses after treatment with lipid solvents such as $KMnO_4$ and OsO_4 [125–127]. (Note that these compounds are also common fixatives used in electron microscopy.) Schmidt believed that the radially positive birefringence was due to lipids arranged in thin layers with the long axes of the lipid molecules oriented radially to the cell surface. To explain the reversal in sign rather than disappearance that occurred after lipid extraction, he said that there were layers of protein adjacent to the lipid layers—the long axes of these protein molecules oriented tangentially to the cell surface. Chinn and Schmitt [19] showed that the remaining negative birefringence after lipid extraction was of the form type and not of the flow type. This indicates molecular orientation since it varies in magnitude when media of different refractive indices are used and approaches zero in one specific medium. Schmitt and Bear [5, 128, 129] developed a quantitative measure of myelin birefringence and their findings agreed with the conclusions of Schmidt.

Schmitt, Bear, and Clark [130] succeeded in obtaining small angle X-ray diffraction patterns from myelin. They found fundamental repeat units in the radial direction of 171 Å for frog peripheral nerve and of 186 Å for mammalian peripheral nerve. They concluded that the myelin repeating unit contained two bimolecular lipid leaflets with monolayers of protein interspersed.

The water content of myelin is important since changes in hydration of the sheath can alter the measurement of repeat units by as much as 50 [47] to 100% [131], which will be reflected in a number of membrane phenomena such as the relative location of reactive groups, permeability properties, and density. Schmitt, Bear, and Palmer [131] estimated that the water in myelin may account for 20–25Å of the fundamental repeat unit. But Finean [43] reduced this figure to only 10 Å after considering the rearrangement of the lipid during drying. Finean's estimate of the water content of the myelin sheath of frog sciatic nerve was between 40 and 50% by weight, which is low compared to the values for water content obtained for other tissues and the 90% water content of the axon contained by the myelin sheath. Furthermore, the greater part of this water may be "organized" water or water bound to the hydrophilic polar groups of the lipid and protein components.

2.5 Supporting Studies—
Electron Microscopy

The cellular membrane can be visualized in high-resolution electron microscopy but this technique requires dehydrated, fixed, and embedded tissues. Therefore, it is only capable of providing static information that must be interpreted in the light of other kinds of observations [33]. In addition to the studies on plasma membranes of cells, there have been many studies of other lamellar structures in cells that have contributed pertinent information about the structure of biological membranes and, equally important, differences in membrane construction comparable to the differences in chemical composition mentioned earlier. Some of the cytoplasmic lamellar structures extensively examined by electron microscopists are photoreceptors, chloroplasts, and mitochondria. Our concerns, however, are primarily with the plasma membrane, although contrasting information derived from these preparations may occasionally appear in later discussions.

One well-studied example of plasma membranes is the myelin sheath of neurons [40, 53, 116, 118]. The embryonic peripheral nerve fibers are embedded within a type of connective tissue cell—the Schwann cell (Fig. 6).

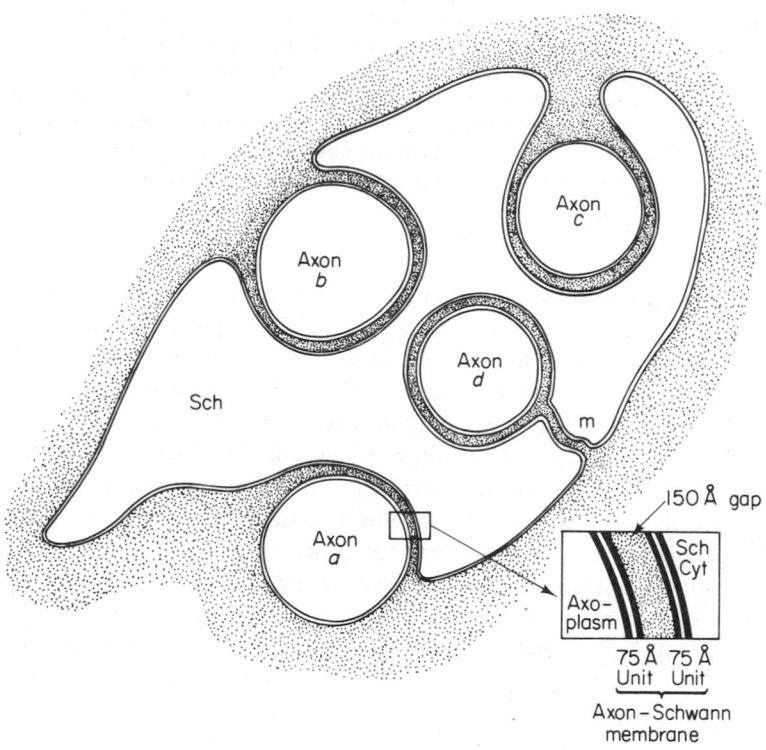

Figure 6. *Diagram of a Schwann cell with four associated axons:* axon *a* is only partially surrounded by the Schwann cell (Sch). Axons *b* and *c* are almost completely surrounded. The membranes of the two lips of Schwann cytoplasm meet to completely envelop axon *d* and to form a paired membrane structure called a mesaxon (m). The insert magnifies the rectangular area indicated at axon *a* to show the unit membrane structures and the intervening 100–150 Å gap that contains intercellular material.

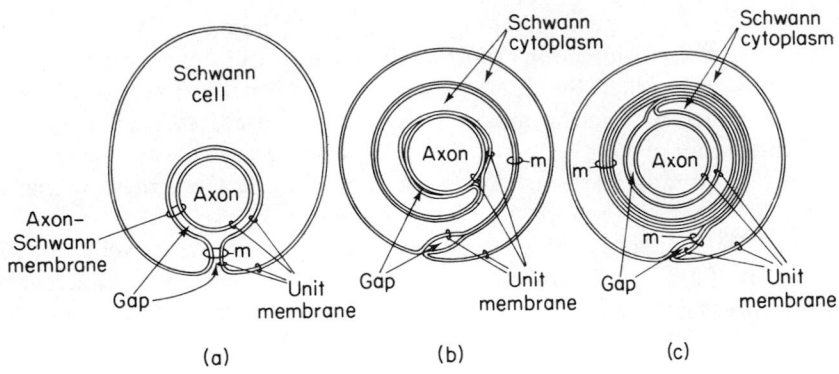

Figure 7. *Diagram of the formation of myelin:* a Schwann cell envelops and spirals around an axon of a peripheral nerve. (a) is the earliest stage in which one Schwann cell is associated with one axon and a short mesaxon (m) is formed. The two unit membranes of the mesaxon fuse so that the intimate apposition of the outer dense lines of the membranes result in the intraperiod line of compact myelin. Note that the gap between the two membranes of the mesaxon is obliterated. (b) shows the mesaxon elongated into a spiral, an intermediate state (compare with Fig. 8). (c) is a further stage in the formation of compact myelin (compare with Fig. 9).

As the cells develop, the Schwann cell rotates around the long axis of the axon, resulting in a spiral wrapping (Figs. 7–9). The spiral wrapping becomes so tight that the intracellular space of the Schwann cell becomes entirely displaced; the internal or cytoplasmic surfaces of the membranes on opposite sides of the same Schwann cell contact each other and make up the major dense line seen in electron micrographs (Figs. 10, 11). The outside surfaces of the Schwann cell membrane also contact each other to exclude the extracellular space as well. Hence, the myelin sheath consists entirely of tightly packed Schwann cell plasma membrane; the tight spiral wrapping results in the regularly repeating radial structure—entirely membrane—that explains the value of this preparation in studies on membrane structure.

The radially repeating unit seen in electron micrographs of myelin correspond to the radially repeating unit detected by X-ray diffraction (Fig. 11). Shrinkage reduces the 170 Å value given by X-ray analysis for the repeating unit in fresh myelin from frog sciatic nerve to values between 100 and 130 Å seen in fixed material in the electron microscope, depending on preparatory procedures. Interpretation of micrographs is complicated by this shrinkage during preparation since these values are significantly smaller than two times the width of a single plasma membrane of Schwann cell that makes up the basic repeating structure. Also, neither the major dense line nor the intra-

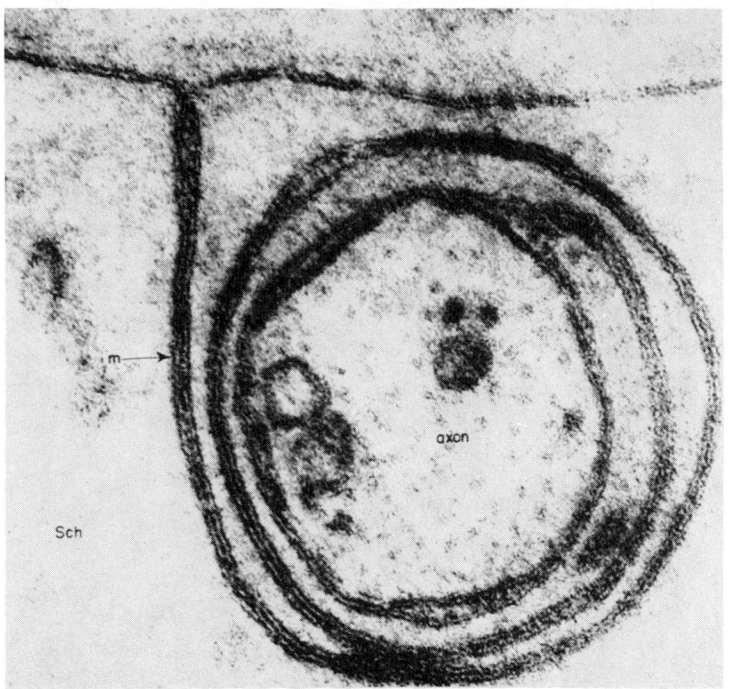

Figure 8. *Electron micrograph of an intermediate myelinating nerve fiber:* the mesaxon (m) is elongated into almost two complete loops about the axon. Note the strata of the surface membrane and mesaxon of the Schwann cell (Sch) (×130,000).

period line seen in micrographs of myelin are wider than one opaque line of Schwann cell membrane although they are both derived from two of the latter. Such difficulties have led to suggestions to reject the hypothesis for myelin structure (see Sec. 2.8). Note that alternate single plasma membranes are reversed: Internal surface contacts only internal surface and external surface contacts only external surface (Fig. 11). This fact explains why double membrane units are described as the repeating unit in the X-ray diffraction studies mentioned earlier—each half of the 170 Å repeating unit contains a complete plasma membrane. Since each of the 85 Å half units show almost identical diffraction patterns, the atomic structure must be nearly symmetrical across the single membrane. Conversely, though, since they are not completely identical, there must be small structural differences between the inside and the outside of a single membrane. To account for at least a part of the different reactivity with fixing agents, Robertson [118] suggested that

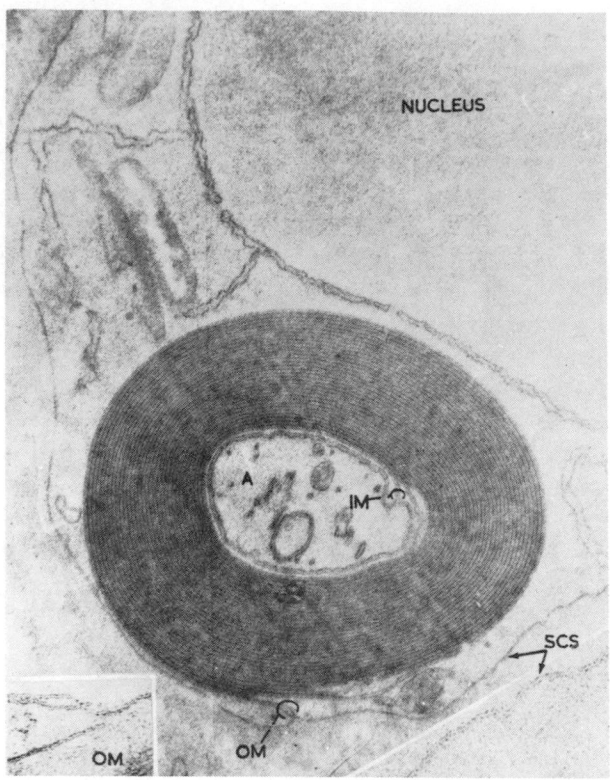

Figure 9. *Electron micrograph of a myelinating nerve fiber:* the Schwann cell surface (SCS) shows a well-defined unit membrane structure that is continuous with the myelin layering at the outer mesaxon (OM). The point where the inner membrane layer (IM) opens out to envelop the axon (A) is also clearly defined. Note that the unit membrane at the surface of the axis cylinder closely resembles the unit membrane of the myelin and of the Schwann cell and that the two unit membranes do not achieve close contact ($\times 31,000$; left insert: $\times 62,500$; right insert: $\times 83,500$).

the outer dense stratum seen in electron micrographs, representing the peripheral surface of membrane, contains a high proportion of mucopolysaccharide or mucoprotein. For the neuronal membrane itself, a protein monolayer on the axoplasmic side is considered to be a functional and structural necessity in order to explain electrophysiological and thermodynamic evidence.

The plasma membrane after $KMnO_4$ fixation has a characteristic internal structure of relatively constant dimensions. It consists of a pair of dense lines measuring 20 Å across separated by a light interzone of about 35 Å in

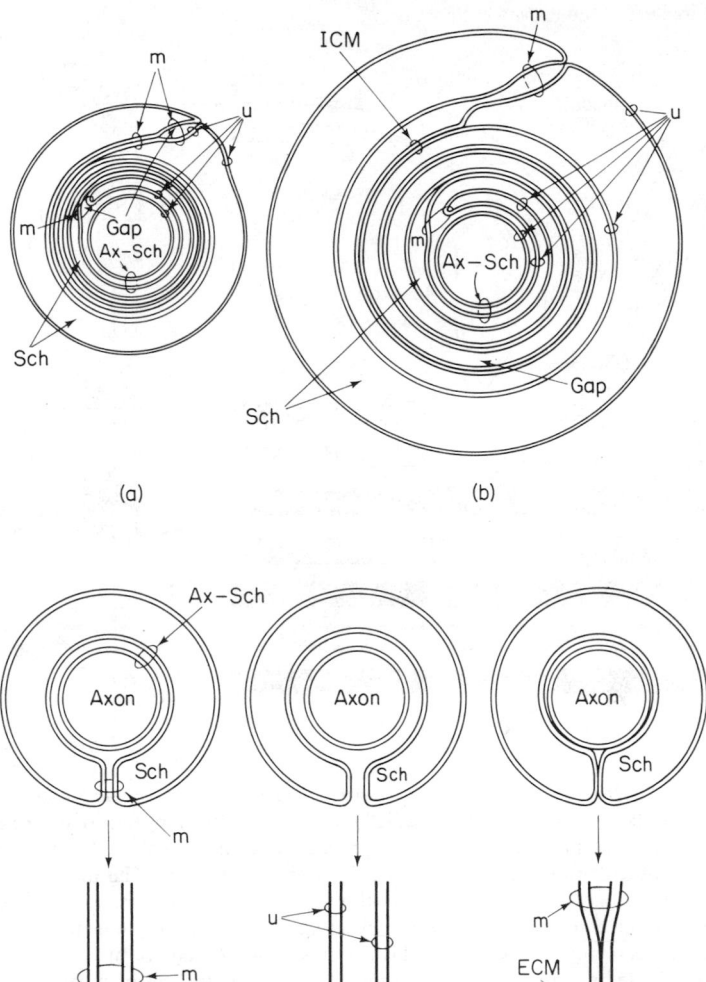

Figure 10. *Diagram of myelin nerve fibers under osmotic stress:* a normal Schwann cell (Sch) swells in hypotonic solutions, such as distilled water, and the gap between the mesaxon (m) unit membrane (u) widens (b). The intraperiod line is split open throughout the sheath but the membranes fail to separate where they are fused along their cytoplasmic surfaces. The paired membrane structures that result are called internal compound membranes (ICM). (c)–(e) show similar effects on unmyelinated nerve fibers. (c) is a normal fiber. In hypotonic solutions (d), the mesaxon gap and the axon-Schwann membrane gap (Ax-Sch) swell. The unit membranes are more widely separated than normal. In hypertonic solutions (e), the gaps are commonly closed, completely in some places, to form external compound membranes (ECM).

23

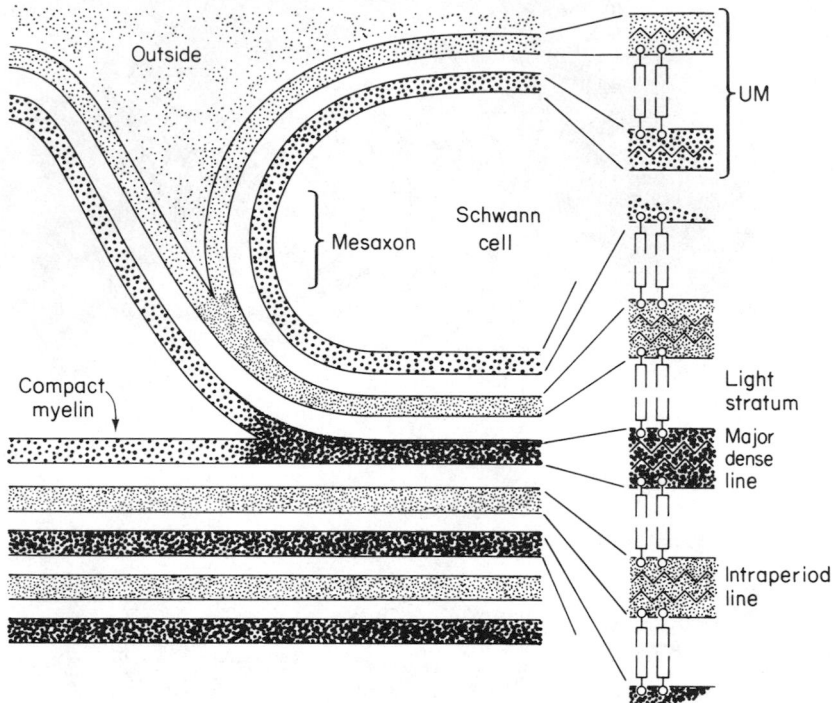

Figure 11. *Robertson's diagram of the ultrastructure of myelin:* Finean's conception of the molecular organization of radially repeating units in myelin (right, see also Fig. 12) can be superimposed on the light and dark lines seen in electron micrographs of a Schwann cell that is forming compact myelin. The diagram (left) shows a mesaxon leaving compact myelin to split into the surface membranes of the Schwann cell. The repeating unit observed in X-ray diffraction studies is the unit between major dense lines and is composed of two cellular membranes. The major dense line results from the fusion of two inner surfaces of membrane and the intraperiod line from fusion of two outer surfaces. The general pattern of the molecular organization of a unit membrane (UM) can then be deduced.

width making a unit structure of some 75 Å across that looks much like a railroad track (Figs. 4, 6). In addition to obvious sources of variation in membrane thickness, such as species and method of preparation, variations appear in membranes taken from different organelles within cells. Mitochondrial membranes, nuclear membranes, Golgi lamellae, and endoplasmic reticulum are all 10–15% thinner than the membranes of the cellular surface synaptic vesicles, capsules and vesicles of multivascular bodies, and Golgi vesicles [178]. Sjöstrand [138–140] restricts the three-layered structure for membrane to the plasma membrane since his later work with mitochondrial and cytoplasmic membranes show 50 Å globular substructures. Robertson

[119, 120] has also seen globular structures participating in synaptic achitec-ture. Robertson emphasizes optical artifacts in electron microscopy and Sjöstrand emphasizes uncertain techniques and labile structures. The point is that biological membranes are certainly not identical and that structures other than the bimolecular leaflet may exist. Some alternate proposals will be considered later.

If all membranes have the same pattern of molecular organization, as Robertson has proposed, then the differences found in properties must result from variation in chemical composition, and much variation has been de-scribed. Indeed, chemical variation will undoubtedly prove important in explaining the functional specificity of the many different types of mem-branes. For example, mechanical stability is a required property of plasma membranes and, indeed, they are found to be among the most stable of cellular structures [88]. Although mechanical stability in membranes is often attributed to the protein present, it is known that cholesterol contributes stability to artificial lipid films [156]. Correlated with this are observations that more cholesterol is found in plasma membranes than in other types of membranes [45] and is especially high in the plasma membranes of eryth-rocytes and nerve myelin (see Table 1, p. 11).

2.6 Molecular Models
of Membrane Based
on Bilayer Theory

Finean [45, 46] has extended Robertson's unit membrane model for plasma membrane on the basis of his studies on myelin. He treated myelin in a variety of ways to obtain a corresponding variety of X-ray diffraction patterns and electron micrographs. He then calculated the sizes of the various layers that he saw. From his data and from the knowledge that 90 % of the lipid in myelin can be accounted for by phospholipid, cholesterol, and cerebroside in a molecular ratio of 2: 2: 1, respectively, as well as knowing the relative size and configuration of each of these molecules, Finean was able to postulate a molecular arrangement of the lipid molecules in the cellular membrane (Figs. 12, 13). To get the lipid molecules to fit into the space between the protein layers in the most probable arrangement, he suggested that the phosphatidyl base moiety of a phospholipid molecule is curled into a J shape with the short arm of the J bent toward an adjacent cholesterol molecule, which then stabilizes this curled configuration [Fig. 13(a)]. This assumption puts the phosphate group into a position that allows it to asso-ciate with the protein's hydrophilic side chains. Thus, he envisages the lipid molecules to be radially oriented in the membrane in a repeating unit made

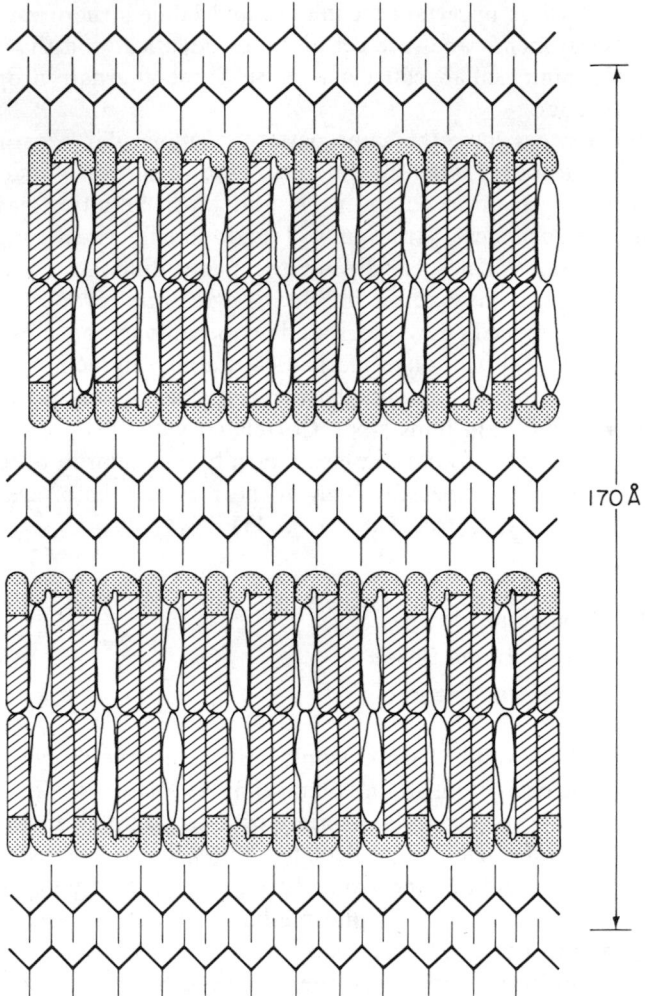

170 Å

Figure 12. *Finean's conception of the molecular organization of the radially repeating units in myelin from frog nerve:* phospholipid molecules are shaded; polar groups are stippled. The clear, irregularly shaped molecules between the phospholipids are cholesterol. This particular model is based upon the polarization studies of Schmidt and the X-ray diffraction studies of Schmitt and Finean. It is an earlier version to that seen in Fig. 13. Basically, it shows myelin to be a layered lipoprotein structure with protein molecules oriented tangentially and lipid molecules oriented radially.

up of a J-shaped phosphatidylserine molecule, a cholesterol molecule fitting beneath the short arm of the J. A second cholesterol molecule appears in a similar relationship to a J-shaped sphingomyelin molecule and, to form the center of the unit, a single cerebroside molecule is fitted between the two phospholipid molecules and adjacent to the two cholesterol molecules [Fig. 13(b)]. A mirror image of this unit makes up the other half of the repeating unit in this leaflet; this bimolecular leaflet fits between layers of protein.

The particular configuration advanced by Finean is only one of several models that can fit the data. While it is a reasonable model and is supported by considerable evidence, it does not represent a complete solution to all the diffraction problems [94] and Finean himself emphasizes its speculative nature. Vandenheuvel [159] criticized the J-shaped phospholipid in Finean's myelin model as sterically and energetically improbable. He proposed a different arrangement in order to maximize London-van der Waals interactions for the lecithin-cholesterol complex in myelin lipid. Vandenheuvel fits cholesterol between the choline moiety of lecithin and the unsaturated chain of the phospholipid, which is curved around in the shape of a C to interact with the cholesterol side chain.

Vandenheuvel [161] also suggested that membrane proteins lie on the surface of the lipid as flattened coils. This enabled him to explain the globular or mosaic structure often seen in electron micrographs and still not contradict the Davson-Danielli-Robertson bimolecular leaflet model. Most of the myelin proteins have a molecular weight near 10,000 daltons [151], which corresponds to a protein chain that is tightly coiled into a disc 60 Å in diameter, a figure close to that of the membrane globular structures seen in some of the electron micrographs.

2.7 Limitations in the Electron Microscopic Evidence

It is not yet clearly resolved just what is stained by the substances used in electron microscopy. Fernández-Morán and Finean [41] extracted myelin with acetone before fixation in OsO_4 and found that the light interzone between the dense lines was greatly reduced in width. They prompty concluded that the light stratum was lipid. Electron micrographs of purified egg cephalin fixed with OsO_4 show a repeating pattern of 20 Å dense lines around 20 Å light zones similar to the pattern seen from myelin. The conclusion from such studies is that the dense lines correspond to the polar ends of the aligned lipid molecules. Korn [72] disagrees. His criticisms are based on his studies that show that OsO_4 reacts with olefinic groups in lipids to form the stable osmic acid esters of glycols; osmium is covalently bound to the hydrocarbon portions of lipids, which are limited in the Davson-Danielli-

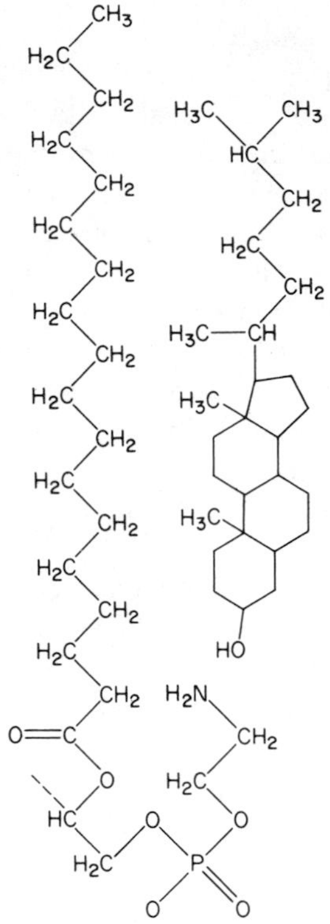

(a)

Figure 13. *Finean's conception of the molecular organization of the radially repeating units in myelin:* (a) indicates a possible relationship of phosphatidyl-ethanolamine to cholesterol in compact myelin. The polar end of the phospholipid molecule is curled around toward the polar end of the cholesterol molecule. (b) *opposite,* shows a more recent version than that in Fig. 12 of the arrangement of the various constituents that Finean considers to be present in the 171 Å repeat unit he described in 1957. The "difference factor" is included to indicate that the structure is not symmetrical since symmetry would demand that the X-ray repeating unit be only half the value observed. The entire region around the "difference factor" line including the adjacent layers of protein and part of the lipid layers corresponds to the intermediate or intraperiod line seen in electron micrographs of compact myelin.

28

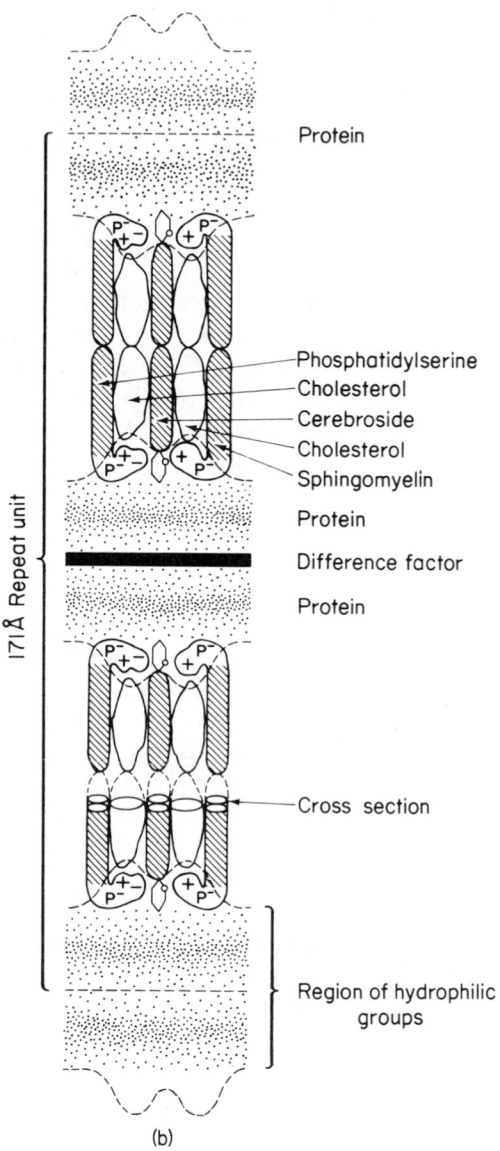

Protein

Phosphatidylserine
Cholesterol
Cerebroside
Cholesterol
Sphingomyelin

Protein

Difference factor

Protein

Cross section

Region of hydrophilic
groups

171Å Repeat unit

(b)

Robertson model to the interior portion of the membrane. Contrary to Korn's interpretation is that of Stoeckenius [149], who based his views on his interpretation of a series of electron micrographs of extracted and fixed lipids. Stoeckenius concludes that the OsO_4 reacts initially with the double bonds of unsaturated fatty acids but the product is unstable. Thus, the partially reduced osmium is permitted to react further with the carboxyl or ester groups located at the hydrophilic ends of the lipid molecules or, in the unit membrane model, at the outer surfaces of membrane. Moreover, according to Stoeckenius, phosphatidylserine is important in the appearance of fixed lipids. For example, a suspension of brain phospholipid contains quantities of phosphatidylserine and the electron micrographs show fine three-layered structures. Mitochondrial lipid contains little or no phosphatidylserine and no layered structures appear in the micrographs although X-ray diffraction patterns show them to be present. This agrees with the difficulties in seeing a railroad track appearance in intact mitochondrial membranes since, according to Stoeckenius, the lipids do not bind enough osmium at the hydrophilic groups to give sufficient contrast. The three-layered structure does appear in micrographs of mitochondria fixed by lead hydroxide in uranyl acetate. Stoeckenius suggests that the appearance of membranes in electron micrographs of OsO_4-fixed tissues is strongly influenced by the lipid composition of the membrane. Korn admits that osmium other than that he finds bound to the hydrocarbon portion of fatty acids may be deposited in tissues after fixation by OsO_4. To what osmium is bound in the membrane to produce the electron image remains unclear despite the vigor of the arguments [114]. Moreover, the polyunsaturated fatty acids of tissues may form polymeric structures linked by diesters of osmic acid. Polymerization would explain why OsO_4 is a good fixative since it would cross-link lipids in the way that gluteraldehyde probably cross-links proteins but this disturbs the view that the original details in the molecular arrangements are maintained. Korn [72] concludes that the dense lines in membranes fixed by OsO_4 reveal nothing about the molecular orientation of the phospholipid in the original membrane.

Phosphotungstic acid (PTA) is known to be capable of fixing the protein filaments of muscle but does not appear to fix purified lipids. PTA-treated myelin shows the dense lines in electron micrographs and this is suggested as evidence that the dense strata of myelin are primarily protein [41]. Frequently, however, OsO_4 will only stain the cytoplasmic dense stratum, while $KMnO_4$ regularly stains both cytoplasmic and peripheral dense strata in myelin figures [118].

Lenard and Singer [75] used circular dichroism and optical rotatory dispersion techniques, which yield information on the helical characteristics of proteins, to demonstrate that the fixatives commonly used in electron microscopy all produce significant changes in all the protein material they tested. Fixing by either potassium permanganate or by the combination of gluteraldehyde plus osmium tetroxide obliterates most of the helical content of proteins, while gluteraldehyde and osmium tetroxide, used separately, are less damaging. The extensive conformational changes produced by $KMnO_4$ are especially disturbing since this fixative has been used extensively to produce the repeatable railroad track figures in electron micrographs of a wide variety of biological membranes. The profound chemical and structural alterations induced by permanganate fixation may leave a structure that is markedly different from that of an unmodified membrane. Because permanganate fixation can leave membrane protein in a random-coil form and can also oxidize some hydrophobic side chains of the lipid into hydrophilic ones, the preferred structure of a protein-lipid system can become one in which the protein lies flat and extended on the surface of a bimolecular leaflet. The lipid-protein structure in the native state may contrast greatly with this. Korn [73] also has questioned the significance of electron micrographs of $KMnO_4$-fixed material. The possibility also exists for significant conformational changes to be induced in the protein during the embedding procedures since most polymers used are nonaqueous liquids.

2.8　Opposition to the
　　　Unit Membrane Theory

Korn [72] has seized the leadership in the criticism of Robertson's unit membrane hypothesis. Korn notes that extensive chemical data are available only for erythrocyte ghosts, and X-ray diffraction data are restricted to model systems and to the myelin sheath, which may not be typical membrane in view of its limited function (although improving techniques for isolating and analyzing membranes are quickly dating these statements). The universality of the unit membrane concept is dependent on the multitude of electron micrographs. Korn places more confidence in the biological significance of studies on model lipid membranes [82, 86, 147, 148] that offer the possibility for phospholipids, especially when mixed with cholesterol, to exist as micelles rather than as bimolecular leaflets. These studies also indicate that small changes in the physical and chemical environment are critical and that osmium may alter molecular structure of phospholipids. Data similar to that summarized in Table 1 (p. 11) show that a range of membrane compositions exist. Myelin is at one end (low protein content and high percentages of unique phospholipids, α-hydroxy fatty acids, and saturated fatty acids) and bacterial membranes are at the other end (high ratio of protein to lipid, no steroids, often only one phospholipid). Korn admits the possibility for membranes with these differences in chemical composition to exist in the same unit membrane structure, but he doubts it. He concludes, therefore, that the unit membrane hypothesis that was derived primarily from studies on myelin—a membrane system he claims to be metabolically inert, chemically atypical, and only an electrical insulator—has little basis for extension to biological membranes in general. Korn believes that membranes differ widely in chemical and enzymatic composition, metabolic activity, biological function, and even in electron microscopic image. Although the greatest similarities occur in their general resemblance in electron micrographs, he cannot interpret this evidence with confidence until he knows more about the chemistry of electron microscopy.

Korn's criticisms seem excessive. While the chemical data are still spotty, some indications of compositions common to membranes have been reported [77, 78]. X-ray diffraction patterns for mitochondrial outer membranes [152] as well as for plasma membranes of erythrocytes, liver, and intestinal epithelia all appear similar, with some variations, to that of myelin and none show any evidence of a subunit structure like that first suggested by Sjöstrand [137, 138]. The chemistry of electron microscopic fixation is really unknown. Conventional preparations extract lipid, so electron micrographs cannot be interpreted confidently in molecular terms. The normal assumption is that the same fixing and staining reactions occur in biological membranes as in known lipid-water systems where the dark regions in electron micrographs correspond to phospholipid polar groups and light areas to hydrocarbon tails [146, 147]. Correlated X-ray diffraction and electron microscopic studies of several lipid-water systems agree in both form and dimension [86] and, when change did occur, as during large variations in temperature and water content in the course of osmium fixation, the change was a loss of structure and not a conversion of one highly ordered structure into another [146]. Variations in membrane thickness, as mentioned by Yamamoto [178] and Sjöstrand [140], are not necessarily suggestive of fundamental differences in structure and, contrary to Korn's view, are not inconsistent with the unit membrane hypothesis. Variations in length and saturation of hydrocarbon tails may account for differences in the light stratum, and variation in protein content may account for differences in the dark line. Also the unit membrane hypothesis does not demand that the ratio between the areas occupied by protein and lipid be similar. While Robertson [121] concludes that the protein layer that interacts with phospholipid probably exists in the extended form, he suggests that much of the protein can be globular and catalytically active. The most important single feature of Robertson's model is that the lipid exists as a bimolecular leaflet. He also wants it to apply universally.

Despite Korn and other critics, the concept of a membrane organized as a protein-bounded bimolecular lipid leaflet is still useful and, indeed, is still trusted by a multitude of competent investigators. There are problems with much of the supporting evidence, however: Some require reinterpretation, some are inconclusive, and some are contradictory. Other structural models have been presented and compromises proposed, including the possibility that membranes may vary in structural organization along their length. Clearly, additional molecular detail and comparative data are needed to clarify the issue.

2.9 Other Models for
Biological Membranes

Prior to the early 1960's and the introduction of a number of new techniques and approaches for the study of both biological membranes and model systems, most investigators were confident of the validity of the Davson-Danielli-Robertson bimolecular leaflet hypothesis. The confidence was premature, however, as many theories and models on aspects of structure and function of lipoprotein membranes have been proposed since. The field has attracted renewed interest and, in fact, has generated strong feelings. Lucy [81] and Hendler [67] have reviewed many of the newer models advanced to describe the molecular structure of plasma membrane. The models of Hechter [66] on membrane depolarization, which is valuable since it emphasized the importance of water, of Watkins [169] on the action of acetylcholine, and of Davies [26] on olfactory mechanisms, are all based on the bimolecular leaflet but each requires some modification, discontinuity, or perturbation in the lipid leaflet to suggest that some of the varied properties of biological membranes are based on other macromolecular organizations of lipid molecules. Other models have included suggestions that the membrane is a sieve or ultrafilter fabricated solely from proteins with pores of molecular dimensions [52], or a mosaic of areas containing purely lipid or purely protein [106], or a mosaic of lipid globular micelles [137, 138]. Others have suggested that the membrane is a complex coacervate or that it is one or another of several kinds of interfacial film. Still others have suggested different ideas or have expressed dissatisfaction with the unit membrane model, for example, Korn's views expressed earlier. Some of the more valuable suggestions have come from physical chemical studies (compare the arguments of Kavanau [70] with those of Chapman and Wallach [17]) that indicate variations in molecular orientations as well as variations in composition. The usual direct staining techniques in electron microscopy reveal the apparent universality of the unit membrane morphology. This implies that, for example, the membrane transport system must be incorporated into the lipoprotein sandwich without any perceptible local alterations of ultrastructure. The less conventional techniques reveal a more complex morphology. It should be emphasized that only the tissue elements that interact with the stains are seen in electron micrographs and these must be interpreted with only minimal knowledge of the chemistry of the interactions of the stains and fixatives with the tissues. Among these alternative suggestions for membrane structure, the micellar models appear important in developing our level of understanding of this problem further. While the alternatives to lamellar lipid structures are exciting, their relevance to real membrane structure is unknown. The lamellar form is the most probable configuration for phospholipid-water mixtures under physiological conditions but phase transitions can be modified by membrane proteins. Alternatives are three-dimensional assemblies, which Maddy [88] thinks are improbable, and the recently described two-dimensional fabric-like structures which consist of hydrophilic channels within a hydrocarbon matrix [85] but which exist under nearly anhydrous or other peculiar conditions. The result is controversy.

2.10 Globular Subunits
in Membranes

In 1963, Sjöstrand [137, 138] published electron micrographs of mitochondrial and other cytoplasmic membranes that showed 50 Å globular micelles in the plane of the membrane lipids, but he did not see any globular substructure in the superficial plasma membrane. Other investigators have since described globular subunits of lipid or lipoprotein using different techniques and a variety of biological membranes, e.g., fixed and positively stained material from mitochondria, frog retina, intestinal microvilli, oranges, and goldfish synaptic discs. Robertson [117, 119] described a granular-fibrillar substructure consisting of 90 Å hexagonal facets as well as a globular substructure in the lipid core of membranes from the outer segment of retinal rods although he is well aware of image-overlap artifacts that can appear in electron micrographs [121]. Negative staining techniques in electron microscopy reveal primarily surface structure and have demonstrated regularly repeating structures in liver cell plasma membranes [6, 8] as well as in mitochondria [42]. Disruption of isolated membranes of Ehrlich ascites tumor cells produces a collection of vesicles of diverse types. Each type of vesicle possesses a different complement and function of the intact surface. Wallach [164] interpreted such data as indicative that this membrane is organized as a mosaic of large and functionally discrete macromolecular assemblies. Membranes of erythrocytes and other cells treated with saponin had regularly arranged pores that led to the suggestion that the saponin extracted the lipid components from these pore areas to leave only the protein framework [32]. Similar patterns were obtained from purified synthetic phospholipids extracted by saponin [3, 54]. Negative stain techniques show such membranes as sheets of closely packed micelles but the image of the same membranes in ordinary electron microscopy reveals structures like bimolecular leaflets. The hexagonal patterns similar to those found by Benedetti and Emmelot [6] in isolated liver membranes are now regularly found in synaptic areas and intercellular junctions associated with intercellular transport and communication functions. Additional evidence for globular subunits have been obtained from bacterial [35, 110] and mitochondrial [9, 61] membranes.

Freeze etching techniques require no previous fixing or staining and, hence, minimize artifacts that may arise from these dehydrating procedures: Tissues are rapidly frozen at liquid nitrogen temperatures and the surface is cut, etched by vacuum sublimation, and shadowed with a heavy metal—the carbon replica is examined in the electron microscope. Freeze etching fractures plasma membranes parallel to the surface to expose the inner hydrophobic faces [12, 13, 29, 95]. Chloroplast lamellae and membranes from yeast and onion root tips treated this way appear to be organized in part as extended bilayers of lipid and in part as 85 Å globular or micellar subunits [14, 171]. Freeze etching of myelin exposes only smooth faces so myelin appears to be entirely organized as a bimolecular leaflet [12, 13].

While it is always difficult to be sure of the conditions of the material used, X-ray diffraction patterns obtained from these membranes show no evidence for subunits [12, 13, 45, 46]. Neither X-ray diffraction nor polarized light studies can show any evidence for subunits if they are few, irregularly spaced, or transient in existence. The suggestive X-ray patterns obtained by Blasie [10] do not distinguish between particles on the membrane surface and particles within it. X-ray diffraction techniques require extensive dehydration of the tissues and long exposure time to the X-ray beam. Unfixed material cannot be easily used and electron microscopists generally experience that even short delays in fixation after a tissue has been removed from its body are deleterious to the fine structure of cells. Hence, the opportunity exists for significant changes to occur in macromolecular structure although only the most dehydrated preparations show a pattern more complex than the more commonly seen lamellar pattern [45]. Clearly, X-ray diffraction data provide no support for a globular substructure in biological membranes but the data must really be considered only uncertain.

2.11 The Green-Benson Hypotheses for Membrane Structure

A considerable body of recently accumulated data on substructure, enzymology, and lipid composition of biological membranes led Green to develop a concept of membrane structure consisting of preformed organized lipoprotein subunits. The electron transfer system, respiratory transformations, and other mitochondrial operations have been associated with macromolecular repeating particles found at the cristae and inner membrane of the external mitochondrial envelope [40, 42, 61]. The repeating lipoprotein unit [Fig. 14(b)], often called the elementary particle, consists of a base piece (40 × 110 Å) and a detachable stalk (50 Å long) and head piece (80 Å in diameter). Isolated base pieces spontaneously associate, side by side, into a membrane sheet, one particle thick. Isolated base pieces that are deficient in lipid associate along all surfaces to form a globular conglomerate. This distinction supports Green's view that the association between base pieces depends on hydrophobic protein-protein interactions and that phospholipid, with outwardly directed polar groups, covers the top and bottom of each base piece to prevent interaction there but not along the four sides, where associations between base pieces can occur and lead to membrane formation.

Benson [9] similarly proposed that membranes are formed by associations of globular lipoprotein subunits [Fig. 14(a)]. The lipid bilayer model is not likely to demand the specificity found in the nature of the hydrocarbon chains of each type of lipid found in membranes. Proteins, in conjunction

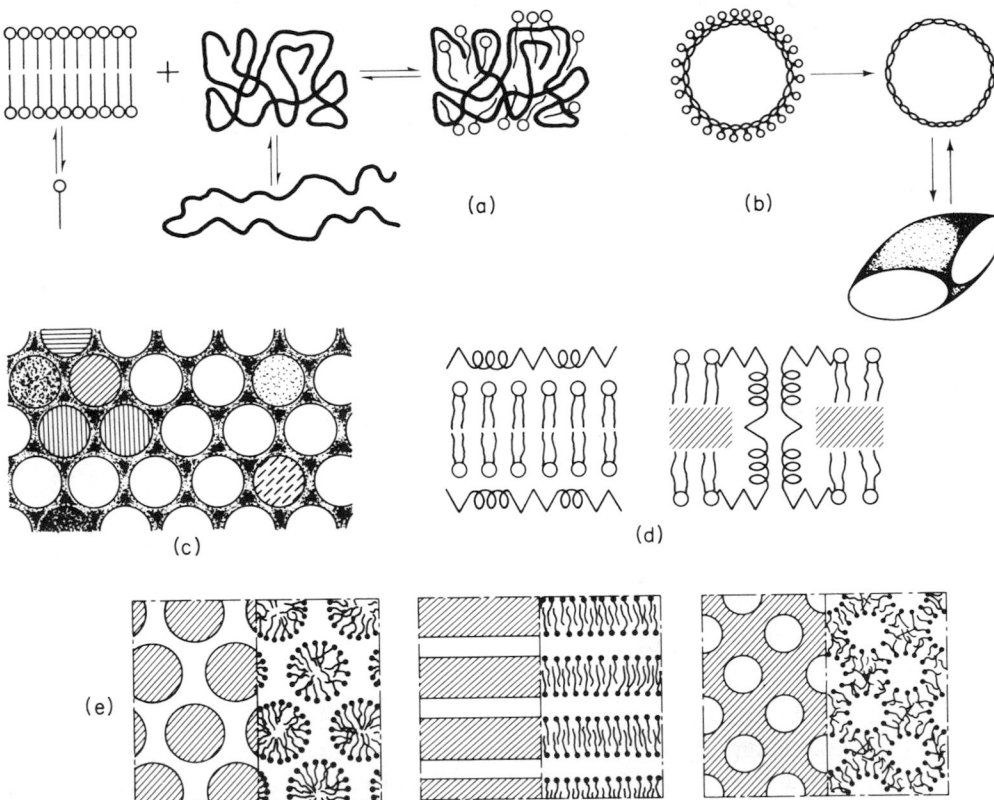

with surfactant lipids, are known to aggregate primarily by hydrophobic interactions to become tightly packed [136]; the interior is largely hydrophobic, while the charged amino acids predominate on the molecule's exterior. From his studies on chloroplast membranes and by extrapolating these relationships, Benson suggested that membrane proteins possess amino acid sequences that can maintain a hydrophobic interior when certain surfactant lipids are available for hydrophobic association. The hydrocarbon chains of the lipids are within the frame of the globular protein molecules and the polar groups of both the lipids and the proteins are located on the outside. While the supporting evidence is primarily circumstantial, electron micrographs of chloroplast lamellae show globular subunits [172, 173]. The idea is diagramed by the equilibrium reactions in Fig. 14(a). The vertical equilibria favor aggregation of individual lipid molecules into a micellar structure and hydrated protein into globular protein, respectively. The

Figure 14. *Mosaic models for membrane structure:* (a) Benson's model for the formation of globular lipoprotein subunits is presented as a series of equilibrium reactions: Lipid molecules aggregate into micelles (left) and hydrated protein aggregates into globular protein (middle). The lipids and protein then associate (right) in such a way that the hydrophobic tails of lipid molecules (thin lines) fit into the hydrophobic interior of a globular protein (heavy line) leaving the polar ends of the lipids (circles) exposed at the surface of the aggregate. This lipoprotein subunit is much like the base piece [(b) right] envisaged by Green to constitute the structural unit of the inner membrane of mitochondria. With lipid present, the base pieces can interact at the four sides to form a membrane sheet. Stalked structures [(b) left] are attached to each base piece and represent functional enzymatic particles. The stalked particles, which cannot form membranes by themselves, can be removed from the base pieces without disrupting membrane structure. Removal of lipid from the base pieces allows them to interact at the top and bottom as well as the sides and to form globular aggregates. (c) diagrams a surface view of the lipid moiety in Lucy's micellar model for a biological membrane. The lipid micelles (unshaded) have diameters near 40 Å. Shaded subunits represent globular proteins, either isolated or organized arrays of enzymes, which replace lipid micelles at certain points in the flexible lattice. Water-filled pores, about 8 Å in diameter, occur between the micelles and are lined by polar groups of either the lipids or the functional proteins. (d) Lenard and Singer modified the Davson-Danielli-Robertson unit membrane (left) to allow for greater hydrophobic interactions between the lipids and proteins and for helical configuration in some of the proteins. They envisage (right) polar groups of the lipid (Ϥ) and random-coiled portions of the protein (ᴧᴧᴧ) to appear at the surface of the membrane and helical portions (ⵡⵡⵡ) of polypeptide chains to be buried in the hydrophobic (shaded) center of the membrane. (e) diagrams the structures proposed by Luzzati for three of the many possible phases that can occur in lipid-water systems. A heavy dot represents the hydrophilic group; a wiggle, the hydrophobic paraffin chain. The hexagonal I phase (left) contains columns of lipid penetrating the water medium, the lamellar phase (center) has bimolecular lipid leaflets separated by layers of water, and the hexagonal II phase (right) has columns of water in a lipid medium.

horizontal equilibrium describes the association of lipid with globular protein to form the membrane lipoprotein subunit, which has a hydrophobic interior and a hydrophilic exterior. This structure is similar to the subunit proposed by Green [Fig. 14(b)]. Globular lipoprotein subunits can then associate to form a two-dimensional membrane with a hydrophobic interior and a hydrophilic surface. The specific lipids are necessary to maintain the proper protein conformation. Such a structure can explain the existence of both railroad track and mosaic figures in different electron micrographs of the same biological membrane.

Vandenheuvel [161] has criticized Benson's model: His space-filling molecular models show that many lipid chains will not fit in the fashion suggested by Benson unless the protein chains form an open structure. If the number of lipid chains present in membranes fill all spaces in which they can fit, the resulting structure will still have many channels open to penetration by smaller molecules. Since the polyunsaturated chains in membranes are not oxidized and since there is no osmophilic reactions in the region that they occupy, Vandenheuvel feels that the membrane is a tight system and, therefore, that Benson's model is not possible.

Green proposes that all membranes are formed from the base pieces of such lipoprotein subunits. Although some variation in size and shape can occur, the base pieces are all morphologically alike. The chemical composition varies to account for differences in functions among membrane types. Green does not believe that structural protein, which is contained in the detached portion of the lipoprotein subunit, can form membranes. He has never observed a reformed membrane develop from a mixture of structural protein and phospholipid as he has when base pieces are mixed with phospholipid. He suggests that the membranes seen in the electron micrographs of Kagawa and Rucker [69], which looks like Green's reformed membranes, are actually phospholipid micelles.

Despite the accumulated evidence supporting subunit structure in biological membranes, the general applicability of the Green-Benson more specific models has been questioned. Electron micrographs of mitochondrial membrane show the usual trilaminar structure [33, 50] but Green's reformed membranes show thick dense lines [59, 92] and may not be equivalent. X-ray diffraction studies do not show regularly arranged subunits in membranes generally including mitochondrial membranes [152]. Many enzymes require lipid to be active [38]. According to Green, the negative charge on phospholipid is not so important as the ability to emulsify large conglomerate enzyme particles to allow substrate molecules to reach the active sites. Without denying this possibility for the electron transfer chain, van Deenen [155] summarized other work that indicates that lipids function as intermediates or provide the correct dielectric properties for the catalytic process. Van Deenen concluded that the restoration of enzymatic activity depends more on the species of phosphatide and the unsaturation of the fatty acyl groups and less on the ability of phospholipid to orient base pieces into sheets rather than conglomerates. Green titrated lipid-extracted and lipid-supplemented electron transfer particles with polyserine. The difference in equivalents of polyserine used in the two titrations was interpreted by Green to be roughly equal to the number of charges on the phosphatides in the particles and, hence, as evidence that polar groups of phosphatides are exposed on the surface of membranes. Alternative explanations to explain these titration data presented by Tourtellotte, McElhaney, and Rader [154] include the possibility that negatively charged protein groups are exposed during the transition from conglomerate to membranous structure, induced by addition of phospholipid, as well as the possibility that phospholipid micelles bound to the electron transfer particles are titrated.

Membranes from various sources dissociate in detergents and re-form after removal of the detergent [6]. Green uses this as support for the existence of subunits in membranes; however, reaggregation of detergent-solubilized membranes cannot by itself be considered adequate experimental support for the existence of subunits. For example, consider the work of Morowitz and his colleagues [35, 110, 150] on isolated membranes of *Mycoplasma laidlawii*, a structurally simple cell so small that its plasma membrane constitutes 40% of its dry mass. These membranes, which are 60% protein and 40% lipid, disaggregate when treated with detergents. The ultracentrifuged material suggests a homogeneous solution of 3.3 S subunits, but turbidity and sedimentation studies show that solubilization of these membranes proceeds through a continuum of states with no defined intermediates to form separate complexes of lipid and protein with detergent. Clearly, a single schlieren peak in the untracentrifuge is an inadequate criterion for homogeneity. The membranes of *M. laidlawii* do not break into homogeneous lipoprotein subunits when exposed to detergent. The disassociated material can be reassembled by dialysis to remove detergent and to add divalent and multivalent cations. Detergent removal alone yields lipoprotein material but the addition of, say, Mg^{2+} yields large lipoprotein aggregates of differing morphologies, as indicated by varying electron microscopic, sedimentation, and centrifugation patterns. Only under a limited set of conditions does the majority of the reaggregated material resemble the original membrane including the railroad track appearance in electron micrographs. The fact that some degree of reformation can occur suggests that synthesis of appropriate proteins and the availability of suitable lipids provide sufficient structure-determining information for synthesis of new membrane. However, these kinds of experiments cannot be interpreted without reservations. It is not known that detergents release subunits, even if they exist *in vivo*, from membranes in their original form. While the reformed membranes of *M. laidlawii* that possess the railroad track appearance in electron micrographs also contain essentially all the NADH oxidase activity, the reformed membranes contain only half of the protein found in the original membranes. Hence, all membrane proteins are not required for structural integrity or physical characteristics, such as electron micrographic image, and single enzymatic activity offers inadequate criteria to establish structural and functional equivalence with the original membrane.

The work of Green and his colleagues on mitochondrial membranes is extensive and he has received support from other workers, for example, that of Benson mentioned above. However, Green has encountered the same difficulty that others have met in trying to extrapolate their ideas on one type of membrane to all membranes. It is much safer to suggest that (1) some membranes contain independent globules of lipid and globules of protein in the plane of the lipid micelles; (2) other membranes contain lipoprotein units; (3) still other membranes contain all three types of globular units; while (4) yet other membranes, like myelin, contain no globular subunit structure at all. Of course, proposing four classes for membrane molecular organization is based as much on ignorance as on evidence.

2.12 Other Proposed
Molecular Arrangements

Lucy [55, 80–82] proposed a compromising equilibrium between a bimolecular leaflet structure and a micellar configuration. The bimolecular leaflet structure can confer to a membrane the stability in structure and the ability to act as a barrier to free diffusion. The micellar configuration, or at least discontinuities or perturbations in the lipid leaflet, is better suited to confer the more dynamic functions [60, 90, 104]. Different organelles, then, may vary in their proportion between extended bilayer and globular subunit to correspond to their respective functions. Figure 14(c) diagrams a surface view of the lipid moiety of Lucy's micellar model for a biological membrane. Lucy feels that the globular lipid micelles, each about 40 Å in diameter, ought to be flexible and in constant motion so that they do not remain in a close hexagonal packing. At certain points in the plane of the flexible lattice, the lipid micelles are replaced by globular proteins, represented by shaded subunits, which form either isolated or organized arrays of molecules with enzymatic or hormonal properties. Water-filled pores, about 8 Å in diameter, penetrate the lipid between the micelles and are lined by polar groups of either the micellar lipids or the globular proteins. Since the micellar arrangement is far too permeable to correspond to experimental data, Lucy suggested that the bimolecular leaflet structure is utilized in portions of the membrane. Alternatively, the pores are plugged by an icelike crystalline structure of water or by polysaccharides associated with the external layers of membrane. Primarily, however, Lucy feels that the adaptability and versatility of membranes result from the utilization of both arrangements for lipid molecules which attain some level of equilibrium between the two configurations which is coincident with function.

It is appropriate to note the large number and variety of structures found by Luzzati [83–87] in pure lipid-water systems. Lipids in biological systems can exist in or, under appropriate conditions, transform from one to another of the several lamellar or rod-like phases. Three of these phases are schematized in Fig. 14(e). The lamellar phase is similar to the bimolecular lipid leaflet required in the Davson-Danielli-Robertson model. The two hexagonal phases resemble lipid structures in the several globular models. The hexagonal II phase is interesting because it provides aqueous channels through the system. Some of Luzzati's rod-like phases, which are not illustrated here, can form two-dimensional fabric-like membranes. The threads of the fabric are formed by lipid rods and the proteins fit either on the lipid framework or between the rods. Such a model, like the Davson-Danielli-Robertson model, provides a continuous nonpolar paraffin matrix coexisting with an ordered two-dimensional structure, which is required by the mosaic models. Luzzati further notes the variety of permeability sites that can occur in fabric-like structures, each site characterized by the local distribution of lipids and proteins, without disrupting the mechanical continuity of the lipid matrix. Since the conditions under which the fabric-like lipid structures exist are unusual in biological systems, i.e., they occur in almost dry preparations, Luzzati suggests that such structures may be limited to only certain regions of membrane and may be transient or traveling. The significant features of Luzzati's work are the demonstration of the number and variety of different structures that can be displayed by lipid-water systems, in contrast to the crystalline structure of paraffin chains that was uncritically presumed by most early workers and equally uncritically rejected by most later workers, and the highly developed long-range order that is compatible with a disordered short-range conformation and a wide chemical heterogeneity of both paraffin chains and polar groups.

Lenard and Singer [74] used circular dichroism and optical rotatory dispersion spectra obtained from aqueous suspensions of human erythrocytes and bacterial membranes to support their conclusion that roughly 30% of membrane protein is in a helical configuration and that the remainder is likely to be in a random coil form. Similar spectra in the regions of the peptide bond absorption bands, which are characteristic of the protein portion of membrane with only a negligible contribution from lipid constituents, were also obtained from membranes of tumor cells, mitochondria, and chloroplasts. In addition to evidence for a helical structure for part of the membrane proteins, this work indicates that membrane components are held together predominantly by noncovalent hydrophobic interactions.

Moreover, phospholipase C, a specific enzyme that catalyzes the hydrolysis of phospholipids to diglycerides and water-soluble phosphorylated amines, rapidly cleaves and releases the major fraction of the ionic heads of membrane phospholipids without disrupting or altering the overall conformation of membrane proteins [76]. Hence, the phosphoester bonds hydrolyzed must be near the surface of the membrane in order to be so readily accessible to the action of phospholipase C. Also, the native structure of membrane is primarily determined more by hydrophobic than by electrostatic interactions between the lipid and the protein. The Davson-Danielli-Robertson model for membrane has most of the protein in an unfolded state on the exterior surface—a situation that does not maximize hydrophobic interactions, demands significant electrostatic interactions between the ionic heads of the phospholipids and the charged groups of the protein monolayer, and exposes a large fraction of the nonpolar amino acids to the water phase. At the same time, a helical protein with sufficient ionic residues to interact with water on one side and with polar head groups of lipid molecules on the other side is an unstable helix. Since these data are not consistent with bilayer models, Lenard and Singer proposed a macromolecular structure for membrane [Fig. 14(d)] to satisfy these arguments. Unfortunately, their concept has little other supporting data. Their model places the ionic and polar head groups of lipid molecules and all the ionic side chains of structural proteins at the exterior surface of the membrane in van der Waals contact with the bulk aqueous medium. The sequences of structural protein that consist predominantly of nonpolar side chains are located in the interior of the membrane along with the hydrocarbon tails of the phospholipids and the relatively nonpolar lipids (e.g., cholesterol). The helical portions of the proteins are specifically located in the nonpolar interior where they can be stabilized by hydrophobic interactions. Although Fig. 14(d) illustrates transmembrane proteins, there is no evidence that a single polypeptide chain traverses the entire thickness of the membrane. Structural proteins possess the unique amino acid sequences that permit them to interact with both the membrane's lipid components and the bulk aqueous medium—these interactions determine the conformations of the protein's structure. Lenard and Singer conclude that such a structure can be organized into subgroups that aggregate into an intact two-dimensional membrane.

Wallach and Zahler [165, 166] described still another model for the molecular organization of membrane, but one which is similar to that proposed by Lenard and Singer. They obtained a partial analysis of protein conformations and lipid-protein interrelationships by using spectroscopic techniques—infrared spectroscopy, fluorescence spectroscopy, and ultraviolet optical rotatory dispersion—on isolated preparations of plasma membranes from Ehrlich ascites carcinoma cells. Their data, similar to those of Lenard and Singer [74], showed little protein in the β-conformation.

Moreover, they found that portions of membrane protein reside within a medium of high refractive index, which indicates extensive hydrophobic interactions between the membrane proteins and lipids. Bilayer models have an extended layer of protein, usually in β-conformation, at the polar surface of a lipid bilayer (or lipid micelles) in an arrangement that prevents nonpolar protein-lipid interactions. Since the bilayer models cannot explain their data, Wallach and Zahler proposed that membrane proteins, both structural and functional, have unique amino acid sequences that impose such tertiary and quaternary structures that two hydrophilic peptide regions are widely separated by a hydrophobic zone. The two hydrophilic sections lie at the membrane surfaces and are connected by hydrophobic rods penetrating the membrane normal to its surface. The length of the hydrophobic units equals the width of the nonpolar membrane core. The hydrophobic units have helical peptide segments packed between hydrocarbon residues of membrane lipids and exist as either single units with only nonpolar side chains or as aggregates similar to microtubules that, depending on primary structure, can have a polar interior. Conjecture permits these tubular structures to be involved in membrane transport phenomena. Helical peptide segments at the hydrated membrane surfaces are in polar interaction with other, including nonpenetrating, peptide chains as well as with head groups of membrane lipids. The complex structure of such membrane proteins depends on their associations, especially hydrophobic bonding, with the appropriate lipids. The specificities of the protein-lipid associations depend strongly on the primary structures of the hydrophobic peptide segments. Wallach and Zahler view the role of membrane lipids to be more than maintenance of membrane permeability and plasticity but they consider the genetic control of membrane structure and the biochemical regulation of membrane function to reside mainly in the membrane proteins.

2.13 Some Additional Data
on Molecular Interactions
in Membranes

The molecular associations in biological membranes are far from being understood and the data presently available are insufficient to specify the protein-protein, the lipid-protein, or the lipid-lipid interactions. Since biological membranes can be easily dissolved by detergents or by sonication and since organic solvents easily extract lipids, strong covalent bonding appears to play only a minor role. The important interactions must be the weaker electrostatic forces, hydrogen bonding, and London-van der Waals dispersion forces [154]. The bilayer and the mosaic theories offer different possibilities for molecular binding. The mosaic structures present a more varied and stronger set of intermolecular binding forces than is possible with the bilayers. If the basic lipid head groups are poorly ionized and are not stabilized in favorable positions for interactions with the protein envelopes in bilayer membranes, only the phosphoryl and carboxyl head groups of the phospholipids can participate in protein binding. If the carboxyl groups of the protein side chains are not able to form direct ion-ion bonds to lipid head groups, then the binding will depend heavily on associations with inorganic cations [70]. Calcium is probably the important cation to consider in maintaining this membrane stability. For example, calcium appears to preserve the trilaminar structure and prevent a transition to globular form [8]. However the relative stability of membranes in solutions of varying salt concentrations argues against ionic stabilization being a major factor. Lenard and Singer [76] found that they could remove 70% of the phosphate from erythrocyte membranes by phospholipase C digestion without altering either the stability of the membranes or their circular dichroism spectra. They argue for extensive hydrophobic interactions between membrane lipids and proteins. Cholesterol, which appears in high concentrations in membranes of red cells and myelin but not in many others, stabilizes sheets of lipids. The stability of lipid bilayers depends on the presence of unsaturated groups; fully unsaturated layers offer no space through which water molecules and small inorganic ions can pass. Hence, optimal porosity, fluidity, and stability in bilayers are obtained with a balance between the saturated and unsaturated groups [159–161].

Much of the current speculation on phase transitions in biological membranes is beyond experimental evidence. Thermodynamic arguments are interesting but insecure. It is generally thought that the hydrocarbon portions of membrane lipid are in a liquid state. The bimolecular leaflet concept of membrane structure corresponds to a more solid lipid interior and the micellar concept to a more liquid interior [105]. The principal evidence for liquid chains comes from X-ray diffraction studies with model lipid systems and some biological membranes [108]. The balance of evidence is against randomly coiled chains of lipid molecules, and Pethica [105] concludes that the interior is more ordered and crystalline than commonly supposed. The role of water is still unexplained although clearly important.

However, some experimental data on lipid phase transitions have been obtained from membranes of living *Mycoplasma laidlawii* by differential calorimetric techniques [111, 143]. These membranes undergo a reversible endothermic phase transition at a temperature that is dependent on the lipid composition of the membrane: The transition temperature is lower with higher unsaturation of the fatty chains of the membrane lipids. The phase transitions in the organisms, in membranes isolated from the organisms, and in lipids extracted from the membranes, all occur at the same temperature. Since these results are similar to the behavior of aqueous dispersions of phospholipids, which exhibit crystalline to liquid-crystal transitions of the hydrocarbon tails within the bulk lamellar array [16, 18], the membrane lipids were suggested to exist in the bilayer conformation. Low-angle X-ray diffraction patterns show spacings below the transition temperature characteristic of a lattice of hexagonally packed hydrocarbon chains within a lamellar array [34]. The enthalpy measurements of Reinert and Steim [111] correspond to the heat of transition for a completely extended bilayer [2]. Any perturbation of a continuous lipid bilayer in a membrane, such as permanent aqueous pores, penetrating proteins, or hydrophobic associations between the lipid and protein, will disrupt the cooperative interaction of fatty acid chains and lower the heat of phase transition. The enthalpy of phase transition is proportional to the amount of lipid in the bilayer state. Hence, the calorimetric data with correlative X-ray analysis are consistent with the presence of an extended lipid bilayer in *M. laidlawii* membranes and are not consistent with appreciable hydrophobic associations between lipid and protein. Reinert and Steim, however, do admit that it is possible for small amounts of protein to penetrate the lipid bilayer without altering the interpretation of their data.

Extensive evidence exists for the cooperation of membrane lipids with protein in several catalytic and antigenic activities. Yet the assembly of these components into a sheet structure remains vague despite clear proposals. The plasma membrane may be a single entity or a mosaic of separate organelles arranged in the form of a limiting envelope. Or plasma membranes may have different organizations in different cells or different combinations of arrangements. Data obtained from one membrane may support one model and data from other membranes may support other models. Difficulties arise when workers try to generalize into an all-embracing theory for membrane structure. Knowledge of chemical composition is necessary but by itself is insufficient for success. Disaggregated membranes show how membrane proteins and lipids may interact but not how they do interact in living cells. Techniques permitting analyses on intact and physiologically active membranes are limited, produce data open to diverse interpretations, and introduce indeterminate artifacts. Electron microscopy has limited value because of the complexities in the mechanisms of fixation and staining. X-ray diffraction patterns cannot distinguish between the two classes of structures especially in the absence of extensive regularity in the arrangement of subunits. Other physical techniques (e.g., optical rotatory dispersion, circular dichroism, infrared spectroscopy, nuclear magnetic resonance, etc.) expose information on the physical state of membrane molecules but not on the interrelationships between protein and lipid. The differences in the characteristics of molecules when they are within membranes and when they are in isolation indicate molecular interactions in membranes. Both theories can accommodate hydrophobic bonding between lipid and protein. Proteins as well as a continuous lipid layer can provide the hydrophobic membrane element necessary to explain the high electrical resistance and the low permeability. Membranes should be considered as dynamic structures whose states change with time. The stability of catalytic factors in membranes may depend on the existence of the precise lipid-protein associations that can persist in a fluctuating matrix. It is generally assumed that the proteins confer strength and mechanical stability to this matrix and that the lipids confer the fluidity and mutability.

2.14 Discontinuities in
Membrane Structures

During the course of earlier discussions on membrane structures, the plasma membrane was often described as lacking a uniform, continuous, and unbroken sheet or film of polymerized material. While obvious in all the mosaic models, even the bilayer models allowed for the occasional appearance of discontinuities, local irregularities, perturbations, and transmembrane molecular structures. The X-ray diffraction studies of Luzzati [83] on pure lipid and phospholipid films show evidence of local irregularities that indicate the existence of less stable lipid crystalline configurations in a hexagonal arrangement in some films. As indicated earlier, Luzzati's later work described numerous phases that can exist in water-lipid systems; the chemical nature of the polar and nonpolar groups on the lipid, their dimensions and charge, the relative amounts of water, and the nature of neighboring molecules, for example, cholesterol or inorganic ions, will determine the existence of one or another of these phases. Luzzati's hexagonal II phase [Fig. 14(e)], which has cylinders of water in a paraffin medium, is interesting because it has water-filled pores that can be altered by these same factors to account for a number of membrane phenomena including membrane permeability properties and the ionic processes associated with excitation and response. The numerous observations of large molecules crossing membranes by themselves suggest "breaks" in the cellular surface. Moreover, discontinuities in cellular membranes have been pictured in electron micrographs from many laboratories.

The plasma membrane cannot be considered fairly as an entirely distinct entity separate from the rest of the cell's elements. The present morphological data are obtained from dead preparations, whereas the living membrane is probably continuously changing in a liquid-crystalline state that is differentiated from the interior of the cell as much by its location at an interfacial boundary as by its molecular composition. The physicochemical organization and special properties that characterize the membrane are then determined by the molecular reorientations that are imposed by this condition.

The point should be made that it is difficult to work with membrane components and to extend this information to living systems. Kavanau [70] illustrates this well in the rigorous development of his theories on membrane structure and function. His extensive review, published in 1965, centers on the physical chemistry and the chemical physics of the components of biological membranes and it is strong in pertinent aspects of intermolecular and interfacial chemistry. His statement that "it is instructive and sobering to be aware of the wide latitude of disagreement that exists over the interpretation of even the most elementary intermolecular phenomena, particular in aqueous media" summarizes the nature of the problem. It is generally thought, however, that biological lipids associate so closely with each other in mixed systems that the properties of the individual components are profoundly altered. The submergence of individual properties in a mixture depends on, among other things, the high miscibility of different hydrocarbon chains. For example, fatty acids, phospholipids, triglycerides, cholesterol, and cholesteryl esters aggregate in aqueous systems to form bimolecular leaflets having a relatively constant thickness. Phosphatides are especially difficult to handle because, in addition to a marked influence on their colloidal behavior by the presence of small amounts of other lipids, they possess a strong tendency for autoxidation and partial hydrolysis [70].

2.15 Pores

One of the difficulties in determining the microstructure of the cell membrane is the thinness of the membrane. Since the macromolecules are probably distorted, stretched, and cross-linked into a planar and dynamically changing structure, the degradation or solution of membrane molecules by conventional analytical techniques cannot be expected to expose the structure or properties of *living* membrane. Good techniques for the study of unaltered membrane are still needed; however, one can infer properties of living structure from indirect studies such as permeation of living membranes by water, ions, or other molecules. One such property of living structure is pore size. Since a bimolecular lipid leaflet has been thought to be a relatively impermeable structure, models for membrane structure based upon a leaflet usually incorporated pores to account for the observed permeability properties [145]. In contrast, the models based upon a micellar configuration for the lipid contain pores as an inherent feature of their structure. The strongest evidence for the existence of pores in membranes comes from the fact that reflection coefficients, which provide a measure of the selectivity that a membrane demonstrates between a solute and solvent, have been derived for a number of permeants of human erythrocyte membrane [56] (see Table 3, p. 71).

The experimental data on permeability of biological membranes have led to the two divergent interpretations of membrane structure. In the first, the permeating molecules enter and diffuse through a hydrophobic environment—through the lattice formed by spaces between the hydrocarbon chains of the lipid. In the second, the permeating molecules diffuse through a hydrophilic environment—through aqueous channels of selected sizes that penetrate the membrane. Each of these views implies a specific molecular structure for the membrane. Stein [144] compared the available experimental data with the models for membrane structure and his present opinion favors a bimolecular lipid leaflet over a membrane composed of water-filled holes. The uniform lamellar structure of Davson, Danielli, and Robertson can easily account for the high resistivity of plasma membrane since this structure should be quite impermeable to ions, but the bimolecular leaflet cannot, in its unmodified form, account for all the functional data. For example, Stein notes the difficulties for this concept to explain the high penetration rates for water and urea, and the penetration of salts is clearly more complex than can be solved in terms of any single model. Other data that suggest a more complicated structure than a lipid film need to be considered. Still, the evidence for pores or their equivalent is indirect and it is difficult to be positive about their existence, but most investigators are.

An estimate of pore size (length and area) requires at least two independent experimental measurements since there are the two-dimensional parameters to be evaluated [134]. Both the rate of diffusion of a substance down a concentration gradient and the rate of flow of a fluid under a mechanical pressure should be greater with a larger area of hole in the membrane and a shorter hole length. Although the rate of transport of water through the cell membrane of erythrocytes is very rapid, both diffusion rate and flow rate can be measured accurately enough to determine the value for average pore diameter in the intact erythrocyte wall [142]. The diffusion rate of water was found by measuring, using a fast flow apparatus, the rate at which radioactively labeled water is picked up by the cells within a few milliseconds of being bathed in the labeled water. The rate of flow into the cell was measured by suddenly changing the osmotic pressure (salt concentration) outside the cell and following the change in cell diameter by means of light-scattering techniques. Analysis of the results gave a pore size of 7 Å diameter—a measure of a physical property of a membrane while it is still living and functional. Limitations in the analysis are the assumptions involved, which may or may not turn out to be correct. Other techniques recently used to measure pore sizes involve differentials in osmotic pressure to measure the leakiness of the membrane to certain ions and the molecular or ion-sieve approach in which large numbers of molecules of various sizes are tested for their penetration; the diameter of the largest molecule that can penetrate is the effective diameter of the pore [144].

Electrophysiological measurements (see Chapter 4) supply evidence for a heterogeneous organization of excitable membrane. The reactive elements responsible for activation, inactivation, and pumping mechanisms act independently. The fundamentally different properties of electrically excitable, electrically inexcitable, and electrogenically inert membrane, often described within the same cell, are indicative of a membrane built as a mosaic of different molecular structures. The action of small molecules on these properties strongly suggests pores: Procaine blocks sodium and potassium activation in squid axon; tetrodotoxin, molecular weight of 319, is a potent blocking agent of sodium activation; the potassium channels, which are smaller than the sodium channels, are clogged by alkali metals, alkaline earth ions, and small organic ions like tetraethylammonium (TEA). Grundfest [62] concludes from such data that there are 2–5 Å openings but that they are rare—channels occupy less than 10^{-4} of the membrane area. Moore, Narahashi, and Shaw [96] estimated only thirteen sodium channels in each $1\mu^2$ of lobster axon membrane.

In Lucy's [80] micellar model for membrane, the polar groups of phospholipids and associated counterions line the pores. His estimate of the effective diameter of a model pore is about 8 Å, which compares well with the experimentally determined values of 7 Å for red cell membrane [56], 8 Å for intestinal mucosal cells [78], and 8.5 Å for resting axolemma of squid [163]. Lucy's micellar model is much more permeable to ions and water than most biological membranes. Less than 1 % of erythrocyte membrane needs to have such a micellar configuration to account for its measured permeability and myelin membrane may have few or no micelles. Or, instead of micelles being rare, an icelike crystalline structure of water can effectively plug pores to explain relatively impermeable membranes [66]. But, Hanai and Haydon [63] argue that water traverses phospholipid films by dissolving into and diffusing through the hydrocarbon phase so that aqueous pores do not need to be postulated at all.

Biological membranes are able to discriminate carefully among cations and anions, and membranes are much more permeable to anions than cations. Speculation concerning the mechanisms for these effects abounds (see Chapter 3). The selectivity is usually attributed to charges along the walls of independent channels that are effected by specific orientations and spacings of polar groups of lipid molecules lining the pores or by a micellar organization of the lipids [81]. Large protein molecules possessing functional groups and enzymatic sites can fit into a flexible lattice of lipid micelles to assist passive and active permeation of the membrane. Many remarkable properties can be explained by long and narrow water-filled channels lined by polar lipid molecules [86].

A continuous monomolecular film of large fatty acids and lipids observed in an electron microscope shows a two-dimensional crystalline structure with many dislocations, which can be interpreted as the precursors of membrane pores [39, 142]. Electron micrographs of red cell ghosts show a network of fine crevices and pores of 30–300 Å size that are filled with microcrystalline structures suggestive of ordered water [40]. From permeability studies, Stein and Danielli [145] proposed pores to explain their observations that hydrogen-bonding molecules, like water, methanol, and formamide, pass through membranes at rates higher than one would calculate. The hypothesis provides hydrogen-bonding spaces through which these molecules can pass to cross the membrane. In restricted portions of membranes, these molecules do not encounter a barrier or restriction to permeation in the transition from aqueous to nonaqueous phases. Small areas of the membrane show non-specific properties with small molecules. With molecules larger than glycerol, the areas are highly specific—as specific as enzymes toward their substrates. Hence, Danielli pictures areas that behave as hydrogen-bonding apertures to extend from one side of the membrane to the other and to be parts of macromolecules or else to depend on some other more complicated, e.g., carrier, mechanism. Kavanau [70] does not believe that the pores are protein lined but that they are lined and crisscrossed by polar groups of the lipids that protrude and interact in the rim of the pore (Fig. 15). Kavanau's pores are located in regions where triads of hexagonally distributed bimolecular discs are formed by localized compression and dilation. The pore, in any case, is not pictured as a blank hole through which fluid can flow according to Poiseuille's law.

Schmitt and Davison [133] have presented a pore model that is capable of a rapid transient existence to explain the dynamic properties of excitable membrane (Fig. 16). Their view of the mechanisms for operating, for example, an axon membrane involves a protein monolayer on the axoplasmic side that is capable of responding to electrical and ionic changes with fast conformational changes, either at the tertiary or quarternary level, of the protein molecules. The linkage between the lipid bilayer phase and the protein monolayer phase is probably accomplished through cationic counter-ions, chiefly calcium, since both phases seem to be negatively charged. As the proteins are induced to change shape by a strong electrical field or by ions driven electrophoretically through the membrane, the lipids in the underlying bilayer phase will change their distribution within the planes of the bilayer, itself a kind of phase transition. This rearrangement of lipid molecules will occur if the change in protein configuration reduces their molecular cross-sectional area and thus the distance between fixed negative charges on the protein that are presented to the undersurface of the lipid bilayer. Each contracting protein molecule, coupled to the negatively charged lipid polar groups by shared counter-ions, compacts the underlying lipid into a condensed-film phase to leave interstitial regions—pores—of less densely packed lipids in a liquid-expanded film phase. These pores can then mediate the crossing of the ions (e.g., Na^+ and K^+) that act as charge-carrier effectors in the regenerative, propagating, action potential mechanism. Reversal of the translocations and transformations follow the removal of the local electric field and the change in ion distribution.

The importance of calcium in membrane models is illustrated by the abundant calcium-dependent phenomena at the cellular level and at the molecular level by phenomena such as the Palmer-Schmitt effect. Palmer and Schmitt [103] described the decrease in the amount of water between leaflets

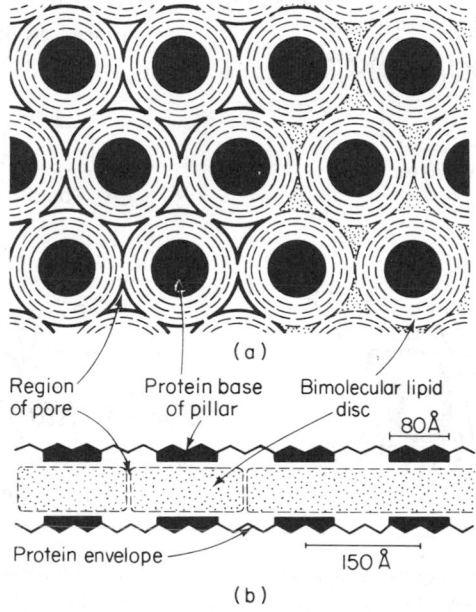

Figure 15. *Kavanau's representation of pores in biological membranes:* (a) is a surface view of a portion of membrane; (b) is a cross section through the four membrane elements just above in (a). The dashed lines show roughly the regions of the lipid phosphoryl groups; the zigzag lines, the positions of the protein envelopes; and solid areas are regions (pillar bases) where the cylindrical lipid micelles are strongly bound to the protein envelopes. The stippled areas in (b) show regions that contain weakly polar and nonionogenic lipids and the lipophilic chains of strongly polar lipids; in (a) the stippled areas show where adjacent bimolecular lipid discs coalesce into a continuous lamella to eliminate pore regions, as diagramed in the right side of both (a) and (b). Pores are formed by the close abutment of the bimolecular lipid discs [left side of (a) and (b)] that fail to obliterate the interstitial spaces. In his monograph, Kavanau develops in greater detail the mechanisms responsible for the transformations into and out of the porous condition to explain membrane functioning.

formed by lipid extracts, especially cephalin, when the ionic strength of the solution was raised with NaCl or KCl. $CaCl_2$ led to the "running out" of almost all the water from between the lamellae, leaving a single mixed lipid system. According to Kavanau [70], this results from the higher valence and charge density of calcium ions compared to potassium and sodium ions.

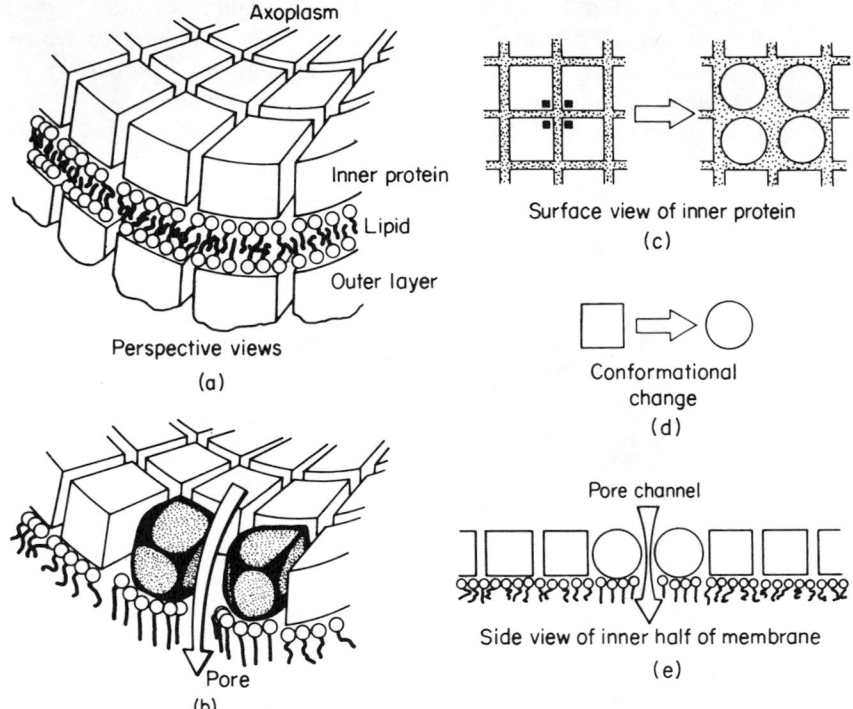

Figure 16. *Diagram of the Schmitt-Davison model for pores in axon membrane:* (a) is a perspective view of axon surface membrane composed of (1) an inner mono-layer of globular protein molecules (shown as cubes) that form a continuous, probably mosaic, sheet; (2) a central, bimolecular, liquid-expanded layer of mixed lipids with polar groups (circles) arranged at both surfaces—the proteins may be bound to the lipid layer through shared cationic counter-ions (e.g., a layer of calcium); and (3) an outer layer of nonlipid molecules of an unspecified nature possibly derived from the intercellular matrix. (b)–(e) show the formation of a pore. In (c), a surface view as seen from the axoplasm, the four inner protein molecules change shape when sites (black squares) are activated. This reversible, contractile conformational change of protein [shown simply in (d)] produces pores [as in (b) and (e)] when the contraction of protein molecules is coupled to a simultaneous local compaction of the lipids in the underlying bilayer, from a liquid-expanded to a liquid-condensed (or solid) state. Schmitt and Davison originally developed this model for the electrogenic protein of axons, which require a rapid (10^{-4} sec) conformational change in response to change in electrical field or other bioelectric parameters, but the model can also be applied to pore formation in other biological membranes.

Calcium can approach the fixed anions more closely to highly polarize and screen the charges on ionized oxygen atoms of the phospholipid head groups and link the phosphoryl and also the carboxyl groups closer together. Calcium ions can reduce the ionization of many cationic head groups as well. Screening of charged groups leads to a closer packing of the charged head groups of molecules, a less abrupt transition of stray electrical fields (i.e., polarity), a reduction in hydrophilic-solvent affinity of the phospholipid head groups and their degree of hydration, and a lowering of the work of adhesion of interfaces for water. When this lowering falls below the high internal cohesion of water, the water will run out from between the smectic leaflets to become a separate bulk phase.

In the final analysis, there is no conclusive evidence to demonstrate the existence of discrete pores in living membranes. The concept of pores, however, is useful since many data can be easily (usually) explained by pores, since one can conjure up an image with which to work, and since something equivalent to pores probably exists although more likely as a continuously changing and meandering brook than as a straight through and walled canal.

2.16 Conclusion

It may seem that the generalized concept of the structure of a biological membrane is, at the same time, clear and unclear. The morphologists can describe a complicated structure that the physiologists call naively simple; the physiologists can use concepts for which there are no morphological equivalents presently described. The molecular biologists are just beginning. The bimolecular leaflet has a maturity and a status that other models do not have. Robertson's unit membrane model is derived from Davson and Danielli's paucimolecular model but both are independently supported by different kinds of data. The bimolecular leaflet model can explain much of the electrical phenomena, the low surface tension, the polarized light data, the X-ray patterns, some electron micrographs, and some of the chemical data. But there are too many minor—and a few major—annoyances, and alternatives have been suggested. The globular structures seen in electron micrographs of negatively stained or freeze-etched preparations and seen in lipid films lend support to a micellar concept for membrane structure. The more cautious workers support a mosaic organization containing both types of structures or dynamic phase transitions. A micellar arrangement provides transmembrane pores more easily than does a lipid leaflet. Either concept offers a basic framework for future studies. But, it appears important, at least for the present, to build in our minds a flexible open-ended concept with which to work. Many problems can be avoided by omitting the word "the" that is often used in front of the word "membrane."

REFERENCES CITED

1. L. A. E. ASHWORTH and C. GREEN, 1965. Plasma membranes: phospholipid and sterol content. *Science* **151**: 210–211.

2. A. D. BANGHAM, 1968. Membrane models with phospholipids. *Prog. Biophys.* **18**: 29–95.

3. A. D. BANGHAM and R. W. HORNE, 1962. Action of saponin on biological cell membranes. *Nature* **196**: 952–953.

4. R. S. BAR, D. W. DEAMER, and D. G. CORNWELL, 1966. Surface area of human erythrocyte lipids: reinvestigation of experiments on plasma membrane. *Science* **153**: 1010–1012.

5. R. S. BEAR and F. O. SCHMITT, 1937. Optical properties of the axon sheaths of crustacean nerves. *J. Cell. Comp. Physiol.* **9**: 275–287.

6. E. L. BENEDETTI and P. EMMELOT, 1965. Electron microscopic observations on negatively stained plasma membranes isolated from rat liver. *J. Cell Biol.* **26**: 299–305.

7. E. L. BENEDETTI and P. EMMELOT, 1967. Studies on plasma membranes. IV. The ultrastructural localization and content of sialic acid in plasma membranes isolated from rat liver and hepatoma. *J. Cell Sci.* **2**: 499–512.

8. E. L. BENEDETTI and P. EMMELOT, 1968. "Structure and function of plasma membranes isolated from liver," pp. 33–120, in: A. J. Dalton and F. Haguenau (eds.), *The Membranes.* New York: Academic Press.

9. A. A. BENSON, 1966. On the orientation of lipids in chloroplast and cell membranes. *J. Am. Oil Chem. Soc.* **43**: 265–270.

10. J. K. BLASIE, M. M. DEWEY, A. E. BLAUROCK, and C. R. WORTHINGTON, 1965. Electron microscope and low angle X-ray diffraction studies on outer segment membranes from the retina of the frog. *J. Mol. Biol.* **14**: 143–152.

11. H. B. BOSMANN, A. HAGOPIAN, and E. H. EYLAR, 1968. Cellular membranes: the isolation and characterization of the plasma and smooth membranes of HeLa cells. *Arch. Biochem. Biophys.* **128**: 51–69.

12. D. BRANTON, 1966. Fracture faces of frozen membranes. *Proc. Natl. Acad. Sci. U.S.* **55**: 1048–1056.

13. D. BRANTON, 1966. Fracture faces of frozen myelin. *Exptl. Cell Res.* **45**: 703–707.

14. D. BRANTON and R. B. PARK, 1967. Subunits in chloroplast lamellae. *J. Ultrastruct. Res.* **19**: 283–303.

15. D. CHAPMAN, 1968. *Biological Membranes—Physical Fact and Function.* London: Academic Press, 438 pp.

16. D. CHAPMAN, P. BYRNE, and G. G. SHIPLEY, 1966. The physical properties of phospholipids. I. Solid state and mesomorphic properties of some 2,3-diacyl-DL-phosphatidylethanolamines. *Proc. Roy. Soc. London A* **290**: 115–142.

17. D. CHAPMAN and D. F. H. WALLACH, 1968. "Recent physical studies of phospholipids and natural membranes," pp. 125–202, in: D. Chapman (ed.), *Biological Membranes—Physical Fact and Function.* London: Academic Press.

18. D. CHAPMAN, R. M. WILLIAMS, and B. D. LADBROOKE, 1967. Physical studies of phospholipids. VI. Thermotropic and lyotropic mesomorphism of some 1,2-diacyl-phosphatidylcholines (lecithins). *Chem. Phys. Lipids* **1**: 445–475.

19. P. CHINN and F. O. SCHMITT, 1937. On the birefringence of nerve sheaths as studied in cross sections. *J. Cell. Comp. Physiol.* **9**: 289–296.

20. K. S. COLE, 1932. Surface forces in the *Arbacia* egg. *J. Cell. Comp. Physiol.* **1**: 1–9.

21. K. S. COLE and R. H. COLE, 1936. Electrical impedence of *Asterias* eggs. *J. Gen. Physiol.* **19**: 609–623.

22. K. S. COLE and R. H. COLE, 1936. Electrical impedence of *Arbacia* eggs. *J. Gen. Physiol.* **19**: 625–632.

23. G. M. W. COOK, D. H. HEARD, and G. V. F. SEAMAN, 1961. Sialic acids and the electrokinetic charge of the human erythrocyte. *Nature* **191**: 44–47.

24. J. F. DANIELLI and H. DAVSON, 1935. A contribution to the theory of permeability of thin films. *J. Cell. Comp. Physiol.* **5**: 495–508.

25. J. F. DANIELLI and E. N. HARVEY, 1935. The tension at the surface of mackeral egg oil with remarks on the nature of the cell surface. *J. Cell. Comp. Physiol.* **5**: 483–494.

26. J. T. DAVIES, 1965. A theory of the quality of odours. *J. Theoret. Biol.* **8**: 1–7.

27. H. DAVSON, 1964. *A Textbook of General Physiology*, 3rd ed. Boston: Little, Brown and Co., 1166 pp.

28. H. DAVSON and J. F. DANIELLI, 1952. *The Permeability of Natural Membranes*, 2nd ed. New York: The Macmillan Co., 365 pp.

29. D. W. DEAMER and D. BRANTON, 1967. Fracture planes in an ice-bilayer model membrane system. *Science* **158**: 655–657.

30. R. B. DEAN, H. J. CURTIS, and K. S. COLE, 1940. Impedence of bimolecular films. *Science* **91**: 50–51.

31. B. J. DOD and G. M. GRAY, 1968. The lipid composition of rat-liver plasma membranes. *Biochim. Biophys. Acta* **150**: 397–404.

32. R. R. DOURMASHKIN, R. M. DOUGHERTY, and R. J. C. HARRIS, 1962. Electron microscopic observations on rous sarcoma virus and cell membranes. *Nature* **194**: 1116–1119.

33. P. F. ELBERS, 1964. The cell membrane: image and interpretation. *Rec. Prog. Surf. Sci.* **2**: 443–503.

34. D. M. ENGLEMAN, 1970. X-ray diffraction studies of phase transitions in the membrane of *Mycoplasma laidlawii*. *J. Mol. Biol.* **47**: 115–117.

35. D. M. ENGLEMAN, T. M. TERRY, and H. J. MOROWITZ, 1967. Characterization of the plasma membrane of *Mycoplasma laidlawii*. I. Sodium dodecyl sulfate solubilization. *Biochim. Biophys. Acta* **135**: 381–390.

36. E. H. EYLAR, M. A. MADOFF, O. V. BRODY, and J. L. ONCLEY, 1962. The contribution of sialic acid to the surface charge of the erythrocyte. *J. Biol. Chem.* **237**: 1992–2000.

37. J. W. FARQUHAR and E. H. AHRENS, 1963. Effects of dietary fats on human erythrocyte fatty acid patterns. *J. Clin. Invest.* **42**: 675–685.

38. L. J. FENSTER and J. H. COPENHAVER, 1967. Phosphatidylserine requirement of (Na^+-K^+)-activated adenosine triphosphatase from rat kidney and brain. *Biochim. Biophys. Acta* **137**: 406–408.

39. H. FERNÁNDEZ-MORÁN, 1962. Cell membrane ultrastructure: low temperature electron microscopy and X-ray diffraction studies of lipoprotein complexes in lamellar systems. *Circulation* **26**: 1039–1065.

40. H. FERNÁNDEZ-MORÁN, 1967. "Membrane ultrastructure in nerve cells," pp. 281–304, in: G. C. Quarton, T. Melnechuk, and F. O. Schmitt (eds.), *The Neurosciences—A Study Program*. New York: Rockefeller Press.

41. H. FERNÁNDEZ-MORÁN and J. B. FINEAN, 1957. Electron microscope and low angle diffraction studies on the nerve myelin sheath. *J. Biophys. Biochem. Cytol.* **3**: 725–748.

42. H. FERNÁNDEZ-MORÁN, T. ODA, P. V. BLAIR, and D. E. GREEN, 1964. A macromolec-

ular repeating unit of mitochondrial structure and function. Correlated electron microscope and biochemical studies of isolated mitochondria and submitochondrial particles in beef heart muscle. *J. Cell Biol.* **22**: 63–100.

43. J. B. FINEAN, 1957. The role of water in the structure of peripheral nerve myelin. *J. Biophys. Biochem. Cytol.* **3**: 95–102.

44. J. B. FINEAN, 1961. The nature and stability of nerve myelin. *Internat. Rev. Cytol.* **12**: 303–336.

45. J. B. FINEAN, 1966. The molecular organization of cell membranes. *Prog. Biophys.* **16**: 143–170.

46. J. B. FINEAN, R. COLEMAN, and W. A. GREEN, 1966. Studies of isolated plasma membrane preparations. *Ann. N.Y. Acad. Sci.* **137**: 414–420.

47. J. B. FINEAN and P. F. MILLINGTON, 1957. Effects of ionic strength of immersion medium on the structure of peripheral nerve myelin. *J. Biophys. Biochem. Cytol.* **3**: 89–94.

48. A. P. FISHMAN (ed.), 1962. *Symposium on the Plasma Membrane.* New York: American Heart Association, pp. 983–1232.

49. D. F. FITZPATRICK, G. R. DAVENPORT, L. FORTE, and E. J. LANDON, 1969. Characterization of plasma membrane proteins. I. Preparation of a membrane fraction and separation of the protein. *J. Biol. Chem.* **244**: 3561–3569.

50. S. FLEISCHER, B. FLEISCHER, and W. STOECKENIUS, 1967. Fine structure of lipid-depleted mitochondria. *J. Cell Biol.* **32**: 193–208.

51. S. FLEISCHER and G. ROUSER, 1965. Lipids of subcellular particles. *J. Am. Oil Chem. Soc.* **42**: 588–607.

52. A. FREY-WYSSLING, 1953. *Submicroscopic Morphology of Protoplasm and Its Derivatives.* Amsterdam: Elsevier Pub. Co., pp. 360–364.

53. B. B. GEREN, 1954. Formation from the Schwann cell membrane of myelin in the peripheral nerve of chick embryos. *Exptl. Cell Res.* **7**: 558–562.

54. A. M. GLAUERT, J. T. DINGLE, and J. A. LUCY, 1962. Action of saponin on biological cell membranes. *Nature* **196**: 952–955.

55. A. M. GLAUERT and J. A. LUCY, 1968. "Globular micelles and the organization of membrane lipids," pp. 1–32, in: A. J. Dalton and F. H. Haguenau (eds.), *The Membranes.* New York: Academic Press.

56. D. A. GOLDSTEIN and A. K. SOLOMON, 1960. Determination of equivalent pore radius for human red cells by osmotic pressure measurement. *J. Gen. Physiol.* **44**: 1–17.

57. E. GORTER and F. GRENDEL, 1925. On bimolecular layers of lipids on the chromocytes of the blood. *J. Exptl. Med.* **41**: 439–443.

58. A. GOTTSCHALK, 1960. Correlation between composition, structure, shape, and function of a salivary mucoprotein. *Nature* **186**: 949–951.

59. D. E. GREEN, D. W. ALLMAN, E. BACHMANN, H. BAUM, K. KOPACZYK, E. F. KORMAN, S. LIPTON, D. H. MACLENNAN, D. G. MCCONNELL, J. F. PERDUE, J. S. RIESKE, and A. TZAGOLOFF, 1967. Formation of membranes by repeating units. *Arch. Biochem. Biophys.* **119**: 312–335.

60. D. E. GREEN and S. FLEISCHER, 1963. The role of lipids in mitochondrial electron transfer and oxidative phosphorylation. *Biochim. Biophys. Acta* **70**: 554–582.

61. D. E. GREEN and J. F. PERDUE, 1966. Membranes as expressions of repeating units. *Proc. Natl. Acad. Sci. U.S.* **55**: 1295–1302.

62. H. GRUNDFEST, 1966. Heterogeneity of excitable membrane: electrophysiological and pharmacological evidence and some consequences. *Ann. N.Y. Acad. Sci.* **137**: 901–949.

63. T. Hanai and D. A. Haydon, 1966. The permeability of bimolecular lipid membranes. *J. Theoret. Biol.* **11**: 370–382.

64. E. N. Harvey and J. F. Danielli, 1938. Properties of the cell surface. *Biol. Rev.* **13**: 319–341.

65. E. N. Harvey and H. Shapiro, 1934. The interfacial tension between oil and protoplasm within living cells. *J. Cell. Comp. Physiol.* **5**: 255–268.

66. O. Hechter, 1965. Role of water in the molecular organization of cell membranes. *Fed. Proc.* **24**: S91–S102.

67. R. W. Hendler, 1971. Biological membrane ultrastructure. *Physiol. Rev.* **51**: 61–97.

68. C. Huang, L. Wheeldon, and T. E. Tompson, 1964. The properties of lipid bilayer membranes separating two aqueous phases: formation of a membrane of simple composition. *J. Mol. Biol.* **8**: 148–160.

69. Y. Kagawa and E. Racker, 1966. Partial resolution of the enzyme catalyzing oxidative phosphorylation. X. Correlation of morphology and function in submitochondrial particles. *J. Biol. Chem.* **241**: 2475–2482.

70. J. L. Kavanau, 1965. *Structure and Function in Biological Membranes.* San Francisco: Holden-Day, Inc., Pub., Vol. 1, 322 pp.

71. E. P. Kennedy, 1967. "Some developments in the biochemistry of membranes," pp. 271–280, in: G. C. Quarton, T. Melnechuk, and F. O. Schmitt (eds.), *The Neurosciences— A Study Program.* New York: Rockefeller Press.

72. E. D. Korn, 1966. Structure of biological membranes. *Science* **153**: 1491–1498.

73. E. D. Korn, 1967. A chromatographic and spectroscopic study of the products of the reaction of osmium tetroxide with unsaturated lipids. *J. Cell Biol.* **34**: 627–638.

74. J. Lenard and S. J. Singer, 1966. Protein conformation in cell membrane preparations as studied by optical rotatory dispersion and circular dichroism. *Proc. Natl. Acad. Sci. U.S.* **56**: 1828–1935.

75. J. Lenard and S. J. Singer, 1968. Alteration of the conformation of proteins in red blood cell membranes and in solution by fixatives used in electron microscopy. *J. Cell Biol.* **37**: 117–121.

76. J. Lenard and S. J. Singer, 1968. Structure of membranes: reaction of red blood cell membranes with phospholipase C. *Science* **159**: 738–739.

77. G. Lenaz, N. F. Haard, H. I. Silman, and D. E. Green, 1968. Studies on mitochondrial structural protein. III. Physical characterization of the structural proteins of beef heart and beef liver mitochondria. *Arch. Biochem. Biophys.* **128**: 293–303.

78. B. Lindemann and A. K. Solomon, 1962. Permeability of luminal surface of intestinal mucosal cells. *J. Gen. Physiol.* **45**: 801–810.

79. W. R. Loewenstein, 1966. Introductory remarks: conference on biological membranes, recent progress. *Ann. N.Y. Acad. Sci.* **137**: 403–1048.

80. J. A. Lucy, 1964. Globular lipid micelles and cell membranes. *J. Theoret. Biol.* **7**: 360–373.

81. J. A. Lucy, 1968. "Theoretical and experimental models for biological membranes," pp. 233–288, in: D. Chapman (ed.), *Biological Membranes—Physical Fact and Function.* London: Academic Press.

82. J. A. Lucy and A. M. Glauert, 1964. Structure and assembly of macromolecular lipid complexes composed of globular micelles. *J. Mol. Biol.* **8**: 727–748.

83. V. Luzzati, 1968. "X-ray diffraction studies of lipid-water systems," pp. 71–123, in: D. Chapman (ed.), *Biological Membranes—Physical Fact and Function.* London: Academic Press.

84. V. Luzzati, T. Gulik-Krzywicki, E. Rivas, F. Reiss-Husson, and R. P. Rand, 1968. X-ray study of model systems: structure of the lipid-water phases in correlation with the chemical composition of the lipids. *J. Gen. Physiol.* **51**: 37s–43s.

85. V. Luzzati, T. Gulik-Krzywicki, A. Tardieu, E. Rivas, and F. Reiss-Husson, 1969. "Lipids and membranes," pp. 79–93, in: D. C. Tosteson (ed.), *The Molecular Basis of Membrane Function.* Englewood Cliffs, N.J.: Prentice-Hall, Inc.

86. V. Luzzati and F. Husson, 1962. The structure of the liquid crystalline phases of lipid-water systems. *J. Cell Biol.* **12**: 207–220.

87. V. Luzzati, F. Reiss-Husson, E. Rivas, and T. Gulik-Krzywicki, 1966. Structure and polymorphism in lipid-water systems and their possible biological implications. *Ann. N.Y. Acad. Sci.* **137**: 409–413.

88. A. H. Maddy, 1969. "Some problems relating to the chemical composition of membranes," pp. 95–108, in: D. C. Tosteson (ed.), *The Molecular Basis of Membrane Functions.* Englewood Cliffs, N. J.: Prentice-Hall, Inc.

89. A. H. Maddy, C. Huang, and T. E. Thompson, 1966. Studies on lipid bilayer membranes: a model for the plasma membrane. *Fed. Proc.* **25**: 933–936.

90. A. H. Maddy and B. R. Malcolm, 1965. Protein conformations in the plasma membrane. *Science* **150**: 1616–1618.

91. G. V. Marinetti and D. Pettit, 1968. The interaction of γ-globulin with lipids. *Chem. Phys. Lipids* **2**: 17–34.

92. D. G. McConnell, A. Tzagoloff, D. H. MacLennan, and D. E. Green, 1966. Studies on the electron transfer system. LXV. Formation of membranes by purified cytochrome oxidase. *J. Biol. Chem.* **241**: 2373–2382.

93. P. Mitchell, 1967. Translocations through natural membranes. *Adv. Enzymol.* **29**: 33–87.

94. M. F. Moody, 1963. X-ray diffraction pattern of nerve myelin: a method for determining the phases. *Science* **142**: 1171–1174.

95. H. Moor and K. Muhlethaler, 1963. Fine structure in frozen-etched yeast cells. *J. Cell Biol.* **17**: 609–628.

96. J. W. Moore, T. Narahashi, and T. I. Shaw, 1967. An upper limit to the number of sodium channels in nerve membrane? *J. Physiol.* **188**: 99–105.

97. W. T. J. Morgam, 1960. A contribution to human biochemical genetics; the chemical basis of blood-group specificity. *Proc. Roy. Soc. London B* **151**: 308–347.

98. A. A. Moscona, 1961. Effect of temperature on adhesion to glass and histogenetic cohesion of dissociated cells. *Nature* **190**: 408–409.

99. E. Mulder, J. de Gier, and L. L. M. van Deenen, 1963. Selective incorporation of fatty acids into phospholipids of mature red cells. *Biochim. Biophys. Acta* **70**: 94–96.

100. D. M. Neville, Jr., 1960. The isolation of a cell membrane fraction from rat liver. *J. Biophys. Biochem. Cytol.* **8**: 413–422.

101. J. S. O'Brien, E. L. Sampson, and M. B. Stern, 1967. Lipid composition of myelin from the peripheral nervous system. *J. Neurochem.* **14**: 357–366.

102. C. H. O'Neill, 1964. Isolation and properties of the cell surface membrane of *Amoeba proteus. Exptl. Cell Res.* **35**: 477–496.

103. K. J. Palmer and F. O. Schmitt, 1941. X-ray diffraction studies of lipide emulsions. *J. Cell. Comp. Physiol.* **17**: 385–394.

104. R. B. Park, 1965. Substructure of chloroplast lamellae. *J. Cell Biol.* **27**: 151–161.

105. B. A. Pethica, 1967. Structure and physical chemistry of membranes. *Protoplasma* **63**: 147–156.

106. E. PONDER, 1961. "The cell membrane and its properties," pp. 1–84, in: J. Brachet and A. E. Mirsky (eds.), *The Cell*, Vol. 2. New York: Academic Press.

107. R. P. RAND and A. C. BURTON, 1964. Mechanical properties of the red cell membrane. I. Membrane stiffness and intracellular pressure. *Biophys. J.* **4**: 115–135.

108. R. P. RAND and V. LUZZATI, 1968. X-ray diffraction study in water of lipids extracted from human erythrocytes. *Biophys. J.* **8**: 125–137.

109. T. K. RAY, V. P. SKIPSKI, M. BARCLAY, E. ESSNER, and F. M. ARCHIBALD, 1969. Lipid composition of rat liver plasma membranes. *J. Biol. Chem.* **244**: 5528–5536.

110. S. RAZIN, H. J. MOROWITZ, and T. M. TERRY, 1965. Membrane subunits of *Mycoplasma laidlawii* and their assembly into membranelike structures. *Proc. Natl. Acad. Sci. U.S.* **54**: 219–225.

111. J. C. REINERT and J. M. STEIM, 1970. Calorimetric detection of a membrane-lipid phase transition in living cells. *Science* **168**: 1580–1582.

112. J.-P. REVEL and S. ITO, 1967. "The surface components of cells," pp. 211–234, in: B. D. Davis and L. Warren (eds.), *The Specificity of Cell Surfaces*. Englewood Cliffs, N.J.: Prentice-Hall, Inc.

113. T. RICHARDSON, A. L. TAPPEL, L. M. SMITH, and C. R. HOULE, 1962. Polyunsaturated fatty acids in mitochondria. *J. Lipid Res.* **3**: 344–350.

114. J. C. RIEMERSMA, 1968. Osmium tetroxide fixation of lipids for electron microscopy, a possible reaction mechanism. *Biochim. Biophys. Acta* **152**: 718–727.

115. J. D. ROBERTSON, 1959. The ultrastructure of cell membranes and their derivatives. *Biochem. Soc. Symp.* **16**: 3–43.

116. J. D. ROBERTSON, 1960. The molecular structure and contact relationships of cell membranes. *Prog. Biophys.* **10**: 343–418.

117. J. D. ROBERTSON, 1963. The occurrence of a subunit pattern in the unit membrane of club endings in Mauthner cell synapses in goldfish brain. *J. Cell Biol.* **19**: 201–221.

118. J. D. ROBERTSON, 1964. "Unit membranes: a review with recent new studies of experimental alterations and a new subunit structure in synaptic membranes," pp. 1–81, in: M. Locke (ed.), *Cellular Membranes in Development*. New York: Academic Press.

119. J. D. ROBERTSON, 1966. Granulo-fibrillar and globular substructure in unit membranes. *Ann. N.Y. Acad. Sci.* **137**: 421–440.

120. J. D. ROBERTSON, 1966. "Current problems of unit membrane structure and contact relationships," pp. 11–48, in: K. Rodahl and B. Issekulz, Jr. (eds.), *Nerve as a Tissue*. New York: Harper & Row, Pub.

121. J. D. ROBERTSON, 1967. Origin of the unit membrane hypothesis. *Protoplasma* **63**: 218–245.

122. B. ROELOFSEN, J. DE GIER, and L. L. M. VAN DEENEN, 1964. Binding of lipids in the red cell membrane. *J. Cell. Comp. Physiol.* **63**: 233–243.

123. S. A. ROSENBERG and G. GUIDOTTI, 1969. Fractionation of the protein components of human erythrocyte membranes. *J. Biol. Chem.* **244**: 5118–5124.

124. G. ROUSER, G. J. NELSON, S. FLEISCHER, and G. SIMON, 1968. "Lipid composition of animal cell membranes, organelles, and organs," pp. 5–69, in: D. Chapman (ed.), *Biological Membranes—Physical Fact and Function*. London: Academic Press.

125. W. J. SCHMIDT, 1936. Der Einfluss von Kaliumpermanganat auf die Doppelbrechung der Markscheide der Nervenfasern und der Aussenglider der Schzellen. *Z. Zellforsch. u. Mikr. Anat.* **23**: 261–267.

126. W. J. SCHMIDT, 1936. Doppelbrechung und Feinbau der Markscheide der Nervenfasern. *Z. Zellforsch. u. Mikr. Anat.* **23**: 657–676.

127. W. J. SCHMIDT, 1938. Polarizationsoptische Analyse eines Eiweiss-Lipoid Systems erlautert am Aussenglied der Schzellen. *Kolloidzscht.* **84**: 137–148.

128. F. O. SCHMITT and R. S. BEAR, 1937. The optical properties of vertebrate nerve axons as related to size. *J. Cell. Comp. Physiol.* **9**: 261–273.

129. F. O. SCHMITT and R. S. BEAR, 1939. The ultrastructure of the nerve axon sheath. *Biol. Rev.* **14**: 27–51.

130. F. O. SCHMITT, R. S. BEAR, and G. L. CLARK, 1935. X-ray diffraction studies on nerve. *Radiology* **25**: 131–151.

131. F. O. SCHMITT, R. S. BEAR, and K. J. PALMER, 1941. X-ray diffraction studies on the structure of the nerve myelin sheath. *J. Cell. Comp. Physiol.* **18**: 31–41.

132. F. O. SCHMITT, R. S. BEAR, and E. PONDER, 1936. Optical properties of the red cell membrane. *J. Cell. Comp. Physiol.* **9**: 89–92.

133. F. O. SCHMITT and P. F. DAVISON, 1965. Role of protein in neural function: an essay. *Neurosci. Res. Prog. Bull.* **3**: 55–76.

134. S. G. SCHULTZ and A. K. SOLOMON, 1961. Determination of the effective hydrodynamic radii of small molecules by viscometry. *J. Gen. Physiol.* **44**: 1189–1199.

135. G. V. F. SEAMAN and D. H. HEARD, 1960. The surface of the washed human erythrocyte as a polyanion. *J. Gen. Physiol.* **44**: 251–268.

136. S. J. SINGER, 1962. The properties of proteins in nonaqueous solvent. *Adv. Protein Chem.* **17**: 1–68.

137. F. S. SJÖSTRAND, 1963. A new ultrastructural element of the membranes in mitochondria and of some cytoplasmic membranes. *J. Ultrastruct. Res.* **9**: 340–361.

138. F. S. SJÖSTRAND, 1963. A comparison of plasma membranes, cytomembranes, and mitochondrial membrane elements with respect to structural features. *J. Ultrastruct. Res.* **9**: 561–580.

139. F. S. SJÖSTRAND, 1967. The structure of cellular membranes. *Protoplasma* **63**: 248–261.

140. F. S. SJÖSTRAND, 1968. "Ultrastructure and function of cellular membranes," pp. 151–210, in: A. J. Dalton and F. Haguenau (eds.), *The Membranes.* New York: Academic Press.

141. H. W. SMITH 1962. The plasma membrane, with notes on the history of botany. *Circulation* **26**: 987–1012.

142. A. K. SOLOMON, 1960. Pores in the cell membrane. *Sci. Am.* **203**: 146–156.

143. J. M. STEIM, M. E. TOURTELLOTTE, J. C. REINERT, R. N. McELHANEY, and R. L. RADER, 1969. Calorimetric evidence for the lipid-crystalline state of lipids in a biomembrane. *Proc. Natl. Acad. Sci. U.S.* **63**: 104–109.

144. W. D. STEIN, 1967. *The Movement of Molecules Across Cell Membranes.* New York: Academic Press, 369 pp.

145. W. D. STEIN and J. F. DANIELLI, 1956. Structure and function in red cell permeability. *Disc. Faraday Soc.* **21**: 238–251.

146 W. STOECKENIUS, 1962. Structure of the plasma membrane. An electron microscope study. *Circulation* **26**: 1066–1069.

147. W. STOECKENIUS, 1962. "The molecular structure of lipid-water systems and cell membrane models studied with the electron microscope," pp. 349–368, in: R. C. J. Harris (ed.), *The Interpretation of Ultrastructure.* New York: Academic Press.

148. W. STOECKENIUS, 1962. Some electron microscopic observations in liquid-cyrstalline phases in lipid-water systems. *J. Cell Biol.* **12**: 221–229.

149. W. STOECKENIUS, 1967. Electron microscopy of fixed lipids. *Protoplasma* **63**: 214–217.

150. T. M. TERRY, D. M. ENGLEMAN, and H. J. MOROWITZ, 1967. Characterization of the plasma membrane of *Mycoplasma laidlawii*. II. Modes of aggregation of solubilized membrane components. *Biochim. Biophys. Acta* **135**: 391–405.

151. E. B. THOMPSON and M. W. KIES, 1965. Current studies on the lipids and proteins of myelin. *Ann. N.Y. Acad. Sci.* **122**: 129–147.

152. J. E. THOMPSON, R. COLEMAN, and B. FINEAN, 1967, Some biochemical and X-ray diffraction studies of mitochondrial outer membrane. *Biochim. Biophys. Acta* **135**: 1074–1078.

153. H. T. TIEN and A. L. DIANA, 1967. Some physical properties of bimolecular lipid membranes produced from new lipid solutions. *Nature* **215**: 1199–1200.

154. M. T. TOURTELLOTTE, R. N. McELHANEY, and R. L. RADER, 1968. Current concepts of membrane structure. *Bull. Inst. Cell Biol. U. Conn.* **9**: 1–14.

155. L. L. M. VAN DEENEN, 1965. Phospholipids and biomembranes. *Prog. Chem. Fats Lipids* **8**: 1–127.

156. L. L. M. VAN DEENEN, 1966. Some structural and dynamic aspects of lipids in membranes. *Ann. N.Y. Acad. Sci.* **137**: 717–730.

157. L. L. M. VAN DEENEN, 1969. "Membrane lipids and lipophilic proteins, pp. 47–78, in: D. C. Tosteson (ed.), *The Molecular Basis of Membrane Function*. Englewood Cliffs, N.J.: Prentice-Hall, Inc.

158. L. L. M. VAN DEENEN and J. DE GIER, 1964. "Chemical composition and metabolism of lipids in red cells of various animal species," pp. 243–307, in: C. Bishop and D. M. Surgenor (eds.), *The Red Blood Cell*. New York: Academic Press.

159. F. A. VENDENHEUVEL, 1963. Study of biological structure at the molecular level with stereomodel projections. I. The lipids in the myelin sheath of nerve. *J. Am. Oil Chem. Soc.* **40**: 455–471.

160. F. A. VANDENHEUVEL, 1965. Structural studies of biological membranes: the structure of myelin. *Ann. N.Y. Acad. Sci.* **122**: 57–76.

161. F. A. VANDENHEUVEL, 1966. Lipid-protein interactions and cohesional forces in the lipoprotein systems of membranes. *J. Am. Oil Chem. Soc.* **43**: 258–264.

162. L. M. G. VAN GOLDE, V. TOMASI, and L. L. M. VAN DEENEN, 1967. Determination of molecular species of lecithin from erythrocytes and plasma. *Chem. Phys. Lipids* **1**: 282–293.

163. R. VILLEGAS and F. V. BARNOLA, 1961. Characterization of the resting axolemma in the giant axon of the squid. *J. Gen. Physiol.* **44**: 963–977.

164. D. F. H. WALLACH, 1967. "Isolation of plasma membranes of animal cells," pp. 129–163, in: B. D. Davis and L. Warren (eds.), *The Specificity of Cell Surfaces*. Englewood Cliffs, N.J.: Prentice-Hall, Inc.

165. D. F. H. WALLACH, 1969. Membrane lipids and the conformations of membrane proteins. *J. Gen. Physiol. Suppl.* **54**: 3s–26s.

166. D. F. H. WALLACH and P. H. ZAHLER, 1966. Protein conformations in cellular membranes. *Proc. Natl. Acad. Sci. U.S.* **56**: 1552–1559.

167. L. WARREN and M. C. GLICK, 1968. Membranes of animal cells. II. THE metabolism and turnover of the surface membrane. *J. Cell Biol.* **37**: 729–746.

168. L. WARREN, M. C. GLICK, and M. K. NASS, 1967. "The isolation of animal cell membranes," pp. 109–127, in: B. D. Davis and L. Warren (eds.), *The Specificity of Cell Surfaces*. Englewood Cliffs, N. J.: Prentice-Hall, Inc.

169. J. C. WATKINS, 1965. Pharmacological receptors and general permeability phenomena of cell membranes. *J. Theoret. Biol.* **9**: 37–50.

170. D. F. Waugh and F. O. Schmitt, 1940. Investigations of the thickness and ultrastructure of cellular membranes by the analytical leptoscope. *Cold Spring Harbor Symp. Quant. Biol.* **8**: 223–241.

171. T. E. Weier and A. A. Benson, 1967. The molecular organization of chloroplast membranes. *Am. J. Bot.* **54**: 389–402.

172. T. E. Weier, T. Bisalputra, and A. Harrison, 1966. Subunits in chloroplast membranes of *Scenedesmus quadricauda. J. Ultrastruct. Res.* **15**: 38–56.

173. T. E. Weier, A. H. P. Engelbracht, A. Harrison, and E. B. Risley, 1965. Subunits in the membranes of chloroplasts of *Phaseolus vulgaris, Pisum sativum,* and *Aspidistra* sp. *J. Ultrastruct. Res.* **13**: 92–111.

174. D. B. Weinstein, J. B. Marsh, M. C. Glick, and L. Warren, 1969. Membranes of animal cells. IV. Lipids of the L cell and its surface membrane. *J. Biol. Chem.* **244**: 4103–4111.

175. L. Weiss, 1965. Studies on cell deformability. II. Effect of surface charge. *J. Cell Biol.* **26**: 735–739.

176. L. Weiss, 1966. Studies on cell deformability. I. Effects of some proteolytic enzymes. *J. Cell Biol.* **30**: 39–43.

177. R. Whittam, 1962. The asymmetrical stimulation of a membrane adenosine triphosphatase in relation to active cation transport. *Biochem. J.* **84**: 110–118.

178. T. Yamamoto, 1963. On the thickness of the unit membrane. *J. Cell Biol.* **17**: 413–422.

PERMEATION OF
BIOLOGICAL MEMBRANES

chapter three

The properties and behavior of the plasma membrane bounding a cell determine to a great extent what that cell can or cannot do. By regulating the rate of flow of materials through the membrane, a cell is able to exert some control over its intracellular composition and activities. The behavior of plasma membranes ranges from insignificant to complete barriers and from passive fences to metabolically linked and highly selective pumping devices. It has proved more difficult to understand how materials cross this structure than to describe these phenomena although the extensive literature on membrane permeability describes numerous mechanisms by which transport is accomplished [11, 22, 45, 52, 69, 108, 118, 125, 127]. Too frequently, however, theory extends beyond the available experimental data.

3.1 Diffusion

The simplest manner by which molecules can cross biological membranes is by diffusion, which is described in standard textbooks as the tendency for any substance to spread evenly throughout the space available to it. Kinetic theory provides the explanation for this behavior. A large number of small elastic particles—molecules—move about in all directions continuously and rapidly. Each molecule flies a straight path until it collides with another or with the walls of the containing vessel (the basis for pressure within a container) and bounces off to continue its flight but in a different direction. An increase in temperature results in a more vigorous movement of the molecules: In fact, the temperature of a gas, for example, is a measure of the average kinetic energy of the moleculus. Motion of the molecules decreases with decreasing temperature until, at absolute zero, molecular motion ceases. From this point, it is not difficult to derive the sundry gas laws. Molecules can move with nearly complete freedom in gases but in liquids and especially in solids the molecules occupy significant amounts of space and interact in more complicated ways than just elastic collisions. Molecules in a liquid are in a crystalline-like array so that relative motion between molecules, e.g., diffusion or viscous flow, can occur only when a molecule possesses enough thermal energy to free itself of the bonds anchoring it in position in the lattice and when it can find an adjacent empty hole in the lattice. Diffusion in a liquid is, therefore, a process in which a molecule moves from hole to hole in a crystal-like lattice. Polar molecules are anchored in the lattice of a polar liquid by effective hydrogen bonds and the associations between these polar molecules dominate the properties of such liquids. Diffusion, then, is exhibited by gases, liquids, and even solids but is most rapid in gases. As long as the molecules are permitted to move (and in random directions) and sufficient time is allowed, it is easy to see, from probability considerations alone, how diffusion can result in the even distribution of molecules throughout the available space. So, in an aqueous medium as in biological systems, diffusion results from the probability that solute molecules will migrate from higher toward lower concentrations until an equilibrium or a steady state is achieved; solvent molecules can be considered similarly.

3.2 Quantitative Approaches

It is difficult—probably impossible—to present a quantitative description of permeation in a small space and still avoid confusion. Yet it is worthwhile to at least define some terms; introduce some components, factors, and relationships; and provide some hints of the nature of the expressions used and the assumptions made. At least two different mathematical approaches have been developed to deal with permeation problems: The older and somewhat more familiar approach is a kinetic one; the newer and somewhat more useful approach is a thermodynamic one. The main purpose of this text is to introduce biological problems and not mathematical ones. Hence, a careful development of the quantitative analyses of permeation data and useful theories will be largely omitted beyond, sorrowfully, the short glimpse that follows and some derivations within the discussions of more specific problems.

Fick's Law

Over a century ago, Adolph Fick described the amount of substance that diffuses through a region of cross-sectional area (A) as being directly proportional to the concentration gradient (dc/dx) across that region:

$$\frac{ds}{dt} = -DA\frac{dc}{dx}.$$

This expression has since been known as Fick's law. ds/dt is the rate of transport or the amount of substance in moles crossing (ds) in unit time (dt); dc/dx is the concentration or, better, the activity gradient (dc) across the point considered (dx); D is the diffusion coefficient or the number of moles transported across unit area per unit time driven by a concentration gradient of unity. The dimensions of D are moles divided by the product of area, time, and concentration gradient, which simplifies to cm²/sec. Since the concentration may vary with time, the expression needs to be integrated over time but, mathematically, this is difficult [61].

The Diffusion Coefficient

The dependence of the rate of diffusion on the nature of the material traversed is represented, although unspecified, by the diffusion constant D. To evaluate D experimentally, the conditions are generally simplified. The term *flux* is restricted to the *rate* of molecular movement in only *one* direction and has often been defined as the number of moles of material diffusing through an area, e.g., of membrane, of 1 cm² in 1 sec and in a given direction. So, if no serious errors result from the assumptions: that the flux in the return direction is insignificant; that molecules bear no net charge to introduce an electromotive gradient; that the concentration gradient is reasonably linear across the membrane, does not change rapidly, and does not alter the nature of the membrane; and that concentrations can substitute for activities, then the expression for the net flux (J) across the membrane becomes

$$J = -D\frac{C_1 - C_2}{x} \text{ moles/cm}^2 \text{ sec.}$$

C_1 and C_2 are the solute concentrations in the bulk solutions on either side of the membrane of thickness x. The expression for the diffusion constant can then be written as

$$D = \frac{-Jx}{C_1 - C_2} \text{ cm}^2/\text{sec.}$$

The diffusion constant D gives an estimate of the rate at which a substance will diffuse through a solution under examination so it will depend on the properties of both solute and solvent and generally it increases with increasing temperature. The temperature coefficient of free diffusion in aqueous solutions is low; Q_{10} is generally less than 1.5. But diffusion across a thin lipid layer from one aqueous solution to another may be slow and show higher Q_{10} values. Hence, Q_{10} values of two or three cannot be used alone as evidence against diffusion as a rate limiting step in material permeating a membrane [11]. Table 2 lists some representative values for diffusion coefficients for solutions of biological interest. One should mention that investigators frequently substitute P, the permeability coefficient, for D, the diffusion coefficient, in the equations when considering substances that penetrate membranes.

TABLE 2

DIFFUSION CONSTANTS (D) FOR SOME SOLUTES
IN FREE AQUEOUS SOLUTION
AND IN THE EXTRACELLULAR FLUID OF VARIOUS VERTEBRATE TISSUES*

K^+	Na^+	*Sucrose*	*Inulin*	*Solvent*
9.9×10^{-4}	7.2×10^{-4}	3.1×10^{-4}	1.0×10^{-4}	Free solution
$4.1–10.9 \times 10^{-5}$	$1.1–3.2 \times 10^{-4}$	3.4×10^{-4}	1.2×10^{-4}	Rat muscle
7.0×10^{-4}	$1.5–3.4 \times 10^{-4}$	—	—	Frog muscle
$3.2–7.2 \times 10^{-5}$	$5.6–14.0 \times 10^{-4}$	2.1×10^{-4}	1.9×10^{-4}	Rat brain
—	9.0×10^{-4}	—	—	Cat nerve
$0.8–5.0 \times 10^{-5}$	$0.7–1.2 \times 10^{-4}$	4.3×10^{-5}	1.5×10^{-5}	Rat nerve
—	1.1×10^{-6}	—	—	Frog nerve
$3.4–4.0 \times 10^{-5}$	$8.3–12.7 \times 10^{-4}$	—	—	Rat liver

* D is given as cm²/min and the values were measured at 20°C. The rates of diffusion for the four solutes can be compared within the same solvent and in different solvents. Note that all the solvents are aqueous and that none contain obvious morphological features that impede diffusion.

This treatment is superficial for biological material since the only consideration mentioned is the random motion of solute that results in a net flux down a concentration gradient.

Gaseous Diffusion

For gases whose free motion is subject only to small intramolecular forces this treatment is satisfactory. The kinetic energy for a gas is given by $\frac{1}{2}NM\bar{c}^2$. N is the number of molecules, M is the molecular weight, $\bar{c}^2$ is the mean square velocity. Thus for any pair of gases at the same temperature, the average energy of the molecules will be the same or

$$\frac{1}{2}NM_1\bar{c}_1^2 = \frac{1}{2}NM_2\bar{c}_2^2.$$

The average velocities of molecules will be inversely proportional to their masses; their velocities will determine their speeds of diffusion. Thus $DM^{1/2}$ is a constant [122]; the diffusion constants of gases are inversely proportional to the square root of their masses.

Einstein's Equation

The molecules are close together in liquids and are within each other's sphere of mutual attraction; hence, any given molecule must break away from its surrounding molecules and push aside other molecules in order to move a significant distance. As a result, we may regard the diffusion coefficient of a solute in a liquid as being related to the viscosity or frictional resistance between molecules of solute and those of solvent. If a molecule is large in comparison with solvent molecules, such as a spherical colloidal particle, the diffusion coefficient D is given by the relationship developed by Einstein:

$$D = \frac{RT}{6\pi r \eta N}.$$

R is the gas constant, T is the absolute temperature, r is the radius of the diffusing molecule, N is Avogadro's number, and η is the viscosity. When the attractions are small, for example, in a homopolar solvent where the forces are mainly of weak van der Waals type, diffusion is rapid. On the other hand, e.g., glycerol diffusing through water, the forces of attraction between the OH groups of glycerol and of water result in the formation of hydrogen bonds between them and lower diffusion coefficients can be expected.

Diffusion Through a Membrane

Migration of molecules across a membrane is a more restrictive event than is simple diffusion in a bulk aqueous medium—a barrier must be crossed. For permeability studies, the Fick equation is generally simplified. The concentration gradient across the membrane is usually considered to remain constant and is simplified to $(C_{out} - C_{in})/x$. Since the membrane is thin, this does not introduce serious errors even if the gradient is not uniform. When V is the volume of the cell, s/V is the intracellular solute concentration. Since the membrane thickness cannot be easily measured, a permeability constant k is used and is equal to $-D/x$ cm/sec. Fick's equation now becomes

$$\frac{dC_{in}}{dt} = \frac{kA(C_{out} - C_{in})}{V}.$$

Among other simplifications involved [21, 118], this relation assumes equal values for the k's in each direction across the membrane, which is not always true. When the cellular dimensions are not known, k' is used and is kA/V cm^{-1}, the transfer constant. For a cell surrounded by an aqueous solution of penetrating molecules, the main obstacle to entrance into the cell is the plasma membrane. Since the rate of diffusion in the external medium is much faster than the rate of diffusion into and out of the membrane, the external concentrations remain effectively constant and the internal concentrations can also be considered to remain constant. If, however, the penetration is not significantly slower than diffusion through the bulk solutions, then the simplification of the concentration gradient is too great and the gradient is less than expressed. Moreover, the ratio A/V determines the rate of change of concentration and equilibration between internal and external concentrations will be reached much sooner for a small cell than for a large cell. Fick's expression can be simplified even more by using a large external volume in comparison to the volume of the cell so that after integrating,

$$k = \frac{V}{At} \ln \frac{(C_{out} - C_{in})}{(C_{out} - C'_{in})}$$

where C_{in} is the concentration before and C'_{in} is the concentration after the time interval t. The force driving water into the cell may be represented as the difference in the osmotic pressures:

$$\frac{dV}{dt} = kA(\pi_{in} - \pi_{out}),$$

π is the osmotic pressure. Customarily, this is measured by the volume of water penetrating unit area in unit time when the difference between osmotic pressures is 1 atm.

Nonequilibrium Thermodynamics

An alternative to a kinetic treatment of diffusion is a thermodynamic treatment. Classical thermodynamics, however, is based upon processes that are completely reversible, a theoretical possibility but a condition difficult to achieve even under precise laboratory conditions. Diffusion and permeation are essentially irreversible processes, are usually not studied under equilibrium conditions, and normally result in an increase in entropy. Hence, the thermodynamic approach used must be concerned with irreversible processes and deal with rates in order to relate the rate of entropy production to the rate of the irreversible process. Most biological phenomena are basically described by the change in entropy (dS) with time (dS/dt). $T(dS/dt)$, the dissipation constant, is given by the sum of all the products of forces (Δ) and corresponding flows (J) or

$$J_1 \Delta_1 + J_2 \Delta_2 + J_3 \Delta_3 + \cdots.$$

The forces are gradients of energy and the units are written in terms of energy per time. Any process can be considered as a net flux produced by the operation of its conjugate force. At low rates, the amount of flux produced is related to the amount of the conjugate force by an appropriate phenomenological coefficient (L). When certain conditions are satisfied [24, 64], the rate of entropy production is given by the sum of the products of these forces and fluxes. Fundamental to the theory of irreversible thermodynamics is the concept of cross-coefficients. Onsager [85] noted that all flows are coupled to all forces so that in a multicomponent system each flow has a component induced by each force, the magnitude of which is indicated by the corresponding phenomenological coefficient. For example, if there are three forces Δ_1, Δ_2, and Δ_3, the corresponding flows are

$$J_1 = L_{11}\,\Delta_1 + L_{12}\,\Delta_2 + L_{13}\,\Delta_3,$$
$$J_2 = L_{21}\,\Delta_1 + L_{22}\,\Delta_2 + L_{23}\,\Delta_3,$$
$$J_3 = L_{31}\,\Delta_1 + L_{32}\,\Delta_2 + L_{33}\,\Delta_3.$$

The dependence of flows on forces is such that the coefficients are symmetrical, i.e.,

$$L_{12} = L_{21}, \qquad L_{13} = L_{31}, \qquad L_{23} = L_{32}.$$

Two net fluxes describe the process of membrane permeation: J_V, the total flow of volume—solute plus solvent—across the membrane and J_D, the relative volocity of solute to solvent or the exchange flow of solute and solvent. The respective conjugate forces are Δ_p, the pressure difference across the membrane, and $RT\,\Delta C_s$, a term related to the solute concentration difference across the membrane. The rate of entropy production is $J_V\,\Delta_p + J_D RT\,\Delta C_s$. The phenomenological coefficient that relates the flow of volume to its conjugate force, the pressure difference that causes this volume flow, is the pressure-filtration coefficient L_p. Therefore, at zero concentration difference, $J_V = L_p\,\Delta_p$. Similarly, the phenomenological coefficient of exchange flow L_D relates the relative velocity of solute to solvent, J_D, to its conjugate force, the concentration difference of solute $RT\,\Delta C_s$. Thus, at zero pressure difference, $J_D = L_D RT\,\Delta C_s$. L_p is related to the permeability coefficient for water P_w and, although the relationship is not simple, L_D is analogous to the permeability coefficient for the solute P_s.

Consider the situation for a semipermeable membrane, which is much more permeable to water than to the solute. A pressure difference Δ_p applied across that membrane results in a net flow of volume J_V through the membrane as determined by the pressure-filtration coefficient L_p. But the applied pressure difference will force relatively more water than solute across the membrane. The different composition of the extruded fluid will result in a change in the relative velocity of solute to solvent J_D. Therefore, the J_D flux will also be determined in part by the applied pressure difference Δ_p; the relationship between the flux and this force is given by the cross-coefficient

L_{Dp}, the ultrafiltration coefficient. Similarly, if a concentration difference is applied across the membrane, a net flow of volume will occur, which is the phenomenon of osmosis. The appropriate cross-coefficient is the osmotic coefficient L_{pD}, the flow of volume per unit concentration difference. These cross-coefficients relate to the solute-solvent interactions. Both the total volume flow J_V and the relative velocity of solute to solvent J_D result from the combined effects of the pressure difference Δ_p and the concentration difference ΔC_s. The general equations that define transport and take into account the multiple effects of solute and solvent flows are

$$J_V = L_p \Delta_p + L_{pD} RT \Delta C_s \quad \text{and} \quad J_D = L_{Dp} \Delta_p + L_D RT \Delta C_s.$$

The permeability of a membrane to a given solute in the presence of water can be characterized by the four phenomenological coefficients L_p, L_D, L_{pD}, and L_{Dp}, but two are identical ($L_{pD} = L_{Dp}$) as predicted by Onsager's general theorem.

It ought to be obvious that the accounts described above are insufficient in many ways but more adequate and sophisticated discussions are beyond the function of this small text. The equations can only describe linear functions that are steady or degrading. They cannot describe oscillatory or feedback phenomena that are also common functions in biological systems. They have proved excellent, however, for membrane phenomena and for other processes that are not far from the equilibrium condition. The equations do, though, show the basic similarity of many transport phenomena and have predicted many of the transport complications that will be described later. Their value is reflected in the abundance of investigators who are currently analyzing their data in terms of irreversible thermodynamics rather than in terms of kinetics. The curious reader will find the necessary background help in the literature. Crank [13], Harris [45], Stein [118], and Teorell [121] present careful analyses of the mathematics of diffusion and permeability, including the many assumptions and limitations involved. The monographs of De Groot and Mazur [24] and Katchalsky and Curran [64] discuss the thermodynamics of irreversible processes and the steady state.

Reflection Coefficient

To describe the relative rates of solute and solvent permeabilities and, therefore, to provide a measure of the selectivity that a particular membrane demonstrates between a solute and its solvent, Staverman [117] defined a reflection coefficient σ. The reflection coefficient is the ratio of the osmotic (ultrafiltration) coefficient to the pressure-filtration coefficient:

$$\sigma = \frac{-L_{pD}}{L_p}.$$

For an ideal semipermeable membrane there is no flow of solute across the membrane and only a flow of solvent. The rates of hydrostatic flow and osmotic flow are necessarily equal [34] and the reflection coefficient equals unity. For a coarse, nonselective, and leaky membrane that can provide no ultrafiltration, the reflection coefficient is zero. For most biological membranes, intermediate values are obtained (Table 3). There are many conditions

TABLE 3

MEASURED VALUES OF THE REFLECTION COEFFICIENT
FOR VARIOUS SOLUTES AND CALCULATED VALUES FOR THE EQUIVALENT PORE RADIUS*

Source of Membrane	Solute	σ	Equivalent Pore Radius (Å)
Human erythrocyte	Glycerol	0.88	
	Propylene glycol	0.85	
	Thiourea	0.85	
	Methylurea	0.80	4.2
	Propionamide	0.80	
	Urea	0.62	
	Acetamine	0.58	
Frog muscle fibers	Mannitol	1.00	
	Sucrose	1.00	
	Glycerol	0.86	4.0
	Urea	0.82	
	Formamide	0.65	
Squid giant axon	Glycerol	0.96	
	Ethylene glycol	0.72	
	Urea	0.70	4.25
	Ethanol	0.63	
	Formamide	0.44	
Necturus kidney tubule	Sucrose	1.00	
	Erythritol	0.89	5.6
	Glycerol	0.77	
	Urea	0.52	

* σ values can range between zero (permeability characteristics show no selectivity between solvent and solute molecules) and one (only solvent molecules can penetrate the membrane). If transport across a membrane takes place through an assembly of capillary channels, the effective pore radius can be estimated from an analysis of solvent flow.

necessary to evaluate a reflection coefficient [43] but the major one is that there is no volume flow across the membrane when the measurement is made. For example, when red cells are suspended in a dilute solution of a penetrating solute, the solute will enter, water will follow osmotically, and the cells will swell. In a very dilute solution, cellular swelling will always occur. If the osmotic effect of the solute is greater than that of the internal solutes of the cell, water will initially leave the cell and the cell will shrink. Later the cell will reswell as permeant enters. By suspending cells in varying concentrations of the permeant, one will be found at which the cells do not swell or shrink at zero time. At this concentration of permeant (C_s), the internal solute concentration (C_i) is exactly equivalent osmotically to the osmotic effect of the external concentration σC_s, or $C_i = \sigma C_s$. By choosing a particular solute that is not able to penetrate the cellular membrane ($\sigma = 1$), C_i can be found and thereafter σ can be determined for all other solutes for which $\sigma \neq 1$.

3.3 Measurement
of Permeability

There are several methods for studying permeability. Some of the older and more qualitative procedures are the following. (1) Dyes such as methanol red enter into cells like *Paramecium*; the cell becomes pink as the dye enters. (2) Dye may be combined with another compound, such as ammonia, and the cell changes from the pink of the methanol red to yellow as ammonia enters. (3) Insoluble precipitates of calcium oxalate crystals are used occasionally as a measure of entry of oxalic acid into *Elodea* cells. Other useful methods include the following. (4) Measures of volume change as in plasmolysis of plant cells or in invertebrate eggs (Fig. 17). Cells are placed in a concentrated solution that causes them to shrink. The recovery from shrinkage is a measure of the penetration of solute molecules (e.g., salts) from this solution into the cells, or cell mass can be estimated by centrifuging before and after immersion of the cells into a hypertonic solution. (5) Hemolysis or osmotic bursting of cells has been used extensively for red blood cells (see Table 4). Hemolysis

TABLE 4

TIME IN SECONDS FOR 75% HEMOLYSIS OF ERYTHROCYTES
FROM VARIOUS MAMMALS IN DIFFERENT SOLUTIONS

Species	0.02M NaCl (control)	Ethylene Glycol +0.02M NaCl	Glycerol +0.02M NaCl
Rat	4.2	6.6	19
Mouse	3.0	8.6	39
Rabbit	3.0	11.3	80
Guinea pig	5.0	15.7	196
Man	8.4	12.6	43
Dog	6.1	28.6	1548
Cat	2.6	18.3	1222
Pig	3.0	16.7	1024
Ox	3.8	35.1	2325
Sheep	1.9	24.1	1623

times are roughly inversely proportional to permeability. (6) Quantitative chemical analyses of the intracellular fluids are rarely performed on single cells except for large cells such as the plants *Valonia* and *Chara* and the giant squid axon. (7) For some small cells, a sample containing many cells is collected and centrifuged, and the packed cells are analyzed. (8) The rate of removal of molecules from a solution can sometimes be determined for small cells. For example, 10^8 yeast cells/ml remove 6 mg of sugar from a solution in 1 hr. In all of these procedures, radioactive tracers are the choice of meticulous investigators.

None of these procedures are without difficulties. For example, permeability coefficients are often evaluated by measuring volume changes of a cell as water enters osmotically following the entrance of a permeant. Errors result from the fact that the volume of the cell changes continuously. Solute enters both by diffusional flow and by bulk flow—being dragged by water molecules. Errors result if the permeant enters rapidly. Normally, the solute needs to enter at a rate some 100 times slower than that of water

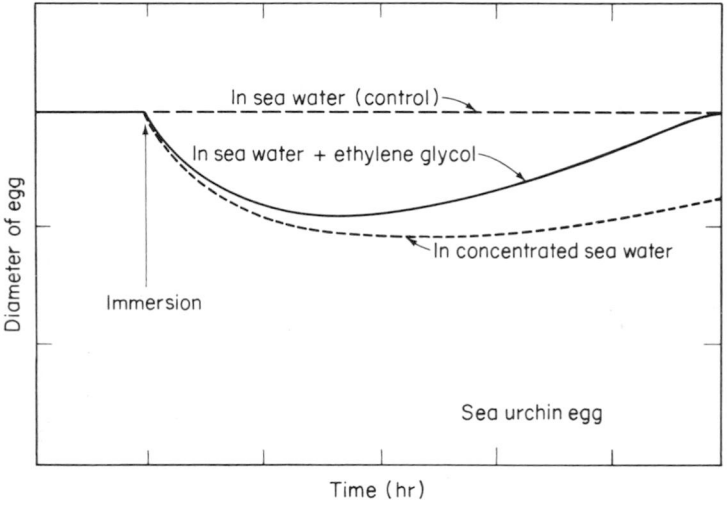

Figure 17. *Volume change as a measure of permeability:* sea urchin eggs shrink when placed in seawater that has been made hypertonic by the addition of, e.g., ethylene glycol. The volume change results from rapid water movement to return the cell to osmotic equilibrium with its environment. The eggs return to normal size as ethylene glycol enters to bring the intracellular fluids into chemical equilibrium with the extracellular fluid. As ethylene glycol enters, so does water, and the cell returns toward normal size. Ethylene glycol enters the egg more rapidly than do salts. Thus, the eggs return to their initial volume sooner in the ethylene glycol solution than in the concentrated seawater. Jacobs [61], among others, presents the basic differential equations for such situations as simultaneous solute and water flow.

for the measurement to be considered accurate. Also, it is incorrect to assume that the permeant exerts its full osmotic effect across the cell membrane. Careful measurements use σC_s (reflection coefficient times permeant concentration).

3.4 Generalizations
on Permeability

Data obtained from permeability measurements indicate that the penetration of biological membranes by molecules can be correlated with three major properties [22]. (1) Lipid solubility enhances penetration. For example, nonpolar compounds, which have a low solubility in water but a high

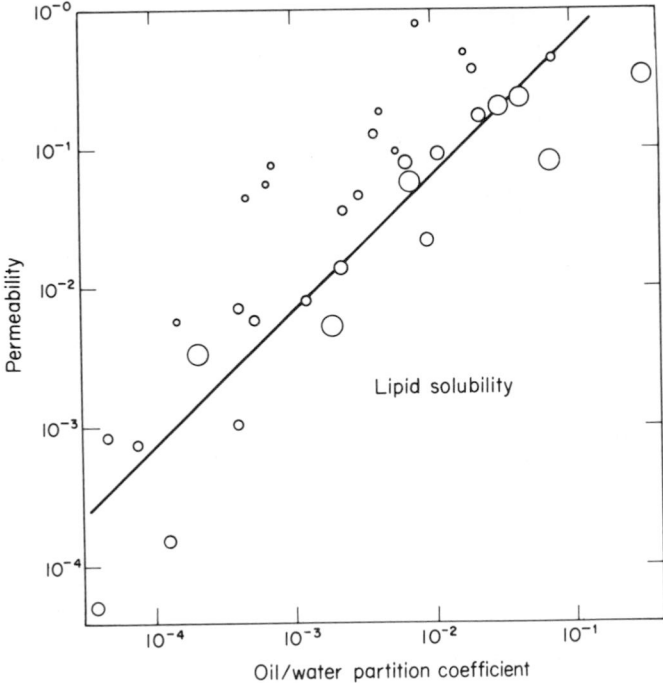

Figure 18. *Permeability as a function of lipid solubility:* when the rate of penetration of various nonelectrolytes into cells of *Chara ceratophylla* is plotted as a function of the olive oil/water partition coefficient for these molecules, a direct correlation is exposed between the permeability of a substance and its solubility in lipids. The diameter of the circles illustrate the relative diameters of the molecules tested. The correlation is improved if $PM^{1/2}$ is used instead of P.

solubility in fats or fat solvents, penetrate faster than polar compounds, so permeability can be expressed in terms of partition coefficients (Fig. 18). (2) Molecular size is important. Large molecules do not enter as rapidly as do small molecules (Fig. 19). For example, the permeability of cells to sucrose is slight; starch, glycogen, and inulin do not pass through many membranes. (3) Permeation depends on the degree of ionization of the solute molecule. The stronger the charge on the permeant, the less rapidly the molecule will penetrate (Fig. 20). For example, increasing the pH of a solution of H_2CO_3 to ionize this compound decreases its penetration into *Valonia* cells.

Permeability values for water require explanation [118]. Each water molecule can form four hydrogen bonds to an aqueous solvent lattice but water is observed to penetrate cells some fifty times faster than this would predict. It is not yet possible to distinguish among the possible explanations for this discrepancy: The number of hydrogen bonds formed may not be four; the penetrating species may be unknown (water is present in a concentration of 55 M and can exist in the form of dimers, tetramers, etc.); the interior of the membrane may be less of a barrier to the diffusion of water than suspected, e.g., holes in the solvent lattice formed by lipid chains may be

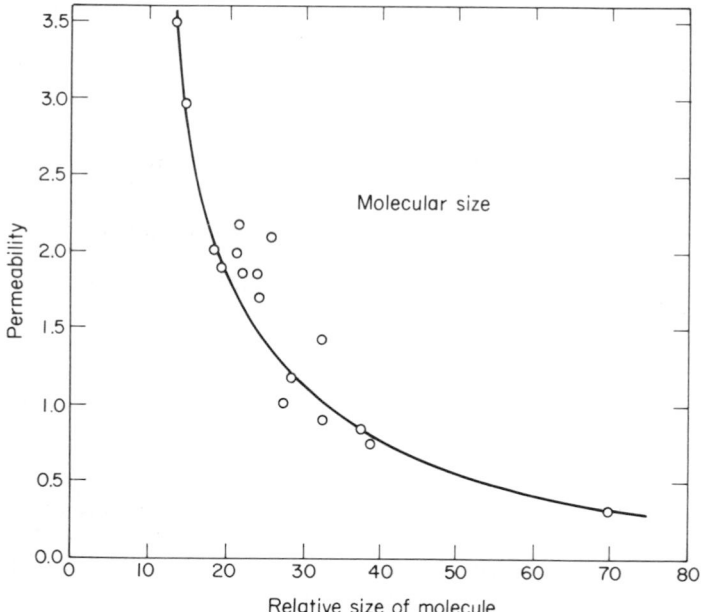

Figure 19. *Correlation between molecular size and permeability in* Beggiatoa, *a large sulfur bacterium:* molecular size was estimated by molecular refraction in yellow light; permeability is given as 10⁴ times threshold plasmolytic concentration in moles per liter.

long enough to allow water to pass but not molecules of greater size; or penetration may occur through water-filled channels.

A collodion or cellophane membrane has been suggested to be a solid skeleton with pores of varying size and, in an aqueous medium, these pores are filled with water. Hence, penetration of molecules through collodion membranes should be like diffusion or, in other words, $DM^{1/2}$ (or $PM^{1/2}$, P is the permeability coefficient) should be tolerably constant for a variety of permeants. For a large-pore collodion membrane, $PM^{1/2}$ is constant but $PM^{1/2}$ does not hold constant for a small-pore membrane. The average velocity of larger molecules is slower than for smaller molecules and only certain pores can be used by the larger molecules. Hence, the magnitude of $PM^{1/2}$ will fall off with increasing size of the penetrating molecules. (For example, see Table 5.)

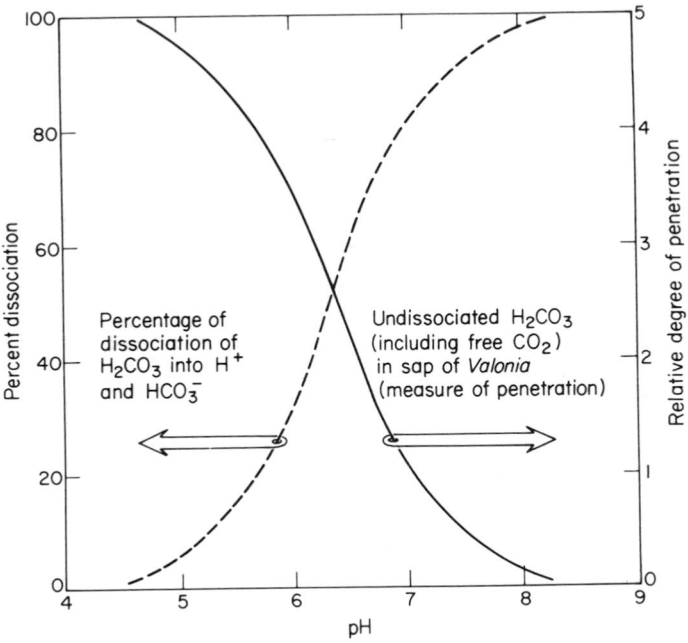

Figure 20. *Correlation between permeability and charge:* the rate of penetration of a weak electrolyte (HCO_3^-) into cells of the alga *Valonia* is plotted as a function of the proportion of charged molecules present, which is a function of pH. Permeation is proportional to the number of undissociated molecules; a change in pH in the direction that increases the proportion of undissociated molecules will enhance penetration. As the proportion of charged molecules (HCO_3^-) in the solution increases, the rate of penetration of these molecules into the cell decreases. The presence of a charge on the molecule decreases its chances for entry.

TABLE 5

VALUES OF $PM^{1/2}$ AND MR_D* FOR TWO COLLODION MEMBRANES

Substance	$PM^{1/2}$ (*very permeable*)	$PM^{1/2}$ (*less permeable*)	MR_D
Methyl alcohol	6.9	5.2	8.2
Ethyl alcohol	7.8	2.0	12.8
Propyl alcohol	7.7	0.8	17.5
Butyl alcohol	7.3	0.7	22.2
Ethylene glycol	6.3	0.2	14.4
Glycerol	7.8	0.21	20.6
Glucose	7.3	0.0	37.5

* MR_D is the molecular refraction that varies approximately with molecular diameter for moderately small molecules.

3.5 Mechanisms
for Penetration

The diffusion of a molecule into a cell is often very low. Permeation rates have been found that are 100,000 times less than the rate computed from a knowledge of diffusion coefficients in water if the membrane is supposed to be aqueous. The plasma membrane is a serious barrier to diffusion. If the membrane is a separate liquid or semiliquid phase between the internal and external media of a cell, the permeating molecule must detach itself from its surrounding solvent molecules and enter into the new phase, the membrane. That molecule must then move across the width of the membrane and detach itself from membrane molecules to jump into the solvent within the cell. Low permeabilities mean that the energy necessary for a molecule to detach itself so completely from water so that it can penetrate the membrane is high and the number of molecules with this energy at ordinary temperatures is small. Within a wide range of size and chemical structure, transport by simple diffusion is possible because there is always a significantly large number of molecules with sufficient energy to make some translational movement through a medium (e.g., Brownian movement).

Danielli [20] developed a potential energy diagram (Fig. 21) to describe kinetically the penetration of molecules into cells. In order to enter a biological membrane, a molecule must possess sufficient potential energy (μ_a) to break free from the bulk exterior medium and to enter the new medium, the membrane. It must then overcome the attractive forces of the membrane molecules (μ_b) and the frictional resistance of the membrane molecules to cross and reach the inner side of the membrane where it can enter the aqueous medium of the cellular interior. Danielli concluded from his analysis that when a molecule penetrates slowly, the difficulty in detaching the permeating molecule from the extracellular water phase overwhelmingly determines the rate of penetration for that molecule. For example, in a homologous series of molecules, Danielli found that increasing the number of OH groups, which increases the possibilities for formation of hydrogen bonds with the water molecules of the medium, decreases their rates of penetration. When the rate of penetration is rapid, the frictional resistance is large compared to the energy jumps for the molecule to pass from water to membrane. Permeability, then, is determined by the partition coefficient for the permeant between the membrane and the medium and by the mass of the permeating molecule.

Molecular weight is still an important factor and, under certain conditions, the factor $PM^{1/2}$ (or $DM^{1/2}$) still holds as a constant for molecules of about the same or of smaller diameter as the distance between centers of the solvent molecules. In water, this corresponds to molecules smaller than about 100 daltons. For larger molecules, $DM^{1/3}$ is found to be constant. This latter constant is also implied by the Einstein equation, which states that the diffusion coefficient D is inversely proportional to the radius of the diffusing particle and, therefore, to the cube root of the molecular weight. But Danielli [20] suggests for liquids that $DM^{1/2}Q_{10}$ holds constant. Q_{10} is the temperature coefficient of diffusion, that is, the ratio of D for values obtained at T and $T + 10°$ Celsius. He argues from his kinetic theory that molecular jumps with large activation energies are affected more by an increase in termperature than are those with low intermolecular attractions. An increase in temperature increases the number of molecules with the necessary activation energy and, empirically, $DM^{1/2}Q_{10}$ is found to be constant.

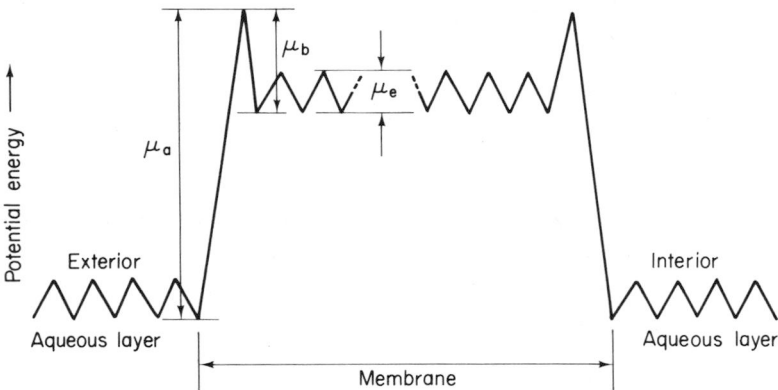

Figure 21. *Potential energy diagram of a cell membrane according to Danielli's analysis of permeation:* the diagram summarizes the potential energy barriers met by a molecule, such as glycerol, as it penetrates a thin membrane of lipoid material.

Moreover, Danielli found that molecular size is more important than inferred by his theory and, as a result, he has repeatedly suggested water-filled pores.

Stein [118] believes that the bimolecular lipid leaflet model for membrane structure is sufficient to give an accurate molecular picture of the unspecialized permeability properties of cellular membranes. The permeant molecules dissolve into the membrane phase and diffuse through the lattice formed by the spaces between the hydrocarbon chains of the lipid molecules. From this model, Stein is able to correlate a large amount of the available data on permeability coefficients. He requires a number of assumptions, however, to account for the high permeability of water and ions in most membranes. Some of the difficulty in understanding water permeation is as much the result of ignorance of the structure of water as it is the result of ignorance of the structure of membrane. Stein also finds difficulties with a membrane structure that is filled with aqueous pores. Everyone finds data for some permeants that require the postulation of specific membrane transport systems.

3.6 Membrane Transport

A body can be considered to be constructed as a number of compartments separated from each other by a series of membranes. Each compartment has a more or less characteristic composition often differing markedly from that of an adjacent compartment so that steep gradients of electrochemical potential can occur across these membranes. The problem is to clarify the means by which various materials cross these membranes and by which the striking gradients are maintained. To some extent, the penetration of various materials can be defined in terms of relatively simple forces (some complications will be mentioned later). In other instances, it is apparent that the movements of certain substances require the utilization of energy derived from cellular metabolism, a process designated as active transport. In no case can it be said that all mechanisms involved have been identified.

Up to now we have considered only the movements of uncharged particles. The presence of charges on ions adds another level of complexity to their transport across membranes. Because of the energy required to separate positive from negative charges across a distance as small, for example, as the thickness of a biological membrane, a requirement for biological systems is that the sum of all the positive charges on one side of the membrane must equal the sum of all the negative charges on that same side. For a cation to penetrate a plasma membrane, it must either exchange positions with another cation or it must be accompanied by an anion. The emphasis of Chapter 4 is on the behavior of charged elements at the membrane.

We may consider the possible ways in which materials cross biological membranes and the forces that effect these movements by following the treatment developed by Üssing [128, 129]. In general, the definable forces that produce movement of material across membranes can be classified as those due to gradients of chemical activity including differences in concentration or activity gradients, those due to gradients of electrical potential, and those exerted upon solutes by the flow of solvent (solvent drag). The rate of transport (ds/dt) resulting from diffusion down an activity gradient is described by Fick's law, as discussed above:

$$\frac{ds}{dt} = -DA\frac{da}{dx}.$$

D is the diffusion coefficient; A, the cross-sectional area; and da/dt, the activity gradient. The electromotive force (E) that propels charged molecules is defined as

$$-ZF\frac{d\psi}{dx}.$$

Z is the valance on the molecule, F is Faraday's constant or 96,519 coul, and $d\psi/dx$ is the gradient of electrical potential across the membrane. The drag force (Y) on solute molecules caused by the flow of solvent through solvent-filled channels is given by the expression

$$\frac{G}{g_w}\frac{d\mu_w}{dx}.$$

G is the frictional coefficient that defines the interaction of solute and water (solvent) or the mobility at unit velocity of material moving in water. G is measured empirically and is constant for any given solute. $1/g_w$ is the frictional coefficient of water through the solute. $d\mu_w/dx$ is the electrochemical potential gradient for water. The general equation derived by Üssing by combining the effects of these forces is unworkable without simplification. One simplification is to disregard a net water movement since it is negligible when the membrane is bathed on both sides by solutions of equal osmotic pressure. This cannot be assumed for cases where isotonic water transport occurs, as demonstrated during secretion and absorption by various internal organs [26]. Üssing [71] showed that the forces that act upon an ion to cause movement across an intervening membrane would produce a flux from one surface to the other given by

$$J = -\frac{da}{dx}\frac{ART}{GN_0 f}\exp\frac{-ZF\psi}{RT}$$

where J is the unidirectional flux of the solute across the membrane, da/dx is the chemical activity gradient across the membrane, A is the area of membrane available for penetration by the solute, G is a frictional coefficient describing the interaction between the membrane and the solute ion, N_0

is Avogadro's number, Z is the valance of the ion, ψ is the electrical potential across the membrane, f is the frictional or activity coefficient, F is Faraday's constant, R is the gas constant, and T is the absolute temperature. This expression has several unknown quantities that are not determinable and, therefore, it is not useful in this form. If attention is directed to the ratio of fluxes from one side of the membrane to the other, the indeterminate quantities cancel out and leave, upon integration, the relatively simple relationship

$$\frac{J_{1-2}}{J_{2-1}} = \frac{f_1 C_1}{f_2 C_2} \exp\left[\frac{ZF}{RT}\psi_1 - \psi_2\right].$$

This equation contains only experimentally determinable quantities except, perhaps, the activity coefficients, which also cancel out when the solutions bathing both sides of the membrane are essentially the same. Another modification of this expression has been given as

$$\frac{J_{1-2}}{J_{2-1}} = \frac{f_1 C_1}{f_2 C_2} \exp\left[\frac{-ZF\Delta\psi}{RT}\right]\left[\frac{a_{w_1}J_{w_1}}{a_{w_2}J_{w_2}}\right]\left[\frac{G}{g_w}\right].$$

In addition to this one of Üssing, other derivations have been described [45, 69, 118, 121].

In addition to the passive forces described above, however, one also finds in practically all biological systems movements of solutes which can be explained by none of these definable forces and which are, as a result, designated as *active transport* [3]. A better definition is that active transport processes are cellular mechanisms that convey material through biological membranes to a higher electrochemical potential [2, 11]. The definition of active transport is quite arbitrary [65, 107, 118]. The first definition mentioned above is not considered adequate by many investigators but it does have the advantage of singling out those solutes that must be considered to be directly involved in the process by which metabolic energy is utilized to perform transport work. Clearly, all concentration and electrical gradients and flows in response to pressure or osmotic gradients found in living material ultimately depend on metabolic work, even though this may be remote from the particular membrane under examination. Hence, the requirement of metabolic work is not sufficiently exclusive for an adequate definition of active transport; however, neither does this definition given above include all processes that have specificities beyond simple diffusion and flow. It also does not include the processes by which certain substances can cross membranes only, say, by combining with a specific carrier, even though the movement is entirely downhill with respect to gradients of electrochemical potential.

The manner in which solutes may cross membranes depends on the nature of the membrane. If it is a continuous lipid layer, substances can pass this layer only by dissolving in it, although passage through such a layer might be facilitated by combination with some component in the lipid layer that would enhance its lipid solubility. Hence, for such membranes, the permeability would be a function of the lipid solubility. Numerous observations, some mentioned earlier, show the extent to which cell permeation is related to lipid solubility. But many features of cellular membranes cannot be explained if they are assumed to consist of a solid lipid layer. It is generally proposed that the lipid layer contains pores through which water forms a continuous phase and through which water-soluble solutes can pass. It is only through such channels that forces produced by the flow of solvent can act. There is evidence that solvent (water) flow does produce movement of solute indicating that solvent-filled pores exist [19]. For the present purposes, it is convenient to accept the view that cellular membranes generally consist of lipid layers penetrated by aqueous pores (see Chapter 2, pp. 47–53, for a better discussion of the evidence), the walls of which may bear an excess of fixed-charged groups that can impede the passage of ionic species of the same sign charge.

In some cases, where membranes consist of more complicated structures than the plasma membrane of a single cell and may consist of an entire layer of cells attached to a common basement membrane, the passage of water-soluble substances may occur between cells rather than through them.

3.7 Complications

Activity beginning in the late 1950's has been centered more on discovering the details of molecular mechanisms for the observed permeation phenomena by rigorous theoretical and empirical description. Much of this breakthrough can be credited to the methods of irreversible thermodynamics. The concepts that must be considered in developing explanations form a long list.

Unstirred Layers

There is always a layer of finite thickness adjacent to a membrane that does not mix easily with the bulk of the bathing medium even with vigorous and efficient agitation [19, 27]. This unstirred layer is in the order of 10–50 μ wide. As a result, calculations of permeability and transport have been in error, some seriously, because the concentration gradients across a membrane are different from those given by the concentrations in the bulk phases. Equilibration between the unstirred layer and the bulk solution must wait for diffusion of the appropriate molecules and, with living membranes especially, may never occur.

Valves

If channels for water flow through cellular membranes have valve-like properties (a rectification effect) or if the membrane structure is deformed by the water flow, there is a consistent but generally nonlinear relation between the osmotic driving force and the water flow; flow is greater in one direction than in the opposite direction for the same osmotic gradient. Flow-induced deformations have been suggested as a partial explanation for nonlinearity in the turtle urinary bladder [5].

Double Membrane Effect

The force-flow relation for an asymmetrical system that consists of several dissimilar membranes arranged in series may not be linear even though the relation may be linear for each membrane taken individually [66]. If, for example, active solute transport takes place into an area between two membranes and the first membrane has narrow channels that are relatively impermeable to the solute but the second membrane has wider channels relatively permeable to the solute, then the solute will exert an effectively greater osmotic pressure difference across the first membrane than across the leaky second [25, 26]. Water will be drawn across the first membrane by osmosis to build up a hydrostatic pressure in the intermembraneous space, which then forces water out of the leaky second membrane. In such a system, active solute transport into the middle space will cause a net movement of water even if the two external bathing solutions are identical; water will move against an external osmotic gradient. Experimental data obtained from artificial membranes agree with theory for such systems [15, 34, 84, 92]. Quantitatively, the effect is more complicated than the description given above since the rates of water and solute movement depend on permeability coefficients, reflection coefficients, and filtration constants for both membranes as well as the active transport rate for the solute.

Direct Effects on the Membrane

Questions have been raised concerning the effects on water flow that a concentration of impermeant molecules may have by exerting a *nonosmotic* effect upon the membrane, e.g., by interacting physically or chemically with the membrane or by increasing the viscosity of the solutions [27]. Such effects can be tested by using a variety of impermeant solutes to alter the osmotic pressures of the bathing solutions, or the concentrations of impermeant solute molecules may exert an *osmotic* effect upon the membrane and alter its permeability [17]. For example, Diamond [27] examined the mechanisms listed above as potential candidates for the explanation of the nonlinear osmotic behavior of rabbit gallbladder. His tight logic and precise experiments permitted him to reject all but the change in water permeability. In this preparation, there is no consistent relation between water permeability and either the direction or the rate of water flow, but there is a decrease in water permeability (decrease in water flow to a given osmotic gradient) with increasing osmolarity of the bathing solutions. The data suggest that channels through the cellular membrane behave as osmometers—an aqueous phase (aqueous pathways through the membrane) is separated by impermeable barriers (membrane-solution interfaces) from the external solutions of impermeant molecules—that can shrink in concentrated solutions of impermeant molecules and thereby increase the membrane resistance to water flow.

Bulk Flow

Water generally passes through biological membranes at rates faster than can be accounted for by diffusion of water alone; osmosis is more complicated than just a special case of diffusion. For example, Mauro [81] estimated the diffusional component of the flux of $^{18}H_2O$ through a collodion membrane as only $\frac{1}{730}$ of the total flux. (He chose collodion membranes for study since he considered the phenomena of living membranes too complex to make any result conclusive or unambiguous.) The pores in the membrane are so large that water can move through them in bulk [82, 99]. The rate of flow is limited by the frictional resistance between the water and the walls of the pore and between the layers of water flowing within the pore. He attributed the difference between his diffusional and total flux values to this "quasilaminar" flux. For membranes with pore diameters near that for a water molecule, the total flux can be adequately described by diffusion. Most biological membranes, however, show a nondiffusional component to the water fluxes. While some investigators have yet to accept this point, bulk flow of water can occur in response to an osmotic gradient. The kinetics of flux during osmosis follow to a significant extent Poiseuille's law of flow in capillaries, suggesting that the flux results primarily from a bulk flow of water through the membrane [70]. Ray [98] and Dainty [16] have extended the idea that osmotic flux is driven by a hydrostatic pressure difference within the membrane and have developed the hypothesis for the origin of the driving force behind osmosis. The concept can be illustrated by considering incompressible water molecules passing through an incompressible pore that is too small for the large solute molecules on either side to penetrate. Such a pore, therefore, is filled only with water molecules. Beyond, say, the inside opening of the pore, there are large solute molecules in addition to water. The number of water molecules diffusing out of the pore into this space will be larger than the number of water molecules diffusing back into the pore since part of the space in the bathing solution is occupied by solute. In effect, a simple concentration gradient is established down which water molecules can move. Beyond the other end of the pore, the outside opening, consider a similar concentration of smaller solute molecules in water. The net result is that more water molecules will be diffusing into the pore at the outside opening than are diffusing out at the inside opening of the pore because of the greater displacement by the larger solute molecules. There will be a net increase in the number of water molecules at the outside end of the pore than at the inside end of the pore or, because of the confined space, a hydrostatic pressure gradient within the pore will be established. When a larger concentration gradient exists between the two bathing solutions, the hydrostatic pressure set up within the connecting pore will also be larger. Note that there are really three concentration gradients—one at each interface and the one across the pore itself—and each is steeper than expected since each exists across only a portion of the width of the membrane, which leads to accelerated transport also. A related effect was mentioned earlier: Water flow through a membrane can produce accumulation of solute; for example, a hydrostatic pressure difference across a membrane that has little resistance to bulk flow (wide pore membrane) can lead to a significant concentration of solutes (e.g., ions) even against an electrochemical gradient [97].

Local Osmosis

If active solute transport sets up a high concentration in a local or restricted space just beyond the cellular membrane, then water can cross in response to a local osmotic gradient. Water transport can be so maintained against an opposing osmotic gradient in the external bathing solutions until this external gradient balances the local osmotic gradient caused by active solute transport [25, 26]. A variation of this phenomonon explains isotonic water transport across the epithelium of rabbit gallbladder [30]. Long extracellular channels provide for an autoregulatory system: Sodium is actively pumped into the channel to make it hypertonic and water enters. Since the channel is long, osmotic equilibration can go on continuously as the solute moves down the extracellular space—isotonicity is ensured regardless of the transport rate. The geometric arrangement also explains the higher water flow during solute transport than can be accounted for by passive osmotic permeability of epithelia. In a long narrow channel, high linear velocities of water can be generated. Indeed, the sweeping effect of the osmotic water flow caused by solute transport would probably be more important than diffusion in carrying solute out of the channel.

Electroosmosis

Living membranes are electrically charged, some more so than others. The passage of cations is accelerated but that of anions is impeded through negatively charged homogeneous membranes or membranes with negatively charged pores. Positive charges on a membrane prevent transport of cations but accelerate anions. Such membranes will produce a flow of water through them. Suppose a negatively charged and permeable membrane separates two salt solutions, identical other than in concentration, then the membrane will permit the passage of cations from, say, the more concentrated side 1 into the more dilute side 2. The anions must remain behind and, as a result, they will prevent the cations from more than just the initial penetration. Thus, the membrane will acquire a double charge, negative on the concentrated side 1 and positive on the dilute side 2. Since the membrane cannot move freely through the aqueous medium toward the positive charge (side 2), as it would if it could behave like a negatively charged body in an electric field, then water passes through it in the opposite direction, which in this case is also the direction of osmotic water movement [6, 18]. While this example implies a complimentarity, electroosmosis is independent of water movement, which is due to differences in osmotic concentrations. Generally, the outer surface of cellular membranes are positively charged, the inner surfaces are negatively charged, and the pores are positively charged. Such a situation results in the transport of water into the cell. Now consider a frog living in fresh water: Water is transported out through his skin as sodium is actively pumped inward. Related phenomena are *streaming potentials*, the electrical potential differences that arise across charged membranes as water flows through them. The aqueous channels through a membrane contain an excess of mobile ions of opposite sign to balance the fixed charges on the walls of the pore [121]. Water, flowing through these channels, carries along the mobile ions to give the side of the membrane toward which the water is moving an electrical potential of sign opposite to that of the fixed charge. For example, pores throughout the rabbit gallbladder contain negative fixed charges, probably organic acids of low pK_a that can be blocked by Ca^{2+} [29]. When water flows through this membrane passively, e.g., in response to a solute concentration gradient in the external bathing solutions, a streaming potential is set up in which the side toward which the water moves goes positive. Diamond [28] has used these streaming potentials as a rapid quantitative measure of water flow. It is possible for such negatively charged fixed sites to be responsible for some discrimination among cations to determine their permeability. The existence of streaming potentials makes it clear that, in the rabbit gallbladder at least, sodium does not diffuse across the plasma membrane by leaving the aqueous phase and moving along a series of fixed charge sites. The existence of streaming potentials implies a frictional transfer of momentum from water to ions and this would be impossible if water and ions had separate routes for passive permeation through the membrane [29].

Facilitated Diffusion

The pattern of migration for many molecules through plasma membranes indicates that simple diffusion cannot be the rate limiting step. This is especially true for the hydrophilic solutes which do not enter a lipid phase easily and which require an opening through the lipid barrier or a modification of their physical properties to permit entrance. Experimentally, the rate of penetration, or flux, of a particular solute reaches a maximum value as the solute concentration continues to rise in the medium; the system becomes saturated (Fig. 22). The addition of structual analogs to compete with the solute for permeation reduces the rate of penetration. Optical enantiomorphs to the solute have different flux values. Specific inhibitors, structurally unrelated to the permeant, can be found and work at low concentrations. Occasionally, a facilitated movement of permeant down an electrochemical gradient can be demonstrated with the movement of a structual analog in the opposite direction against the gradient (see counterflow below). Because of similarities to other phenomena, the solute is believed to react with a limited number of some other molecular structure or chemical site; the supply of this structure limits the transport rate at the higher concentrations. The system controls the rate of entrance but not the final concentration of the permeant. The transport system operates passively down an existing electrochemical gradient to promote its disappearance; it is called *facilitated diffusion*, a term preferred over the more ambiguous "mediated transport" [11]. No other input of energy is required other than the electro-

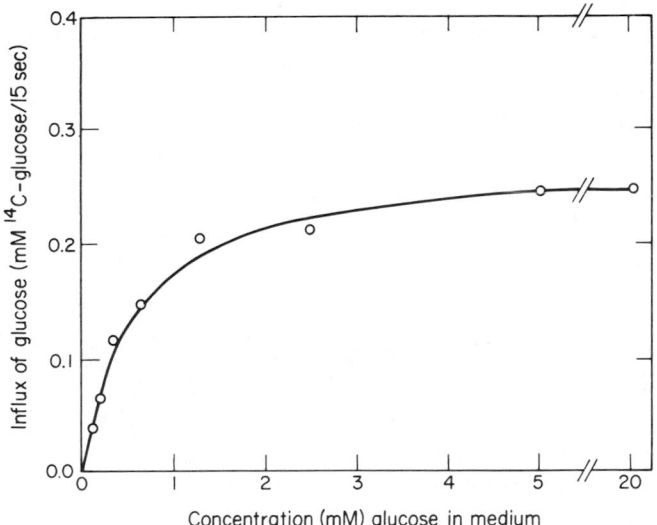

Concentration (mM) glucose in medium

Figure 22. *An example of facilitated diffusion:* the influx of radioactively marked glucose into human erythrocytes is measured as a function of the concentration of glucose in the plasma bathing the red cells. The solid line is computed from kinetic theory. The key feature shown by the system is the saturation effect at the higher substrate concentrations.

chemical gradient and the energy necessary to maintain cellular membrane structure.

Restricted Diffusion

Migration rates for solutes that are diffusing through pores only slightly larger than the size of the molecules will fall off with increasing concentration as more and more collisions occur unfavorable to passage [11]. Restricted diffusion can be distinguished from facilitated diffusion for, in restricted diffusion, (1) D and L isomers have similar transport rates; (2) a flux in one direction tends to be depressed by a flux in the opposite direction; and (3) the analogs that inhibit transport of a solute do not pass as easily themselves, whereas they are transported more rapidly in mediated diffusion.

Flow Driven by Counterflow

In some cases after a red blood cell, for example, has reached a 1:1 equilibrium distribution with a given monosaccharide, the addition of a second competing sugar to the bathing solution will cause the first sugar to migrate out of the cell, creating a concentration gradient, while the second sugar is entering in the direction of its concentration gradient [101]. Wilbrandt and Rosenberg [137] interpreted such data as evidence for two different sites of action, one for entry and one for exit. The sites can be either fixed or mobile within the membrane, but most likely mobile. The addition of the second solute causes the entry site to become, in effect, an exit site so that the second solute competes with the original solute for the entry site and thus slows its entrance, without having a corresponding effect on its exit, which continues at its original rate; the exit rate becomes effectively larger than the entry rate for the original solute.

Exchange Diffusion

The chemical structure that mediates transport can work in both directions so that most of the flux in one direction will be canceled out by the flux in the opposite direction; the net flux is only a fraction of the total flux. Üssing [128] warns against deriving conclusions, e.g., about the active transport energy requirements, from single flux measurements, as is commonly the case in experiments using isotopes. If part of the movement of a particular solute occurs by an exchange of a molecule from the low-energy side (more dilute) of the barrier for another of the same species from the concentrated or high-energy side, which can occur without energy cost, then measurements of the rate at which radioactive tracers enter (or exit) that cell will be higher than what actually occurs by active (or passive) transport. The phenomenon is more important in experimental design than it is in the physiology of the cell.

Cotransport

Some transport processes are coupled so that the transport of one species is required for the transport of another. Coupling between sodium and potassium movements is well-known. Another important example is the associated transport of cations, especially sodium, and nonelectrolytes, primarily organic solutes such as sugars [109]. In all these cases, the organic solute moves against an electrochemical gradient and the dependence on the cotransport of the cation (e.g., sodium) is quite specific. Although the mechanisms are yet to be clarified, the energy for the transport of the organic solute seems to be provided by the asymmetry in the distribution of sodium across the membrane and not directly provided by metabolic reactions. Once the sodium gradient is established, the organic solute moves; nonelectrolyte movement is directly coupled to sodium movement.

The basic similarity in these last three cases of countertransport, exchange diffusion, and cotransport is revealed when the systems are expressed in the quantitative terms of nonequilibrium thermodynamics. Indeed, the mathematics predicts the existence of multiple coupled-transport systems.

3.8 Active Transport

The term *active transport* is reserved for those processes by which materials cross membranes against an electrochemical gradient (uphill) *and* which utilize metabolic energy. It is helpful to know that the process in question cannot be explained by any of the passive processes mentioned above, some of which are uphill and some of which depend on metabolic energy. Two pieces of evidence that are convincing, if they are available, are (1) that the gradients established are beyond the plausible limits for binding agents, and (2) that water movements are in approximately direct relationship to solute uptake. Investigators have found examples of active transport for a large variety of molecules but, so far at least, not for water; water transport, when examined carefully, is described only by passive processes. A description of only a couple of the more popular preparations should adequately illustrate current thinking.

3.9 Frog Skin

One of the most studied biological membrane systems is the abdominal skin of a frog. Üssing [130] in particular has led much of its extensive study including the theoretical considerations. Over 100 years ago, du Bois Reymond discovered that an isolated frog skin can maintain its outer, or epithelial, surface at a negative potential with respect to its inner, or corium, surface. Huf [60] found that an isolated frog skin can, when bathed on both sides with Ringer's solution, move sodium and chloride from the outside to the inside bathing solution. Krogh [72] demonstrated that an intact, but salt depleted, frog can take up salt through its skin from a NaCl solution as dilute as 10^{-5} M or against a 10,000-fold concentration gradient. Krogh found this uptake to be specific for sodium since he could not get either potassium or calcium to substitute for the sodium.

The skin from the abdomen of a frog can be easily stripped, trimmed, and stretched across the opening between two chambers as a diaphragm. It survives for many hours *in vitro* performing metabolic operations including active ion transport [133]. The frog skin is structurally complicated (Fig. 23): It consists of a sheet of tissue containing a loose layer of connective tissue, a continuous basement membrane, and several layers of epithelial cells. Outward from the basement membrane, the cells become progressively flattened and keratinized and, presumably, lose their metabolic activities. Only the basal layer of cells is believed to be involved in the transport processes [58] and the skin is, therefore, treated as a single membrane comparable to the rabbit gallbladder [29] or the toad urinary bladder, which has only a single layer of cells [24]. Actually, the work in Üssing's [132] laboratory has demonstrated that while more than one cell layer is involved in transporting sodium or in providing energy for the transport, only a few cell layers are necessary for the transport; their evidence excludes the need for the glands, the stratum germinativum, and the basement membrane in the frog's ability to transport sodium across the skin. As a result, the evidence presented to locate the pumps to specific layers in the skin is contradictory. The evidence cited by Cereijido and Rotunno [10] show pumps located generally throughout the skin. The differences are probably as much related to the different concepts used to explain the sodium transport phenomenon as to different techniques or species of frog used in the individual studies.

When mounted between two chambers containing oxygenated and stirred Ringer's solutions, the isolated frog skin generates an electrical potential difference across itself [133]. The corium side (inside) is electrically positive with respect to the epithelial side (outside). The value of the potential varies greatly in individual skins from a few millivolts in some preparations to over 100 mV in others. The usual value cited is about 50 mV. The conditions are important since the potentials can vary not only in magnitude but even in sign [79]. Curiously, the system is sensitive to mechanical stimulation. A small bump will alter the potential and current measurements for several minutes [79]. Since, in this system, ions are transported and ions carry charge, the transport can be conveniently followed by measuring the electrical currents that flow and the potentials that are generated.

The electrical potential difference ($\Delta \mu$) of the total potential energy difference per mole of ion is the sum of the electrical and concentrational energy differences across the membrane for the ion being considered. The electrical potential energy difference of $1\ M$ of, e.g., K^+ ions (or any other since the same considerations hold), is the work that must be done solely against electrical forces to carry $1\ M$ of K^+ ions across the membrane from outside to

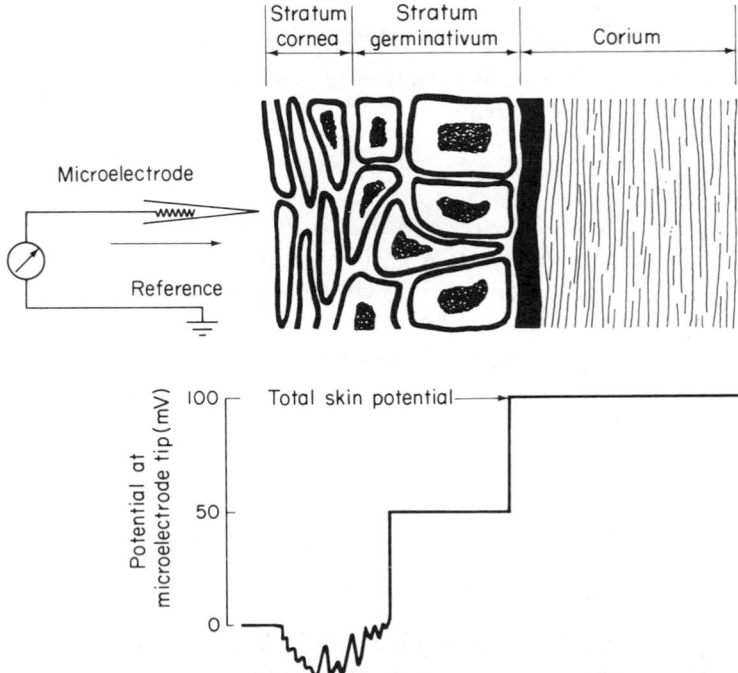

Figure 23. *Diagrammatic correlation of potential profile through frog skin with skin histology:* a micropipette, used as an electrode, is passed through the skin. The potential difference measured between its tip and a reference electrode placed in the outer solution is plotted as a function of the distance through the skin (the location of the microelectrode tip) in the lower figure. Upon puncturing the skin, the first cells encountered are the dying cornified cells and the potential recorded is an unstable one, negative with respect to the outside solution. The first distinct positive potential is recorded when, presumably, the microelectrode crosses through the outer border of a cell in the basal epithelial layer (stratum germinativum). This potential remains steady as the electrode is advanced further until a sudden jump to the full skin potential occurs when the corium face of the basal cell is pierced. Thus the skin potential is the sum of two potentials, as predicted by Üssing's model illustrated in **Fig. 27.**

inside while holding the transmembrane potential at its original value. The electrical work (W_E) is given by

$$W_E = Z_K F E_m.$$

Z_K is the valance of the potassium ion (1), F is Faraday's constant (96,519 coul/mole), and E_m is the transmembrane voltage (joules per coulomb). The concentration potential energy difference (W_C) is the work required to carry 1 M of K$^+$ ions from outside to inside solely against a concentration gradient while holding the internal and the external potassium concentrations constant:

$$W_C = RT(\ln [K^+]_i - \ln [K^+]_o).$$

R is the gas constant and T is the absolute temperature. The electrochemical difference for potassium then is

$$\Delta \mu_K = W_E + W_C \quad \text{or} \quad \Delta \mu_K = Z_K F E_m + RT \ln \frac{[K^+]_i}{[K^+]_o}.$$

When E_m, $[K^+]_i$ and $[K^+]_o$ are such that $\Delta \mu_K = 0$, the K$^+$ ions are equilibrated across the membrane. If $\Delta \mu_K$ is not zero, it is a measure of the net tendency for K$^+$ ions to diffuse through the membrane. The larger the value of $\Delta \mu_K$ is, the greater the net flux of K$^+$ becomes.

The condition for ionic equilibrium is that the electrochemical potential of an ion is zero. Setting $\Delta \mu_K = 0$ and replacing E_m by E_K and solving for E_K gives the familiar Nernst equation:

$$E_K = -\frac{RT}{FZ_K} \ln \frac{[K^+]_i}{[K^+]_o}$$

or the more general form,

$$E_B = \frac{RT}{FZ_B} \ln \frac{a_{B_o}}{a_{B_i}}.$$

E_B is the electrical potential in millivolts, R is the gas constant (8.314 joules/degree mole), T is the absolute temperature (°K), Z_B is the valance of ion B (i.e., the number of Faradays involved, 1 for a monovalent ion), F is Faraday's constant (96,519 coul/mole), a_{B_i} is the activity of ion B in the inside solution, and a_{B_o} is the activity of ion B on the opposite side (outside) of the interface. E_K determines the value E_m must have if the K$^+$ ions are to be in equilibrium and is called the *potassium equilibrium potential*. Substituting the values into the expression for a potassium case at room temperature (20°C) and converting the logarithm to the base 10, a useful form of the Nernst equation results:

$$E_K = -58 \log \frac{[K^+]_i}{[K^+]_o}.$$

E_K is now given in millivolts (1 volt = 1 joule/coul). Or, for a tenfold difference in concentration (e.g., log 10 = 1) for a given ion, the potential determined at equilibrium is -58 mV. If this relationship is applied to the frog skin with the assumption that it is a simple physical-chemical system, the relationship obviously fails. Inserting the concentration values for the major ions in Ringer's frog solution, which bathes both sides, into the Nernst expression provides the following result (the values are expressed as meq/l):

$$E_{skin} = -58 \log \frac{[111]Na_i^+ + [117]Cl_o^- + [2]K_i^+ + [2]Ca_i^{2+}}{[111]Na_o^+ + [117]Cl_i^- + [2]K_o^+ + [2]Ca_o^{2+}}$$

$$= -58 \log(1) = 0 \text{ mV}.$$

Under these conditions, however, the skin potential usually has a value of 50 mV. Some more explaining is in order.

Consider a membrane separating two solutions of equal concentration of ions (e.g., the frog skin under the conditions just specified). Under these initial conditions, the number of positive charges are equal to the number of negative charges in each solution. If positive charges migrate by some means from, say, left (or outside) to right (or inside) or if negative charges migrate in the opposite direction, the movement will give rise to an electrical potential. Under these circumstances, the movement of ions is related to the potential by this relationship:

$$\frac{J_i}{J_o} = e^{(E-E_s)F/RT}$$

where J_i is the inward flux (mole/cm^{2}sec); J_o, the outward flux; E the observed potential; E_s, the potential determined by the concentration gradient of the ions in question; and e, R, T, and F have their common connotations. For conditions of passive movement of a given ion, $J_i/J_o = 1$. If active transport is involved, the ratio of influx to outflux will be greater or less than unity. The general expression for the flux ratios that is given here does not hold when such phenomena as exchange diffusion or solvent drag occur in the system. The inward and outward fluxes of ions have been measured by tracer techniques for both Na^+ and Cl^- in the frog skin [133]. Such experiments show that sodium is actively transported across the frog skin but chloride moves passively across the skin. The flux ratio for Cl^- is unity but

$$\frac{J_{22Na_i}}{J_{24Na_o}} > 1.$$

If the skin is poisoned (e.g., with CN^-) or the oxygenation of the bathing solutions is stopped, the potential across the skin becomes zero and the flux ratios become unity.

Üssing has developed a simple technique for measuring the active transport of sodium across the frog skin. The method involves short-circuiting the skin battery with an external potential of opposite polarity and adjusting the current that flows through the external circuit so that the observed skin potential becomes zero [133]. An electrical model of the skin and short-circuit current technique is illustrated in Fig. 24. Studies of the short-circuited frog skin serve to show that only sodium is actively transported by the frog skin and that when the potential across the skin is kept at zero, there is a remarkable correspondence between the current flowing through the skin and the net movement of sodium ions. These observations are compatible with the hypothesis that the only active process involved in the frog skin is the transport of sodium ions from outside to inside (this statement will be modified in a moment). No other ion can substitute for sodium, except lithium to a small extent; potassium particularly cannot, although the com-

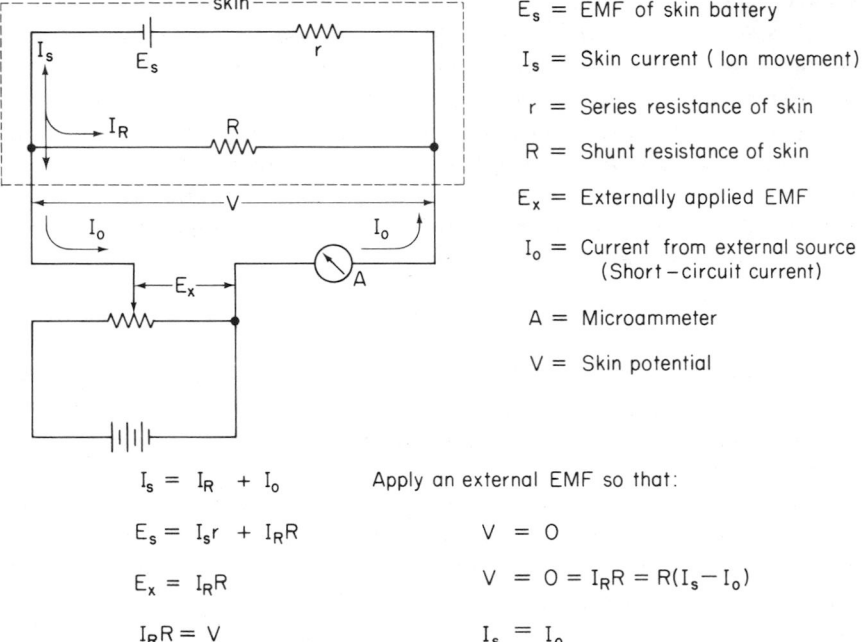

E_s = EMF of skin battery

I_s = Skin current (Ion movement)

r = Series resistance of skin

R = Shunt resistance of skin

E_x = Externally applied EMF

I_o = Current from external source (Short–circuit current)

A = Microammeter

V = Skin potential

$$I_s = I_R + I_o$$

$$E_s = I_s r + I_R R$$

$$E_x = I_R R$$

$$I_R R = V$$

Apply an external EMF so that:

$$V = 0$$

$$V = 0 = I_R R = R(I_s - I_o)$$

$$I_s = I_o$$

Figure 24. *An electrical analog model for the frog skin preparation and the basic circuitry used to measure its electrical parameters:* the lower expressions summarize the argument used to arrive at a measure of the short circuit current across the frog skin. One side of the frog skin, which lies between two chambers both containing Ringer's solution, is connected to the other side through a circuit containing a battery, voltmeter, and ammeter. The ions moving through the skin carry current (I_s), some of which passes through the skin (I_R) and some through the measuring circuit (I_o). The voltage measured across the skin (V) is the IR drop across the shunt resistance of the skin ($I_R R$). If the battery is varied until the voltage across the skin is adjusted to zero, the current passing through the external circuit (I_o) and measured by the ammeter (A) is equal to the current passing through the skin (I_s). The quantity of charge moved is measured and, hence, the quantity of ions transported by the pump mechanism becomes known under conditions of no electrochemical gradient across the preparation.

plete absence of potassium in the bathing solutions reduces markedly the transport of sodium [4]. The chloride fluxes appear to be entirely passive and the sodium influx is so much greater than that of chloride that there is no question of bulk movements of elements from the outside solution to the inside. Generally, the potential across the skin is higher when the sodium influx is higher or when the chloride influx is lower, as happens when the skin is treated with 10^{-5} M Cu^{2+} solutions. The potential difference across the skin is, therefore, derived from the transport of sodium. As seen later (i.e., Fig. 27, p. 98), the pumped sodium exchanges for potassium in the body fluids of the frog.

Summarizing, in 1949 Üssing developed a hypothesis to describe the relationship between the electrical potential across the frog skin and its ability to transport sodium. He showed that sodium could be transported by frog skins against both a concentration gradient and an electrical potential gradient. He showed that this transport must involve the expenditure of energy by the tissue. He suggested that (1) primarily sodium ions are transported actively from the outside to the inside surfaces of the skin; (2) the skin potential is clearly associated with the active transport of sodium; and (3) this potential is responsible for the net inward movement of chloride. Since that time, many investigators have basically confirmed Üssing's hypothesis although several inadequacies have appeared. Similarly, the model developed by Üssing and extended by many other workers has proved useful in understanding the underlying mechanisms but, as discussed later, it is currently being challenged [10, 118].

The sodium transport system appears to be fundamental and there are suggestions that the same basic system that operates in frog skins also works in many different cells in a variety of organisms. Yet the basic question of the nature of the transport mechanism remains open. One of the less popular theories of ion transport, which attempts to explain the movement of sodium across the membranes of the frog skin, is the redox or electron-transport pump. Variations have been presented by several investigators. These hypotheses assume that oxidation and reduction of the cytochrome ion are spatially separated and that hydrogen ions formed in the reductive step are available for exchange for some cations outside the cell. Electrons are then passed along the chain; the cations that are taken up are possibly moving with them. The electrons are finally donated to oxygen to yield hydroxyl ions. Such a scheme has a stoichiometric limitation however: Only four electrons and four univalent cations can be transported per oxygen molecule consumed. Careful measurement of oxygen consumption and current flow (net sodium transport) in the frog skin has shown that close to seven sodium ions are transported for every oxygen molecule consumed [76, 77]. If one divides the *change* in transport by the *increase* in oxygen consumption when sodium transport is increased in one of several ways, ratios in excess of 20 can be obtained [80]. Thus, even if all the oxygen consumed were utilized in ion transport, the ratio of transport to oxygen consumption would be too great to be attributed to a redox pump mechanism.

The observations of Koefoed-Johnsen and Üssing [71] have suggested an alternative mechanism for active sodium transport across the frog skin. The effective anion permeability of frog skins is reduced to a minimum by treatment with copper or by replacing the chloride ions in the outside solution with sulfate ions (to which the skin has a low permeability). Such preparations yield maximal electrical potentials. By varying systematically the concentrations of sodium and potassium on each side of the skin, it can be seen that the inside of the skin behaves as if it were a potassium electrode (the potential follows the expression for the Nernst equilibrium potential). The potential across the skin drops 60 mV for every tenfold increase in potassium concentration in the inside solution (Fig. 25). Changing the potassium concentration on the outside surface of the skin has no effect but changing the outside sodium concentration does—the potential increases as the sodium concentration increases (Fig. 26). This observation suggests that the two surfaces of the cells that make up the skin have quite different and specific properties, a fact also suggested from many other kinds of measurements. The mechanism suggested, illustrated in Fig. 27, involves the cycling of potassium at the inner surface and the net movement of sodium from outside to inside. A Na ion is extruded from the inner cell surface by a forced exchange for K; a charged

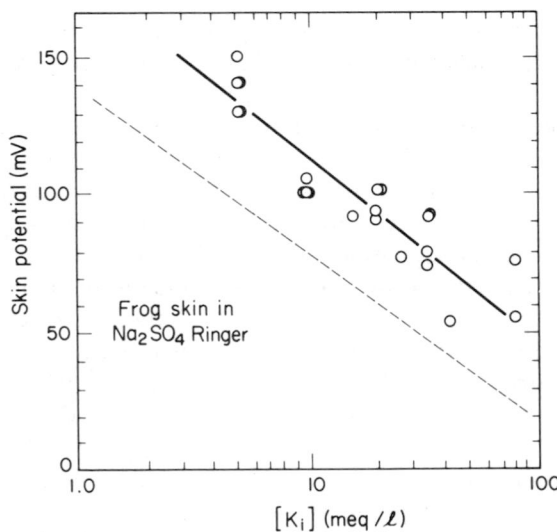

Figure 25. *Dependence of the frog skin potential on the potassium concentration in the inside bathing solution:* the dashed line shows the theoretical slope expected from a simple potassium electrode.

carrier drops off Na and takes up K. At the inner surface, by some exergonic reaction, the specificity of the carrier is changed and it gives up K for Na. As a result, the Na concentration inside the cell decreases, while the K concentration increases—a process that does not alter the electrical potential. Since the inner surface membrane is highly permeable to K, K diffuses out and this leads to a diffusion potential with the cell contents negative. The negativity of the cell contents tends to produce the uptake of a cation through the outer surface and since this surface is permeable only to Na, it is a Na ion that enters. The net effect is a cycling of K at the inner surface of the cellular membrane and the net movement of Na ions from the outer bathing solution, through the cell, to the inside of the skin. In the presence of Cl and a normal permeability to Cl, the potential is maintained at a lower level by the diffusion of the Cl ion from outside to inside. In the normal state, the net movement of Na and Cl must be nearly equal, although most investigators, to free their experiments from a number of complications, rarely duplicate the conditions that are faced by a living frog in nature. This model is also supported by the work of Hoshiko [58] who found potential changes to occur in two jumps when micropipette electrodes are progressively passed through the

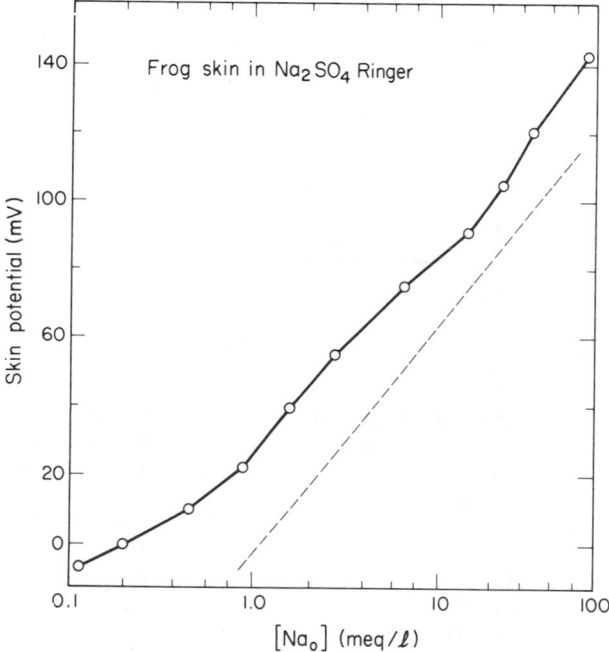

Figure 26. *Dependence of the frog skin potential on the sodium concentration in the outside bathing solution:* the dashed line shows the theoretical slope expected from a simple sodium electrode.

skin (Fig. 23). These two jumps can be interpreted as representing the changes that occur at the two surfaces of the active layer of cells. The model is not without defects, however. It predicts that zero potassium at the inner skin surface should stop the system. This happens but the reason is that the epithelial cells have shrunk; if cell shrinkage is prevented in some way, the active salt transport continues even in the absence of potassium [4]. Snell and Chowdhury [115] found more complicated effects than the model is able to predict when they simultaneously changed the composition of both bathing solutions. Cereijido and Curran [7] carefully measured the skin potentials with a microelectrode and found that changes in the external concentration of sodium altered the potential difference across the outer but not the inner boundary—as predicted by the model—but changes in the inner potassium levels altered the potential difference across the inner (predicted) and the outer (not predicted) boundaries. Despite problems, the model remains useful.

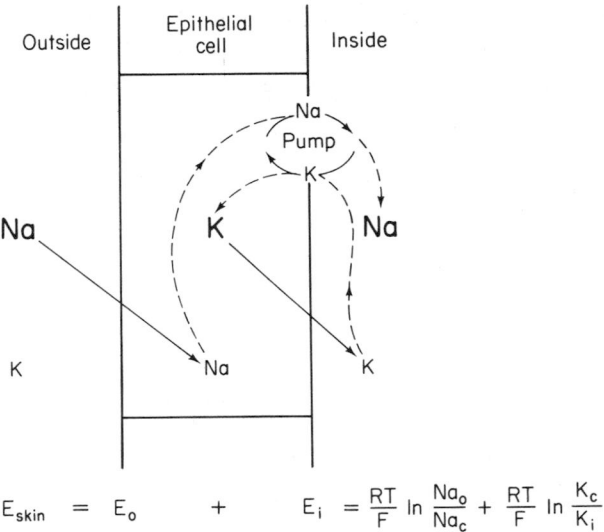

$$E_{\text{skin}} \quad = \quad E_o \quad + \quad E_i \ = \frac{RT}{F} \ \ln \ \frac{Na_o}{Na_c} + \frac{RT}{F} \ \ln \ \frac{K_c}{K_i}$$

Figure 27. *Üssing's model for active sodium transport across the frog skin:* the plasma membrane of the basal epithelial cell represented that is in contact with the outer bathing solution is presumed to be selectively permeable to sodium; the membrane facing the inside solution is presumed to be selectively permeable to potassium. In the absence of anion penetration (e.g., in sulfate-Ringer's), the skin electromotive force (E_{skin}) is the sum of the sodium diffusional potential (E_o) across the outer border and the potassium diffusional potential (E_i) across the inner border. The subscripts o, c, and i refer to the respective ion concentrations in the outside bathing solution, the cytoplasm, and the inner bathing solution. The size of the symbols for Na and K illustrate their relative concentrations. The solid arrows represent diffusion through membrane down a concentration gradient and the dashed arrows refer to diffusional paths into and out of the active transport pump.

Calcium and antidiuretic hormone (ADH), a polypeptide hormone of the posterior pituitary, are known to alter permeability properties of biological membranes. The effects of these two agents are interrelated [47]. Calcium added to the solution bathing the outside surface of an isolated frog skin causes a net decrease in the sodium transport across the skin. ADH, which works only when added to the solution bathing the inside surface of the skin, increases the sodium transport. The decrease in Na transport as a result of adding Ca is the same both before and after treating the skin with ADH. The increase in Na transport due to ADH is the same both before and after treating the skin with Ca. Hence Ca and ADH do not compete against each other but act independently at two different sites. The sites are in parallel, however; they are located on the same barrier to Na movement in the skin. Also, Ca causes a decrease in Cl influx across the skin but the ADH has no effect on Cl. The action of these agents is independent with respect to Cl movement. The effects on Na transport produced by Ca (added to the outside solution) and by ADH (added to the inside solution) can be explained in terms of changes in Na permeability of the outer membrane of the cells involved [14, 47]. Figure 28 gives the average values for calculated unidirec-

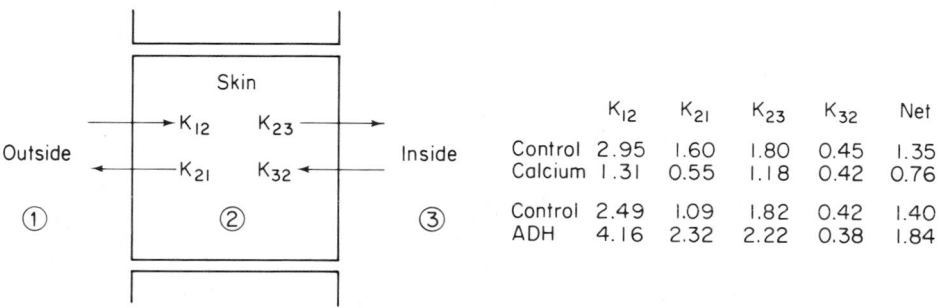

		K_{12}	K_{21}	K_{23}	K_{32}	Net
Control		2.95	1.60	1.80	0.45	1.35
Calcium		1.31	0.55	1.18	0.42	0.76
Control		2.49	1.09	1.82	0.42	1.40
ADH		4.16	2.32	2.22	0.38	1.84

Figure 28. *Model of frog skin used for kinetic analysis:* compartment 2 is not meant to refer to a particular structure in the frog skin but merely to represent that portion of the skin between the two major barriers to sodium movement. The arrows indicate the directions for the rate constants of the unidirectional sodium movements. The table contains the average values (microequivalents of Na per square centimeter hour) for calculated unidirectional sodium flux across the two barriers and the measured net sodium flux across the skin under the conditions given. Calcium decreases the sodium influx across the outer membrane as well as the net flux, whereas ADH increases both the influx and the net flux for sodium. There is no significant change in the sodium flux from the inside solution toward the cell compartment but the flux from cell compartment toward inside solution, assumed to represent the active transport step, is increased by the presence of ADH and decreased by Ca—both can be explained solely by changes in the Na pool inside the cell.

tional Na fluxes, given as microequivalents of sodium per square centimeter hour across the two barriers to Na movement in the frog skin. The average net Na flux across the skin, determined by short circuit current, is included. Both Ca and ADH act on the Na permeability of the outward facing membrane—Ca decreasing it and ADH increasing it—but neither affect directly the Na transport system located within the inner barrier. The change in the size of the Na pool within the cells alters the rate of active Na transport; permeability properties of the outer membrane can play an important role in the control of the active transport rate.

A frog or toad skin also shows water transport especially when the animal has been dehydrated. The active accumulation of salt leads to an osmotic gradient and the subsequent entry of water. The rate of water transport is under hormonal control mediated by the antidiuretic hormone (ADH). The osmotic water permeability of frog skin is low in the absence but high in the presence of ADH and appears to act only by increasing the permeability of the tissues to water [47, 79]. A frog sitting in water has a net uptake of water several times higher than predicted from measurements of the influx of radioactively tagged water and from the assumption that the flux in each direction across the skin should be proportional to the water activity in the solution of origin, which, in this case, is the lymph or extracellular fluid of the frog and the water in which the frog is immersed. Some investigators have suggested that this indicates an active transport of water. This is not a necessary conclusion, however, and the movements of water can usually be explained to result from the operation of osmotic effects produced by a variety of other phenomena.

Cereijido and Rotunno [9, 10] distinguish between transport across the membranes of individual cells and across more complex structures represented by epithelial membranes. They have proposed a quite different mechanism for the passage of ions through epithelial membranes from their work with the abdominal skin of the South American frog *Leptodactylus ocellatus*. Data assembled from many laboratories including their own are inconsistent with the view that ions permeate epithelial membranes by passing through the intracellular compartments of cells: (1) The major barrier to the passage of ions across biological membranes is presented by the layer of lipid components in the plasma membrane as clearly illustrated by the studies documenting the very poor ability of sodium ions to cross artificial lipid bilayers. Yet published data show that the sodium fluxes measured in the active direction across epithelial membranes are uniformly much greater than similar fluxes measured across single plasma membranes of individual cells. According to the transcellular models, such as discussed above, a flux through an epithelial membrane requires passage through two plasma membranes, not just one. (2) These same transcellular models predict that sodium enters the cellular compartment by a passive mechanism or that the concentration of sodium within the cells must be lower than in the outer bathing medium, which, in the case of frogs, is pond water. Measured values for sodium concentration within frog skin vary but most range around 100 mM. The magnitude of the electrical potential difference across the skin cannot explain sodium entry so an active sodium transport step must be present in the system. Ouabain-sensitive Na-K-ATPase has always been found associated with the sodium pump mechanism but this enzyme has not been found at the outer border of the skin [37]. Instead it has been found over more than half of the thickness of the epithelium. Moreover, calculations show that at least two layers of cells are required to supply sufficient Na-K-ATPase to pump the amount of sodium that is measured to pass across the skin, which is, in the case of *Leptodactylus ocellatus*, near 50 μM/hr cm². (3) Other studies have shown that only about 10% of the total amount of sodium present within the skin is participating in the transport process [8] and that more than 60% of the sodium present within the skin is bound and not present as free ions [104]. (4) Making the outside bathing solution hypertonic can produce a net influx through the frog skin of several small solutes that otherwise behave passively. While the transfer of these solutes can be demonstrated to occur through the extracellular space, the phenomenon is dependent on the maintenance of the active sodium transport mechanism [131]. Since no transcellular model could explain such data, Cereijido and Rotunno were forced to consider alternative models that would provide a route for sodium: (a) to enter passively an epithelium of high sodium content, (b) not to cross two lipid bilayers, and (c) to reach an active step located deep in the epithelium.

The essential feature of the Cereijido and Rotunno model is the location of the sodium-pumping route in the plasma membrane of the epithelial cells rather than through their cytoplasm; the sodium that is transported is not passing through the epithelial cells but is passing around them. Sodium is selectively adsorbed onto the surface of the outward facing cells. They assume that the sodium selectivity of the outer facing membrane is due to a particular ionic strength of polar groups of the outer layer of lipid molecules of the bimolecular lipid leaflet of cellular membranes. The ability of any ion to penetrate the hydrophobic lipid bilayer is so poor that it is energetically easier for it to travel tangentially by jumping from fixed polar group to fixed polar group than to cross the hydrophobic compartment of the membrane. The ion goes around the cell rather than penetrate it. Moving as counterions to the fixed membrane polar groups, the ions can pass through the regions of zonulae occludens, the regions of tight contact between cells in epithelia that prevent mass exit of interior fluids out of bodies and entrance of exterior fluids. Once passed through this region, an ion has reached the sodium-selective groups within the inner facing membrane of an epithelial cell. In addition to the membrane lipid bilayer, there is a second barrier of low permeability to sodium. This second barrier occurs only on the inner facing membranes and is located between the polar groups where sodium is migrating and the intercellular (or extracellular) space. This second barrier facing the intercellular space is covered with sites of ATPase that are also sites of the sodium pump. This second barrier is missing from the membrane facing the outer bathing solution; hence, sodium is pumped from sodium-selective polar groups into the extracellular space between the cells by these sodium pumps and is now effectively in the inner bathing solution. Each empty site left by pumping a sodium ion cannot be reached by sodium from either the intercellular space or the intracellular space. The site can only be refilled by sodium, acting as a counterion, jumping from neighboring polar group sites. Effectively, then, the empty site is moved sidewise toward the outward facing membrane where another sodium ion can be adsorbed. The mechanism is independent of the one used by the cell to maintain sodium balance, which helps explain evidence that metabolic inhibitors and diuretics stop net sodium transport without disturbing the electrolyte balance of the individual cells in the epithelium.

3.10 Sodium Transport
Across the Toad Bladder

Morphologically, the urinary bladder of a toad is much simpler than the abdominal skin of a frog. Toad bladder consists of but a single layer of mucosal cells plus a small amount of connective tissue and a noncellular serosal membrane [93]. Toads conserve water by reabsorbing it from urine in its bladder. Water follows osmotically the active transport of salt by this single layer of cells. The net flux of sodium equals the short circuit current [74], which means that sodium is basically the only ion transported. The potential difference across the bladder averages near 50 mV; the inside (mucosal) surface is negative with respect to the outside (serosal) surface [38]. Sodium is pumped from the inside of the bladder to the outside or into the body fluids of the toad. Frazier also found that the potential occurred in two steps when he slowly pushed a microelectrode through the wall. ADH increases the short circuit current by increasing the passive permeability of the mucosal side to sodium and thereby increasing the sodium pool that is available for active transport [73]. The toad bladder pumps between sixteen and nineteen sodium molecules for each molecule of oxygen consumed [75]— values that compare well with the sixteen to twenty found for frog skin [76, 77, 133]; hence, the toad bladder preparation is much the same as the frog skin.

3.11 Active Sodium Transport
by Excitable Cells

Since much will be told about excitable cells in Chapter 4, a quick summary will suffice here to indicate that active transport has been well studied in muscle and nerve cells. Table 6 summarizes the ionic distribution

TABLE 6

APPROXIMATE STEADY STATE ION CONCENTRATIONS AND
POTENTIALS IN MAMMALIAN MUSCLE CELLS AND INTERSTITIAL FLUID

Interstitial Fluid [ION] ($\mu M/cm^3$)			*Intracellular Fluid* [ION] ($\mu M/cm^3$)		$\dfrac{[ION]_o}{[ION]_i}$	$E_{ion} = \dfrac{61}{Z} \log \dfrac{[ION]_o}{[ION]_i}$ mV
Cations:		P L	Cations:			
Na$^+$	145	A S	Na$^+$	12	12.1	66
K$^+$	4	M	K$^+$	155	1/39	−97
H$^+$	3.8×10^{-5}	A	H$^+$	13×10^{-5}	1/3.4	−32
pH	7.43	M E	pH	6.9		
		M				
Anions:		B R	Anions:			
Cl$^-$	120	A	Cl$^-$	4*	30	−90
HCO$_3^-$	27	N E	HCO$_3^-$	8	3.4	−32
Others	7		A$^-$†	155		
Potential	0		−90 mV			−90

* Calculated from membrane potential using the Nernst equation for a univalent anion.
† A$^-$ refers to largely unknown organic anions.

and relevant electrical potentials for a mammalian muscle cell. The steady transmembrane potential (E_m) is defined as the potential of the inside solution minus the potential of the outside solution. This is the potential measured across the plasma membrane using an intracellular microelectrode. Since the cellular interior is negatively charged, E_m is a negative number. The concentration values given in the table are the measured amounts of substance per volume of water in that tissue sample. The right-hand column gives the calculated equilibrium potentials for each ion species as calculated from the Nernst expression using mammalian body temperatures.

From the values calculated for E_{Cl} and E_K, it can be seen that these ions are distributed across the membrane in near equilibrium with the membrane potential E_m. There are two possible interpretations: (1) E_m is generated by existing concentration gradients of Cl^- and K^+, the mechanisms establishing these gradients not specified; or (2) E_m is maintained by unspecified mechanisms and the K^+ and Cl^- distribute themselves in equilibrium with that potential. At present, it has proved difficult to distinguish between these two possibilities.

The distribution of Na^+ is far from equilibrium with the resting membrane potential; in the example listed in Table 6, $E_{Na} = +66$ mV contrasts with $E_m = -90$ mV. Two explanations have been proposed to explain this discrepancy. During the early part of this century, it was thought that Na^+ ions were not able to penetrate the membrane so that the disequilibrium was only apparent and persisted indefinitely, but the experimental data obtained since then clearly show that radioactive isotopes of sodium do penetrate the plasma membrane of muscle cells although not as readily as do potassium or chloride. The measured sodium influx in a frog sartorius muscle is 10^{-5} μeq/cm^2 sec. The passive outflux is less than 1% of the influx. Therefore, some force other than concentration and potential differences must exist to expel sodium at an average rate equal to the rate of passive entry. This force prevents the sodium influx from raising the internal concentration of sodium at a rate of, for example, 14 μeq/cm^3 hr in a muscle fiber 100 μ in diameter that contains only 12 μeq/cm^3. The power to eject sodium is derived from the cell's metabolic energy and the process is informally called the sodium pump.

For a frog skeletal muscle, Keynes and Maisel [68] obtained these values: $E_m = -88$ mV, $[Na^+]_o = 115$ μeq/cm^3, $[Na^+]_i = 20$ μeq/cm^3, and $J_{Na_o} = 8.7$ μeq/hr g of muscle. (This is not the best unit for outflux figures but is desirable here since oxygen consumption is given in cubic millimeters per hour gram). Energy production, calculated from oxygen consumption, was 0.17 cal/hr g of muscle tissue. They calculated the power requirements for sodium transport and found it to be 0.027 cal/hr g of muscle, assuming 100% efficiency for the process; a minimum of 15% of the oxygen consumption of a noncontracting muscle is used to pump sodium. Since the pumping efficiency is unlikely to be greater than 50%, the cell is probably using a third of its resting oxygen consumption to pump sodium. In any case, the energy demands of the sodium pump are not excessive and the postulate of active sodium transport is possible energetically.

3.12 Red Blood
Cell Ghosts

Another useful preparation for membrane studies has been the red blood cell ghost—the isolated sacs of plasma membrane that remains after erythrocytes are hemolyzed in hypotonic solutions. The plasma membrane does not break during hemolysis but it does become sufficiently permeable to allow hemoglobin and the like to leak out. The properties possessed by any particular type of ghost will depend on the manner of isolation, but after hemolysis a ghost can recover its impermeability to hemoglobin and regain its normal permeabilities to cations [51]. The isolated ghost preparation is valuable because of its relative simplicity and the great latitude it allows for experimental manipulations.

Electron micrographs of a ghost show a sheet of plaques, short cylinders each about 40 Å thick and 250 Å in diameter, packed in a hexagonal array over the entire surface and overlying a sheet of fibers, each about 20 Å in diameter and 100–400 Å long [48, 50]. Several methods estimate the thickness of a dried ghost membrane at 60–70 Å. The structure can be interpreted in terms of the paucimolecular theory of Danielli, Davson, and Robertson if the lipid contained in the ghost membrane is radially oriented and arranged as a bimolecular leaflet between the sheets of plaques and fibers. The structure can also be interpreted in terms of the pore hypothesis of Parpart and Ballentine [90] if some of the lipid is contained in the sheet of plaques and oriented tangentially similar to the structures in Fig. 15, p. 51.

The lipid components of the red cell membrane, which vary [23], do not appear to be important to normal cation transport in those cases examined. Hoffman [50] examined the permeability to Na and K in red cells obtained from rats raised on diets deficient in essential fatty acids and from people suffering from acanthocytosis. These rat cells are almost completely devoid of certain phospholipids, contain new phospholipids, and look much like normal cells in size and shape. Acanthocytotic cells, which have a characteristic thorny appearance, contain normal amounts of phospholipid but less lecithin and more sphingomyelin than normal. Both types of cells are deficient in linoleic acid. These two types of red cells present specific but different alterations in their lipid compositions. Hoffman's measurements of the influx and the outflux of both K and Na under a number of conditions on both types of red cells yielded the same values as did his measurements on normal cells. This suggests that these lipids, at least the altered ones, are either not involved in the various pathways of normal cation transport or, if they are, that these types of alterations are not important for the transport processes.

A different series of experiments gave the same conclusion. Mammals can be separated into three distinct groups on the basis of the concentrations of Na and K in their red blood cells [123]. The erythrocytes of most mammals, such as man, rabbit, and rat, have intracellular concentrations high in K and low in Na, much the same as for most body cells of all mammals. The reverse is true in some carnivores, such as cats and some dogs, whose red cells are high in Na and low in K. The third group is variable: Animals in the order Artiodactyla, such as pig, ox, or sheep, have red cells that are high either in K or in Na. One type of sheep has red cells high in K but low in Na, as in man, while another type of sheep has cells high in Na and low in K. The two types of sheep represent the difference in a single gene [36] and the low K type is inherited as a Mendelian dominant trait [123]. Both types of sheep have the same lipid composition in their red cells [51]; therefore, there is no correlation between lipid composition and Na or K content of the red cells nor is there any correlation with the K or Na permeability.

The relationship between lipid composition and permeability of membranes to inorganic ions is more complicated than implied by the last experiment. For example, membranes constructed of lecithin extracted from rats raised on diets deficient in essential fatty acids show a faster sodium efflux at 25°C but a slower efflux at 50°C than control membranes constructed from lecithin extracted from egg, fish, or normal rat material [83]. Such data suggest that rats grown on diets deficient in essential fatty acids have a more inefficient use of calories and a higher metabolic rate than normal rats partly because of the higher level of active transport of ions required to maintain a concentration difference across a more permeable membrane.

3.13 Cation Movements
in Red Blood Cells

The movements of cations across membranes can be classified into three pathways [126] as summarized in Fig. 29. The first is passive diffusion represented by the term *leak*—ions moving in the direction of their electrochemical gradients. Exchange diffusion, a special type of diffusion that can be observed only with radioactive tracers, is the obligatory exchange of an ion in the

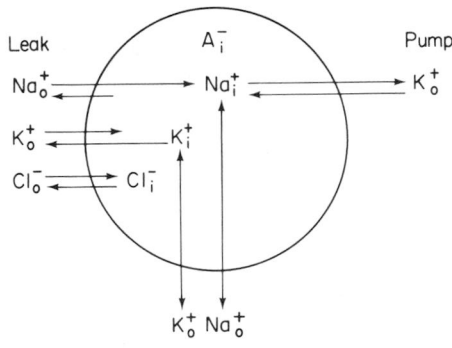

Figure 29. *Diagram of a cell characterizing the three major modes of cation transport.* Pump refers to the active transport mechanisms; leak and exchange diffusion are passive processes; A_i^- illustrates the quantity of large and impermeable anions, primarily protein, inside the cell.

cytoplasm for another of the same kind in the environment. No net transport can be accomplished so its importance is more in the analysis of the experimental data than in the discussion here. Active transport is represented by the term *pump*—ions are moving against their electrochemical gradients requiring the expenditure of metabolic energy. Net movements can occur only by the leak or the pump: the pump involves the exchange of Na inside for K outside; exchange diffusion involves the exchange of Na for Na and K for K; the leak is the combined movement of cation with anion.

The intracellular cation composition can be controlled by a pump that operates to compensate for the changes brought about by diffusion leak. In red cells, there is ample evidence to indicate that the pump flux of Na is tightly linked to the pump flux of K [126] and the idea of a coupled Na-K pump in a 1:1 ratio in a number of systems is not new [39]. Tosteson and Hoffman derived expressions to predict the membrane permeabilities that are needed to maintain the internal concentrations of cations or to predict the steady state Na and K concentrations from known permeabilities, given extracellular conditions. For any given cell, the pertinent membrane parameters turn out to be (1) the ratio of the two leaks, the influx of Na to that of K; (2) the ratio of the permeabilities, pump to leak; and (3) the coupling ratio of the pump, in this case, unity. Their predictions concerning differences in membrane parameters between the two types of sheep erythrocytes were confirmed by experiment. Both cell types have an active transport mechanism to exchange an internal Na for an external K that operates in parallel to the diffusion pathway for both Na and K. The red cells high in potassium (HK cells) have higher ratios for both pump/leak and J_{Na_i}/J_{K_i} than do the low potassium (LK) cells. Both ratios are decisive in determining the difference in steady state ionic composition. For example, consider a red cell living in plasma that has a certain leak permeability to sodium and another—but higher—leak to potassium; the 1:1 coupled Na-K pump will raise the intracellular concentration of potassium the faster the pump turns over. The pumps for each cell type, both LK and HK cells, differ only in a quantitative way. The Na-K pump is not absent from the LK cells but it operates at about one-fourth that rate observed in HK cells; hence, both cell types control their steady state composition in essentially the same manner.

This conclusion and the observation that the differences between LK and the HK cells represent the genetic activities of a single pair of alleles [34] provoke interesting speculation about the evolution of cellular transport. The distinctions between the three modes of ion transport cannot be as independent as it may appear. A common step in early cellular differentiation can occur to allow one process to determine the characteristics of the other two. A less emphasized but perhaps more primitive cellular function of active cation transport is to control the total intracellular solute content of the cell and thus its volume. Indeed, red blood cells are in osmotic equilibrium with their environment. It is not difficult to imagine that the first animal cells used active sodium and potassium transport to control their total cation content and, as a result, their volume much like the function subserved in LK sheep red cells. A later evolutionary development may have been the development of cells with high concentrations of potassium and low concentrations of sodium, which would permit the development of cells capable of generating action potentials and the like characteristic of excitable cells [123].

3.14 ATP and Active Cation
Transport in Erythrocytes

Ghost preparations of erythrocytes have yielded important information about the energy source for the cation pump and its correlation with an enzyme. The pump properties of ghosts are similar to those of intact cells but known amounts and kinds of compounds can be incorporated into ghosts at the time of hemolysis without difficulty [120], an experimental flexibility not permitted with intact cells. Intact cells are impermeable to all the phosphorylated intermediates in the metabolic pathways and there always exists the possibility for other interrelated reactions to interfere with the interpretation of the experiments. The study of single reactions is possible in ghosts because interference can be eliminated either by dilution or by removal of substrates. Any intracellular composition that is wanted can be created so that the properties of the cellular membrane can be studied in isolation.

Another important development that permitted studies on the energy source for the transport pump was the discovery by Schatzmann [105] that cardiac glycosides, such as ouabain or strophanthin, completely inhibit sodium extrusion and potassium accumulation by erythrocytes without interfering with glycolysis, the energy supplying metabolic activity of mammalian red blood cells. The inhibitory action of these compounds depends on competition for a carrier molecule in the pump mechanism. Whittam [136] noted that they spared the ATP supply of the cell while inhibiting active cation transport, which is strong evidence for the direct participation of high-energy phosphate in the active transport process. The action of the glycosides, which act only at the outer surface of the membrane [49], was confirmed by Glynn [40] when he found that an increase in the outside concentration of potassium partially reverses the inhibition, increasing the outward flux of sodium.

Hoffman's studies, using these two developments, deserve further comment. Hoffman [50] measured sodium outflux from human red blood cells by hemolyzing them with ten volumes of water containing a trace amount of ^{24}Na and a test substance. The resultant labeled ghosts were washed with $MgCl_2$ to remove as much excess ^{24}Na as possible. The cells were then suspended in thirty volumes of an isotonic medium, pH 7.4, and incubated at 37°C. The loss of ^{24}Na was measured over several hours. Such ghosts contain only about 9% of their original hemoglobin and enzymatic activity. Two types of media were used: (1) a sodium-free medium in which the sodium was replaced by magnesium and (2) a sodium-containing medium. 5 mM K was maintained in both media. In some of the experiments, inosine was added directly to the incubation medium, since the membrane is permeable to inosine, and served as an energy source. Inosine enters the Emden-Meyerhof pathway through the hexose monophosphate shunt and thus can stimulate glycolysis. Strophanthin was added to some experiments as a tool to assay the activity of the cation pump. Hoffman thusly concluded that ghosts behave as intact cells since all three modes of transport can be demonstrated (Fig. 30). The difference between curves A and C is due to the action of strophanthin and, therefore, represents the activity of the pump. The difference between curves A and B is due to the presence of Na and, hence, represents the Na exchange diffusion. This also applies to the difference between curves C and D. Curve D represents the leak alone since, in the

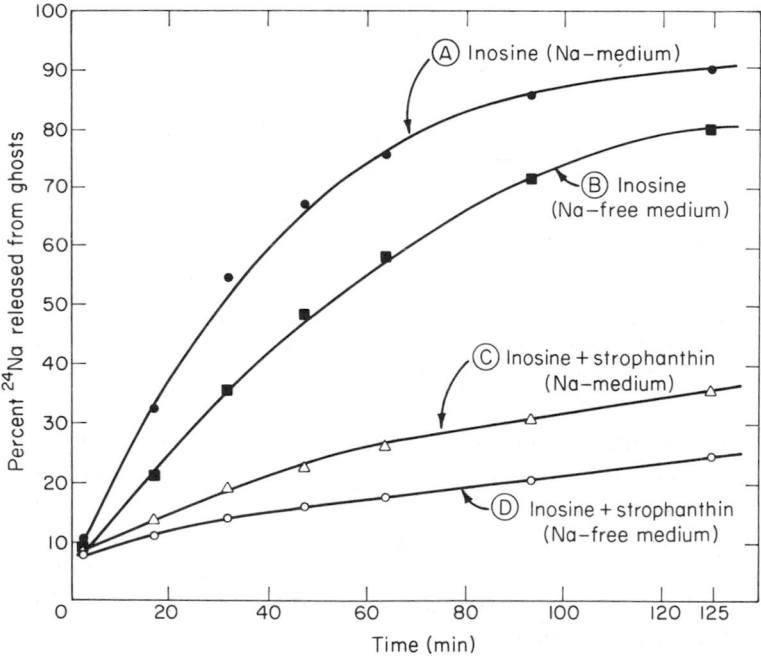

Figure 30. *The rate of loss of ^{24}Na from red cell ghosts suspended in different media.* Inosine supplies energy for the pump. Strophanthin inhibits pump activity.

presence of strophanthin, there is no active transport and there can be no exchange diffusion in a Na-free medium.

Hoffman also demonstrated that the pump can be activated by extracellular potassium in reconstituted ghosts. Figure 31 shows that the addition

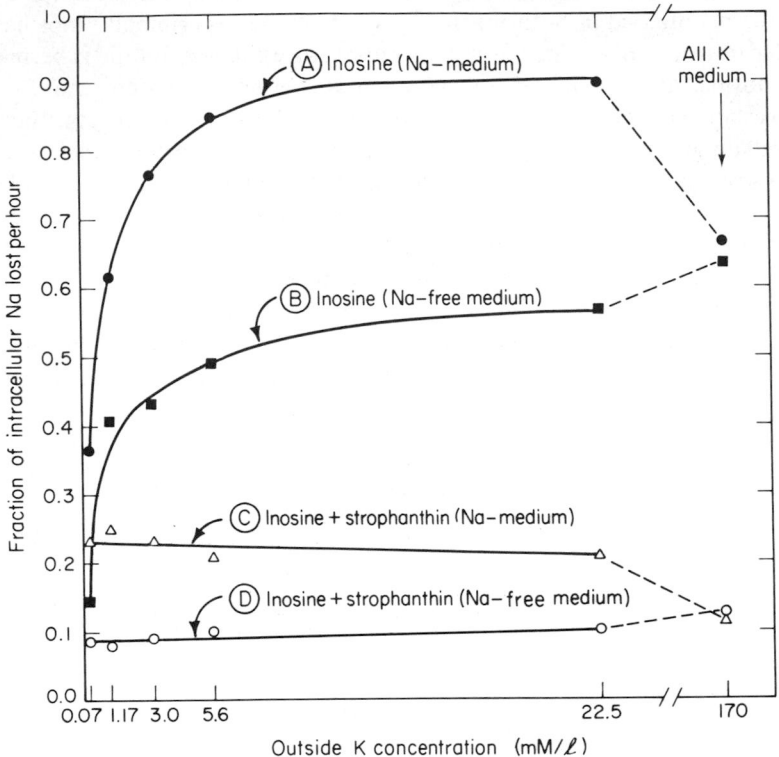

Figure 31. *The nature of the influence of extracellular K concentration on the rate of loss of* ^{24}Na *from ghosts:* the increased loss of Na is only effected by K when the pump is operable (no strophanthin present in the system).

of K to the outside medium increases the Na outflux until saturation is reached. This occurs in the presence or the absence of any Na exchange diffusion. This K activation of the pump can be inhibited by the addition of strophanthin. Steinbach [119] in muscle and Harris and Maizels [46] in red cells first observed this kind of activation of Na outflux by extracellular K. Harris [44] and Glynn [39] used these observations as the basis for a coupled pump since the dependence indicates that K must exchange for Na for the pump to operate. The addition of strophanthin or the removal of K, hence, results in the inhibition of the pump and both procedures have been used to assay the pump in ghost preparations. Hoffman used ghosts to confirm Glynn's experiments on intact cells, mentioned above, that showed strophanthin and K competing for the pump site. Activities are similar in both intact red cells and in ghosts.

Several experiments provide clues to the identity of the specific energy source or substrate for the pump. Fresh ghost preparations, i.e., ghosts prepared immediately after withdrawal of the blood from the donor, can use either adenosine or inosine as a substrate in the activation of the sodium pump mechanism. But only adenosine—not inosine—can activate the pump if the cells have been incubated for 18–24 hr in the Mg-K medium, a treatment that depletes ATP, ADP, hexose diphosphate, and diphosphoglyceric acid. The depleted control ghosts produce no lactate but the addition of either adenosine or inosine stimulates glycolysis to evoke large but essentially similar increases in lactate production. Adenosine and inosine are about equally effective in promoting a partial replenishment of all the phosphorylated glycolytic intermediates. Hoffman [51] confirmed these experiments and also demonstrated the synthesis of inosine triphosphate from inosine and adenosine triphosphate from adenosine. Obviously, the differences produced by these two compounds cannot be attributed to the intermediates but must be due to the different nucleotides formed. When ATP or ADP are incorporated into reconstituted ghosts, the pump outflux of sodium is activated. This activation is unaffected by the addition of adenosine or inosine to the system. Strophanthin inhibits the ATP activation, which indicates that ATP, by itself, can stimulate the pump. It is likely that the apparent ADP activation of the pump results from the formation of ATP by the enzyme adenylic kinase known to be present in the system. The pump is specific for ATP as other nucleotides—guanosine triphosphate (GTP), uridine triphosphate (UTP), and inosine triphosphate (ITP)—are not effective. In another series of experiments, Hoffman demonstrated that the apparent activation by phosphoenolpyruvate (PEP) is due to a trace amount of ADP. The ADP serves as the acceptor for the phosphate removed from PEP by the enzyme pyruvate kinase in the formation of pryuvic acid. The ATP that is formed drives the pump. In fact, any reaction that will generate ATP will activate the pump. The addition of glucose to the system even in the presence of iodoacetate poisoning inhibits ATP activation of the pump [32]. Natural hexosekinase is denatured during hemolysis so hexosekinase must be added to the medium. ATP is degraded to ADP during the conversion of glucose to glucose-6-phosphate by this enzyme. Apparently, the hexosekinase reaction preferentially uses all the available ATP leaving none to react with the pump mechanism. The conclusion from such studies is that ATP is the immediate source of energy for the pump; the link between metabolism and transport is ATP.

An ATPase displaying characteristics of the pump has been found in crab nerve and attached to red cell membrane [33, 94]. There are now many preparations of this enzyme and they are all attached to the membrane and not soluble [2, 88, 102, 103]. Ghost systems used in the ATPase studies differ from the ghost systems used to examine the cation pump. Hoffman [51] used ghosts permeable to ATP, Mg, Na, and K to characterize the red cell ATPase. Strophanthin added to ghosts incubated with Mg, ATP, Na, and K inhibits a sizable fraction of the total amount of inorganic phosphate liberated, a measure of the enzymatic activity of the ATPase. The strophanthin-inhibited ATPase activity is located at the inside of the membrane and the remainder—the apyrase activity—is at the outside. The characteristics of the ghost membrane ATPase—the requirement of both Na and K for activation, the inhibition by strophanthin, and the specificity for ATP (ADP, ITP, UTP, and GTP are all inactive)—parallel those for the pump. Calcium inhibits the ATPase in the presence of Mg but activates the apyrase fraction, which requires Mg. There is no appreciable breakdown of ATP in the absence of Mg. Hoffman demonstrated similar activity in the pump by saturating the binding capacity of the ATP and substituting Mg for Ca at low Ca concentrations: The total outflux of Na decreased as the calcium level rose. Calcium inhibits the pump only when it is located on the inside of the ghost preparation. Calcium has no effect when it is in the plasma nor does it have any effect upon the leak component. Summarizing the common properties of pump and ATPase, (1) both Na and K are needed for activation; (2) ATP is the substrate specifically; (3) pH sensitivities are alike; (4) both systems are inhibited by Ca, F, Cu, and strophanthin; and (5) both systems are unaltered by iodoacetate, sodium nitrate, arsenic, dinitrophenol, cyanide, or iodine. Hoffman concludes from these similarities that the ATPase is an intimate component of the cation pump in red cells.

Tosteson [123] found that ATPase activity, like pump activity, is four times greater in HK cells than in LK cells of sheep ghost preparations. The ratio of the K-activated ATPase in intact red cells to (Na + K)-activated ATPase in ghosts is unity, or equal, for both LK and HK cells. This observation is strong and independent evidence for the involvement of the membrane-bound ATPase in the active transport mechanism. Also, the Na plus K stimulation of ATPase is present in LK membranes at a quarter the activity observed in HK cells rather than complete absence. Kinetic analysis produced evidence that ATP is split by essentially the same enzyme molecule in both sheep genotypes although there is a substantial number of catalytic sites in the HK membrane that require both Na and K to be present to unmask them, contrary to the case for LK membranes. As Tosteson points out, such data are not consistent with the view that the alkali metals form intermediate complexes with substrate-enzyme molecules, as proposed in most models for active transport. Rather, the data suggest a more direct action of the alkali metals on some component of the enzyme system itself.

Hoffman and Kregenow [53] identified a second pump in red cells that extrudes sodium. The first pump, discussed above, is a ouabain- or strophanthin-sensitive, linked Na-K active transport system that displays a strong correlation between the level of the potassium in the cell and the activity of transport. The second pump is insensitive to ouabain or strophanthin, is inhibited by ethacrynic acid (a diuretic), has a higher affinity for sodium, is less dependent on metabolic depletion, and predominates in cells with low intracellular concentrations of sodium or in metabolically depleted cells.

3.15 Models of Active Transport

Many different hypotheses have been advanced to explain transport across biological membranes and no attempt will be made to summarize them all. Most models can, however, be conveniently classified as being either carrier mechanisms with mobile units or noncarrier mechanisms with fixed or anchored units. At present, it is not possible to distinguish unambiguously between them. There does not now exist any experimental basis for characterizing an actual membrane process as carrier or noncarrier although the work on model systems is providing help [35]. Patlak [91] gives theoretical arguments that emphasize the insufficiency of kinetic data per se to distinguish between the two types of processes. Again, as in the passive transport case, the advent of good thermodynamic analyses have permitted much progress in our understanding [59]. Detailed discussion of most of the proposed models appears almost annually in good reviews [2, 11, 31, 52, 63, 96, 103, 118, 135].

Osterhout [87] carried out the initial and now classical experiments relating transport to a carrier mechanism. He used liquid guaiacol as an artificial membrane separating two aqueous phases to show that it could selectively transport cations from one aqueous phase to the other and that the transport depended on the formation of a dissociable guaiacol-cation complex. More recently, others have used artificial membranes that mimic the behavior of biological membranes to study transport properties, and many examples of specific carriers for the permeating molecules have been documented for these artificial systems (see Chapter 5). While many of these substances possess the desirable feature of being constitutents of cellular membranes, however, their involvement in the process of active transport remains to be demonstrated.

Figure 32 illustrates a carrier-type scheme that is sufficient to describe most transport systems [41]. X and Y are carrier molecules confined to the membrane phase; X is specific for K^+ and Y is specific for Na^+. The carriers can cross the membrane only when they are combined with the ions, which can be adequately explained on the basis of concentration gradients for the complexes formed alone. K carriers (X) are converted to Na carriers at the

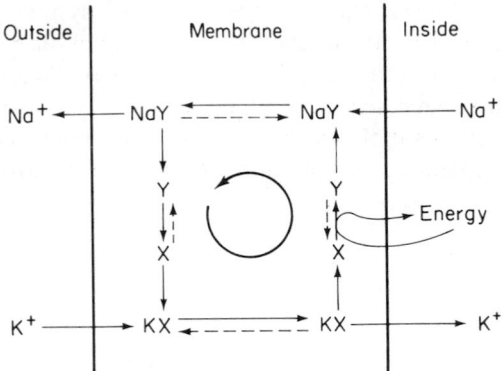

Figure 32. *An example of a carrier model for active transport:* X, a carrier molecule specific for K, can be converted to Y, a carrier for Na, by the addition of energy. Y reverts back to X spontaneously. Continuous cycling of the carriers results in the transport of K inward and Na outward.

inside surface of the membrane by metabolism. The Na carriers are re-converted to K carriers again spontaneously at the outer surface of the membrane. Active transport as well as the K-Na linkage can be satisfactorily explained by this model. If the system cycles in a counterclockwise direction, as indicated, a net movement of Na will occur toward the left and a net K movement will occur toward the right. Red cell transport appears to operate by a system like this one. For example, the removal of K from the outside bathing solution will result in the accumulation of only K carriers at the outer surface of the membrane and the sequence will come to a halt, as observed empirically. But this model is not very specific and, hence, not very useful.

Another carrier-type model with more specific features has been suggested by Hokin and Hokin [54–57] to explain data obtained from the salt-secreting glands of the marine bird, the albatross. This bird possesses salt-secreting glands, stimulated by cholinergic nerves, in its beak and is able to maintain its blood hypoosmotic to the seawater it drinks by actively pumping salt out of its body through this gland. The Hokins discovered a phosphatidic acid turnover correlated with NaCl transport. In the model (Fig. 33), diglyceride (DG) and phosphate transferred from ATP are converted to phosphatidic acid (PA) by the enzyme diglyceride kinase. As an aside, this enzyme is also implicated in transport in human red cells by Parker and Hoffman [89]. Na on the inside combines with the PA and the complex moves to the other side of the membrane. Here an enzyme, phosphatidic acid phosphatase, splits the complex to liberate the Na, which passes out; the DG, which returns to begin the cycle again; and the phosphate. The phosphate is somehow prevented from escaping and it returns to the inside compartment to await

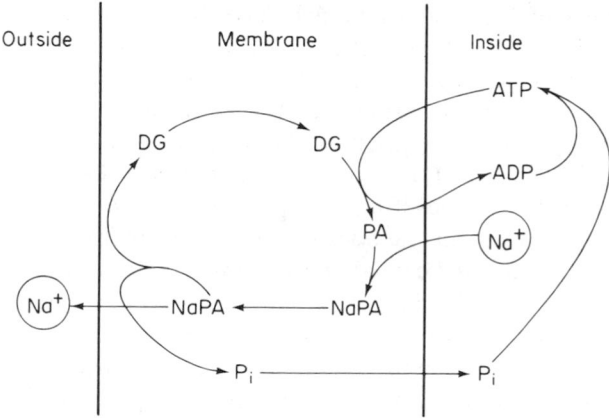

Figure 33. *Another carrier model for active transport of sodium:* the enzyme diglyceride kinase catalyses the reaction between diglyceride (DG) and ATP to form phosphatidic acid (PA), which binds Na. The complex is dissociated at the outer surface by the enzyme phosphatidic acid phosphatase to release the Na and permit inorganic phosphate to return to the inside compartment.

incorporation into ATP. The Hokins feel that the phosphatidic acid cycle reactions can produce conformational changes in a carrier lipoprotein that could bring about active transport in a number of ways—the phosphatidic acid is not the carrier but some lipoprotein is. The cyclic phosphorylation and dephosphorylation of a molecule of phosphatidic acid that forms part of a carrier lipoprotein is considered to be a mechanism by which the chemical energy of ATP is fed into the transport system [55]. In this way then, the phosphatidic acid cycle reactions act as a transducer to convert chemical energy into conformational changes of specific lipoprotein carriers and the conformational changes lead to the performance of osmotic work or secretion. One of the desirable aspects of Hokins' model is the specific use of ATP, but the model leaves unexplained the pump flux of K and, more important, the linkage between K and Na. Also, the experiments correlating transport and phosphatidic acid synthesis have not yet been accomplished. Acetylcholine added to slices of salt glands stimulates the incorporation of inorganic phosphate into phospholipids, especially into phosphatidic acid [56]. But phosphatidic acid is a central intermediate in the biosynthesis of phospholipids and it is difficult to separate effects on phospholipid metabolism from those on sodium transport [57–67]. The actual intermediate in the Na-K-activated ATPase has the properties more of an acyl phosphate [113, 135] than of a phosphatidic acid, as required in scheme presented above. There is more current interest in the acyl phosphate intermediate than in the phosphatidic acid cycle as compounds in transport systems.

Other types of basic carrier models have been proposed, for example, the fixed-carrier model. An ion is attached to the carrier and without any translational movement by the carrier the complex transports the ion across a thin but high potential barrier and then releases the ion. Transport is accomplished by either a rotation of the complex or by attachment of the substrate to the "free" end of a long carrier molecule that then contracts if initially expanded or extends if initially contracted [100]. Another model is a generalization of this with the addition of several more carrier molecules. The ion joins onto the first carrier and is transferred to a second carrier and so on. In this manner, the ion is transported across the region [78]. Cation transport is accomplished by the exchange of H^+ ions for the cations. The source of energy is the drop in potential of the electrons.

Many types of noncarrier mechanisms have been proposed [86]. Goldacre [42] envisages the folding of protein molecules that have picked up the material at the cell surface, a contractile protein molecule attached to the membrane extending and contracting and coiling. Uncoiling involves the breaking of intramolecular bonds and so forming affinities that can be satisfied only by absorption of foreign molecules or ions. When the protein recoils, the original bonds are reformed and the foreign matter is released or desorbed inside the cell. The protein molecules may also act as enzymes capable of liberating energy from, e.g., high-energy phosphate compounds to get the transport work accomplished. The idea presented by Abrahamson and Moyer [1] is basically a form of electroendosmosis or the mass movement of fluid through a fixed membrane as a result of a difference in electrical potential on both sides of the membrane. It is difficult to see how such a process could be highly selective and it has not received much attention from current investigators.

Skou's model [112] has received much attention since it does correlate with much of the pump evidence. Skou's model uses ATPase. The enzyme E combines with ATP, Mg^+, Na^+, and K^+ to form a complex. Phosphorylation of the enzyme frees Mg^{2+} and ADP from the complex. The transport reaction is the breakdown of the phosphorylated complex to yield again the enzyme and inorganic phosphate; Na and K are moved in the appropriate direction as they are released. In summary,

$$E + ATP + Mg^{2+} + Na^+ + K^+ \rightleftharpoons \overset{\displaystyle K}{\underset{\displaystyle Na}{E}}\text{--Mg--ATP} \rightleftharpoons \overset{\displaystyle K}{\underset{\displaystyle Na}{E}} \sim P + Mg^{2+} + ADP$$

$$Na^+ + E + P_i + K^+.$$

One possibility for this to occur is through the operation of a gate-type or rotator mechanism. Skou's formulation does not show that the enzyme mol-

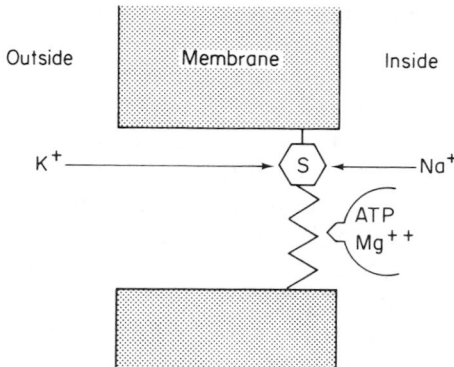

Figure 34. *An example of a noncarrier model for active transport:* S indicates the site on the ATPase enzyme that reacts with Na at the outer surface, rotates inward to give up the Na and become a K site. The site now spontaneously returns to its original position yielding K to the inside and again becoming a Na site.

ecule—the ATPase itself—is the pump, as is diagramed in Fig. 34. The enzyme is structurally fixed and is situated at the end of a pore in the membrane. The enzyme is therefore able to control passage through that pore. S indicates the position on the enzyme of the site that can be occupied by Na or by K sequentially but not by both concurrently. In the unenergized state, the site is occupied by Na and faces toward the inside of the cell. As enzymatic breakdown of ATP occurs, this site is rotated or shifted in such a way that it now faces toward the outside. Not only is the site displaced spatially but its specificity is changed: The site changes from a Na one to a K one. After the reaction is completed, the enzyme returns to its original state and the K site returns to a Na site facing the inside of the cell. The reaction requires both Na and K and, hence, the proper ion must occupy the site during both phases of operation.

It is possible that there are two sites, one Na and one K, as in Skou's original formulation. Moreover, the mechanism must depend on the ability of the enzyme to show a little molecular flexibility; that is, the enzyme changes shape during its reaction with the substrate. One characteristic of this induced fit mechanism is that the curve relating the rate of enzymatic reaction to the reciprocal of the absolute temperature (an Arrhenius plot) is nonlinear, as is true for the red cell ATPase [49]. Although this is consistent with this model, it is not a convincing piece of evidence.

Skou [114] summarized his later work in which he has accumulated data on the enzyme system and correlated it with known properties of ion-transport mechanisms to support the essential features of this model. The enzyme system that he has isolated—the enzyme, the sites with affinities for monovalent cations, and the structural components that are necessary for their function—appears much like the transport system and, indeed, it seems that Skou has isolated the transport system in his preparations. The affinity of the ATPase for Na is about three to four times the affinity for K. The hydrolysis of ATP is a two-step reaction: A Na-dependent phosphorylation of a compound in the enzymatic particle is followed by a K-dependent dephosphorylation [95]. Such evidence implies that there are two separate sites—one for Na and one for K—rather than one site that changes in form. Other evidence is more suggestive than conclusive that the cations which activate the ATPase are also the cations which are transported. For example, the same inhibition phenomena of the transport system and the isolated ATPase system by strophanthin are found. The result is a model that helps explain how the chemical energy from ATP is transformed by the transport system into a movement of cations against an electrochemical gradient. Another aspect offered by this type of model is that the pump can be considered to convert the chemical energy of ATP into mechanical energy of unidirectional movement—a transducer role that relates the pump to mechanisms for muscular contraction.

Pardee [88] summarized the work on isolation and characterization of proteins that appear to be parts of membrane transport systems. One isolation method is illustrated by the work of Stein and others [118]. These workers isolated a protein specific for the transport of β-galactosides in bacteria by growing normal cells (those capable of transporting β-galactoside) in ^{14}C-arginine and abnormal cells (cells incapable of transporting β-galactoside) in ^{3}H-arginine. When both types of E. coli cells are mixed, broken, and fractionated, the protein fraction with the highest ratio of ^{14}C to ^{3}H should contain the transport protein; the transport protein should be labeled by ^{14}C but not by ^{3}H, whereas all other proteins should have both kinds of labels. Other experimenters have modified this double labeling technique to yield a cleanly isolated transport protein. Another approach used to identify a protein associated with a transport process is to measure specific substrate binding by cellular fractions. The equilibrium between two solutions will be altered by the amount of substrate bound to the transport protein. In this way, Pardee, Roseman, and Kundig, among others [62, 88], were able to isolate and crystallize a sulfate-binding protein from Salmonella typhimurium. Most of the transport protein identified in these ways are similar but not identical. They all have about the same size (molecular weight near 30,000 daltons) and one specific binding site each and are associated with the membrane fractions. Other transport proteins have been identified by reactions associated with the energy supply for the active transport process. In some cases, the energy donor changes the substrates and, in other cases, the transport protein is changed. For example, ATP appears to alter the transport protein in the translocation of sodium and potassium across membranes [113].

Most of the evidence for each protein is incomplete. Critical experiments, for example, that try to add the protein to cells incapable of transport in order to reconstitute the transport system are not conclusive. No evidence is presently available to distinguish among the translocation mechanisms proposed. Of the carrier systems, the permease model suggests that the permease enzyme, the recognition protein, catalyzes bond formation between the substrate and a low-molecular-weight transporter molecule in the membrane, which is assumed to be soluble in membrane lipids and to diffuse across to the other side, dissociate, and release the substrate. Another hypothesis suggests that the recognition molecule itself carries the substrate across the membrane by some conformational change. Hopefully, work with the isolated proteins will produce more conclusive evidence concerning transport mechanisms. Presently, it seems that the transport systems generally have few specific parts because few genes appear to be involved in their construction.

3.16 Conclusions

In the words of K. S. Cole [12], "we find lively, noisy problems in the traffic of molecules and ions across a bilayer." Some of the ideas of lipid solubility have persisted since those of Overton in 1899. Exceptions in expected behavior have led to descriptions of molecular sieving, charge interaction, solvent drag, bulk flow, direct effects on membrane molecular architecture, shrinking pores, local osmosis, restricted and exchange diffusion, unstirred layers with some of the ingenious methods for their measurement [111], carriers or protein-lined pores for both passive and pumped transport across a small fraction of the bilayer, asymmetrical membranes in series, co-transport, counterflow, electroosmosis, and streaming potentials. We are not often able to tell which came first—the potential or the ionic distribution—so that the Nernst-Planck model becomes more of a description than an explanation. Nonequilibrium thermodynamics has joined kinetics as the method for analysis of data but we still have more data than conclusions. The current status of our understanding of the physical basis for ion and non-electrolyte selectivity shown by biological membranes has been effectively summarized by Diamond and Wright [31].

A strange variety of preparations have proved their worth in these studies. Any list would include frog skins; rabbit gallbladders; squid axons; toad urinary bladders; sheep brains, guts, and blood (and in that order with respect to decreasing pump ATPase activity); and the albatross and seaweed from the oceans.

Red blood cells have contributed much to our model building [116] and to our knowledge of the source of energy to drive the active transport pumps. An ATPase enzyme system that is dependent on both Na and K is membrane bound and found in most animal cells [114, 135]. Enzyme activation occurs only when Na is at the inner and K at the outer surface of red cell membranes [134]. The cardiac glycosides, especially ouabain, inhibit this enzyme system specifically as they do cation transport [106], with similar kinetics at physiological conditions [86]. Tosteson implied that the HK type of sheep red cell is a more recent evolutionary development than the LK type of cell but he also found indications that the HK-type membrane appears earlier in developing sheep red cells than does the LK-type membrane [124]. Curiously, cats do not maintain much of a gradient to either K or Na across their red cell membranes and, correlated with this, cat erythrocytes lack an ATPase activated by Na and K and show no inhibition by ouabain [110].

These few comments merely point up the observation that this traffic problem has not yet been solved either.

REFERENCES CITED

1. H. A. ABRAHAMSON and L. S. MOYER, 1937. Some recent developments in electrokinetic methods and their applications to biology and medicine. *Trans. Electrochem. Soc.* **71**: 135–152.

2. R. W. ALBERS, 1967. Biochemical aspects of active transport. *Ann. Rev. Biochem.* **36**: 727–756.

3. R. W. BERLINER, 1959. Membrane transport. *Rev. Mod. Phys.* **31**: 342–348.

4. N. S. BRICKER, T. BIBER, and H. H. ÜSSING, 1963. Exposure of the isolated frog skin to high potassium concentrations at the internal surface. I. Bioelectric phenomena and sodium transport. *J. Clin. Invest.* **42**: 88–99.

5. W. A. BRODSKY and T. P. SCHILB, 1965. Osmotic properties of isolated turtle bladder. *Am. J. Physiol.* **208**: 46–57.

6. C. W. CARR and K. SOLLNER, 1964. The electroosmotic effects arising from the interaction of the selectively anion and selectively cation permeable parts of mosaic membranes. *Biophys. J.* **4**: 189–201.

7. M. CEREIJIDO and P. CURRAN, 1965. Intracellular electrical potentials in frog skin. *J. Gen. Physiol.* **48**: 543–557.

8. M. CEREIJIDO and C. A. ROTUNNO, 1967. Transport and distribution of sodium across frog skin. *J. Physiol.* **190**: 481–497.

9. M. CEREIJIDO and C. A. ROTUNNO, 1968. Fluxes and distribution of sodium in frog skin: a new model. *J. Gen. Physiol.* **51**: 280s–289s.

10. M. CEREIJIDO and C. A. ROTUNNO, 1970. *Introduction to the Study of Biological Membranes.* New York: Gordon and Breach, Science Pub., Inc., 261 pp.

11. H. N. CRISTENSEN, 1962. *Biological Transport.* New York: W. A. Benjamin, Inc., 133 pp.

12. K. S. COLE, 1966. The melding of membrane models. *Ann. N.Y. Acad. Sci.* **137**: 405–408.

13. J. CRANK, 1956. *The Mathematics of Diffusion*. Oxford: Clarendon Press, 347 pp.

14. P. F. CURRAN, F. C. HERRERA, and W. J. FLANIGAN, 1963. The effect of Ca and anti-diuretic hormone on Na transport across frog skin. II. Sites and mechanisms of action. *J. Gen. Physiol.* **46**: 1011–1027.

15. P. F. CURRAN and J. R. MacINTOSH, 1962. A model system for biological water transport. *Nature* **193**: 347–348.

16. J. DAINTY, 1963. Water relations of plant cells. *Adv. Bot. Res.* **1**: 279–326.

17. J. DAINTY and B. Z. GINZBURG, 1964. The premeability of the cell membranes of *Nitella translucens* to urea and the effect of high concentrations of sucrose on this permeability. *Biochim. Biophys. Acta* **79**: 112–121.

18. J. DAINTY, P. C. GROGHAM, and D. S. FENSOM, 1963. Electro-osmosis, with some applications to plant physiology. *Can. J. Botany* **41**: 953–966.

19. J. DAINTY and C. R. HOUSE, 1966. 'Unstirred layers' in frog skin. *J. Physiol.* **182**: 66–78.

20. J. F. DANIELLI, 1952. "Theory of penetration of a thin membrane, Appendix A," pp. 324–335, in: H. Davson and J. F. Danielli (eds.), *The Permeability of Natural Membranes*. Cambridge: University Press.

21. H. DAVSON, 1964. *A Textbook of General Physiology*. Boston: Little, Brown and Co., pp. 251–480.

22. H. DAVSON and J. F. DANIELLI, 1952. *The Permeability of Natural Membranes*. Cambridge: University Press, 365 pp.

23, R. M. C. DAWSON, N. HEMINGTON, and D. B. LINDSAY, 1960. The phospholipids of the erythrocyte "ghosts" of various species. *Biochem. J.* **77**: 226–230.

24. S. R. DE GROOT and P. MAZUR, 1962. *Non-Equilibrium Thermodynamics*. Amsterdam: North-Holland Pub. Co., 510 pp.

25. J. M. DIAMOND, 1964. Transport of salt and water in rabbit and guinea pig gallbladder. *J. Gen. Physiol.* **48**: 1–14.

26. J. M. DIAMOND, 1964. The mechanism of isotonic water transport. *J. Gen. Physiol.* **48**: 15–42.

27. J. M. DIAMOND, 1966. Non-linear osmosis. *J. Physiol.* **183**: 58–82.

28. J. M. DIAMOND, 1966. A rapid method for determining voltage-concentration relations across membranes. *J. Physiol.* **183**: 83–100.

29. J. M. DIAMOND and S. C. HARRISON, 1966. The effect of membrane fixed charges on diffusion potentials and streaming potentials. *J. Physiol.* **183**: 37–57.

30. J. M. DIAMOND and J. McD. TORMEY, 1966. Role of long extracellular channels in fluid transport across epithelia. *Nature* **210**: 817–820.

31. J. M. DIAMOND and E. M. WRIGHT, 1969. Biological membranes: the physical basis of ion and nonelectrolyte selectivity. *Ann. Rev. Physiol.* **31**: 581–646.

32. E. T. DUNHAM, 1957. Linkage of active cation transport to ATP utilization. *The Physiologist* **1**: 23.

33. E. T. DUNHAM and I. M. GLYNN, 1961. Adenosine triphosphatase activity and the active movements of alkali metal cations. *J. Physiol.* **156**: 274–293.

34. R. P. DURBIN, 1960. Osmotic flow of water across permeable cellulose membranes. *J. Gen. Physiol.* **44**: 315–326.

35. G. EISENMAN, J. P. SANDBLOM, and J. L. WALKER, Jr., 1967. Membrane structure and ion permeation. *Science* **155**: 965–974.

36. J. V. EVANS, J. W. B. KING, B. L. COHEN, H. HARRIS, and F. L. WARREN, 1956. Genetics of hemoglobin and blood potassium difference in sheep. *Nature* **178**: 849.

37. M. G. FARQUAHAR and G. E. PALADE, 1966. Adenosine triphosphatase in amphibian epidermis. *J. Cell Biol.* **30**: 359–379.

38. H. S. FRAZIER, 1962. The electrical potential profile of the isolated toad bladder. *J. Gen. Physiol.* **45**: 515–528.

39. I. M. GLYNN, 1956. Sodium and potassium movements in human red cells. *J. Physiol.* **134**: 278–310.

40. I. M. GLYNN, 1957. The action of cardiac glycosides on sodium and potassium movements in human red cells. *J. Physiol.* **136**: 148–173.

41. I. M. GLYNN, 1957. The ionic permeability of the red cell membrane. *Prog. Biophys.* **8**: 242–307.

42. R. J. GOLDACRE, 1952. The folding and unfolding of protein molecules as a basis of osmotic work. *Internat. Rev. Cytol.* **1**: 135–164.

43. D. A. GOLDSTEIN and A. K. SOLOMON, 1960. Determination of equivalent pore radius for human red cells by osmotic pressure measurements. *J. Gen. Physiol.* **44**: 1–17.

44. E. J. HARRIS, 1954. Linkage of sodium- and potassium-active transport in human erythrocytes. *Soc. Exptl. Biol. Symp.* **8**: 228–241.

45. E. J. HARRIS, 1956. *Transport and Accumulation in Biological Systems.* London: Butterworth, & Co., Ltd., Pub., 291 pp.

46. E. J. HARRIS and M. MAIZELS, 1951. The permeability of human red cells to sodium. *J. Physiol.* **113**: 506–524.

47. F. C. HERRERA and P. F. CURRAN, 1963. The effect of Ca and antidiuretic hormone on Na transport across frog skin. I. Examination of interrelationships between Ca and hormone. *J. Gen. Physiol.* **46**: 999–1010.

48. J. HILLIER and J. F. HOFFMAN, 1953. On the ultrastructure of the plasma membrane as determined by the electron microscope. *J. Cell. Comp. Physiol.* **42**: 203–247.

49. J. F. HOFFMAN, 1961. "Molecular mechanisms of active cation transport," pp. 3–17, in: A. M. Shanes (ed.), *Biophysics of Physiological and Pharmacological Actions.* Washington: AAAS.

50. J. F. HOFFMAN, 1962. The active transport of sodium by ghosts of human red cells. *J. Gen. Physiol.* **45**: 837–859.

51. J. F. HOFFMAN, 1952. Cation transport and structure of the red-cell plasma membrane. *Circulation* **26**: 1201–1213.

52. J. F. HOFFMAN (ed.), 1964. *The Cellular Functions of Membrane Transport.* Englewood Cliffs, N.J.: Prentice-Hall, Inc., 291 pp.

53. J. F. HOFFMAN and F. W. KREGENOW, 1966. The characterization of new energy dependent cation transport processes in red blood cells. *Ann. N. Y. Acad. Sci.* **137**: 566–576.

54. L. E. HOKIN and M. R. HOKIN, 1960. Studies on the carrier function of phosphatidic acid in sodium transport. I. The turnover of phosphatidic acid and phosphoinositide in the avian salt gland on stimulation of secretion. *J. Gen. Physiol.* **44**: 61–85.

55. L. E. HOKIN and M. R. HOKIN, 1963. Phosphatidic acid metabolism and active transport of sodium. *Fed. Proc.* **22**: 8–18.

56. M. R. HOKIN and L. E. HOKIN, 1964. The synthesis of phosphatidic acid and protein-bound phosphoryl-serine in salt gland homogenates. *J. Biol. Chem.* **237**: 2116–2122.

57. M. R. HOKIN and L. E. HOKIN, 1967. The formation and continuous turnover of a fraction of phosphatidic acid on stimulation of NaCl secretion by acetylcholine in the salt gland. *J. Gen. Physiol.* **50**: 793–811.

58. T. HOSHIKO, 1961. "Electrogenesis in frog skin," pp. 31–47, in: A. M. Shanes (ed.), *Biophysics of Physiological and Pharmacological Actions.* Washington: AAAS.

59. T. HOSHIKO and B. D. LINDLEY, 1967. Phenomenological description of active transport of salt and water. *J. Gen. Physiol.* **50**: 729–758.

60. E. HUF, 1935. Versuche über den Zusammenhand zwischen Stoffwechsel, Potentialbildung und Funktion der Froschhaut. *Pflüger's Arch. Ges. Physiol.* **235**: 655–673.

61. M. H. JACOBS, 1935. Diffusion processes. *Ergebn. Biol.* **12**: 1–160.

62. H. R. KABACK, 1970. Transport. *Ann. Rev. Biochem.* **39**: 561–598.

63. H. R. KABACK, 1970. "The transport of sugars across isolated bacterial membranes," pp. 36–99, in: F. Bronner and A. Kleinzeller (eds.), *Current Topics in Membranes and Transport.* New York: Academic Press.

64. A. KATCHALSKY and P. F. CURRAN, 1965. *Nonequilibrium Thermodynamics in Biophysics.* Cambridge: Harvard University Press, 248 pp.

65. O. KEDEM, 1961. "Criteria of active transport," pp. 87–93, in: A. Kleinzeller and A. Kotyk (eds.), *Membrane Transport and Metabolism.* New York: Academic Press.

66. O. KEDEM and A. KATCHALSKY, 1963. Permeability of composite membranes. Part 3—Series array of elements. *Trans. Faraday Soc.* **59**: 1941–1953.

67. E. P. KENNEDY, 1967. "Some recent developments in the biochemistry of membranes," pp. 271–280, in: G. C. Quarton, T. Melnechuk, and F.O. Schmitt (eds.), *The Neurosciences—A Study Program.* New York: Rockefeller Press.

68. R. D. KEYNES and G.W. MAISEL, 1954. The energy requirement for sodium extrusion from a frog muscle. *Proc. Roy. Soc. London B* **142**: 383–392.

69. A. KLEINZELLER and A. KOTYK (eds.), 1961. *Membrane Transport and Metabolism.* New York: Academic Press, 608 pp.

70. V. KOEFOED-JOHNSEN and H. H. ÜSSING, 1953. The contributions of diffusion and flow to the passage of D_2O through living membranes. *Acta Physiol. Scand.* **28**: 60–76.

71. V. KOEFOED-JOHNSEN and H. H. ÜSSING, 1958. The nature of the frog skin potential. *Acta Physiol. Scand.* **42**: 298–308.

72. A. KROGH, 1937. Osmotic regulation in the frog (*R. esculenta*) by active absorption of chloride ions. *Scand. Arch. Physiol.* **76**: 60–74.

73. A. LEAF, 1960. Some actions of neurohypophyseal hormones on a living membrane. *J. Gen. Physiol. Suppl.* **43**: 175–189.

74. A. LEAF, J. ANDERSON, and L. B. PAGE, 1958. Active sodium transport by the isolated toad bladder. *J. Gen. Physiol.* **41**: 657–668.

75. A. LEAF and E. DEMPSEY, 1960. Some effects of mammalian neurohypophyseal hormones on metabolism and active transport of sodium by the isolated toad bladder. *J. Biol. Chem.* **235**: 2160–2163.

76. A. LEAF and A. RENSHAW, 1957. Ion transport and respiration of isolated frog skin. *Biochem. J.* **65**: 82–89.

77. A. LEAF and A. RENSHAW, 1957. The anaerobic active ion transport by isolated frog skin. *Biochem. J.* **65**: 90–93.

78. H. LUNDEGARDH, 1954. Anion respiration. The experimental basis of a theory of absorption, transport, and exudation of electrolytes by living cells and tissues. *Symp. Soc. Exptl. Biol.* **8**: 262–296.

79. D. W. MARTIN, 1964. Reversed potentials of isolated frog skin. *J. Cell. Comp. Physiol.* **63**: 245–251.

80. D. W. MARTIN and J. M. DIAMOND, 1966. Energetics of coupled active transport of sodium and chloride. *J. Gen. Physiol.* **50**: 295–315.

81. A. MAURO, 1957. Nature of solvent transfer in osmosis. *Science* **126**: 252–253.

82. A. Mauro, 1960. Some properties of ionic and nonionic semipermeable membranes. *Circulation* **21**: 845–854.

83. J. L. Moore, T. Richardson, and H. F. Deluca, 1969. Essential fatty acids and ionic permeability of lecithin membranes. *Chem. Phys. Lipids* **3**: 39–58.

84. J. T. Ogilvie, J. R. MacIntosh, and P. F. Curran, 1963. Volume flow in a series-membrane system. *Biochim. Biophys. Acta* **66**: 441–444.

85. L. Onsager, 1931. Reciprocal relations in irreversible processes, I. *Phys. Rev.* **37**: 405–426; II. *Phys. Rev.* **38**: 2265–2279.

86. L. J. Opit and J. S. Charnock, 1965. A molecular model for a sodium pump. *Nature* **209**: 471–474.

87. W. J. V. Osterhout, 1940. Some models of protoplasmic surfaces. *Cold Spring Harbor Symp. Quant. Biol.* **8**: 51–62.

88. A. B. Pardee, 1968. Membrane transport proteins. *Science* **162**: 632–637.

89. J. C. Parker and J. F. Hoffman, 1967. The role of membrane phosphoglycerate kinase in the control of glycolytic rate by active cation transport in human red blood cells. *J. Gen. Physiol.* **50**: 893–916.

90. A. K. Parpart and B. Ballantine, 1952. "Molecular anatomy of the red cell plasma membrane," pp. 135–148, in: D. H. Barron (ed.), *Modern Trends in Physiology and Biochemistry*. New York: Academic Press.

91. C. S. Patlak, 1957. Contributions to the theory of active transport. II. The gate type non-carrier mechanism and generalizations concerning tracer flow, efficiency, and measurement of energy expenditure. *Bull. Math. Biophys.* **19**: 209–235.

92. C. S. Patlak, D. A. Goldstein, and J. F. Hoffman, 1963. The flow of solute and solvent across a two-membrane system. *J. Theoret. Biol.* **5**: 426–442.

93. L. D. Peachey and H. Rasmussen, 1961. Structure of the toad's urinary bladder as related to its physiology. *J. Biophys. Biochem. Cytol.* **10**: 529–553.

94. R. L. Post, C. R. Merritt, C. R. Kinsolving, and C. D. Albright, 1960. Membrane adenosine triphosphate as a participant in the active transport of sodium and potassium in human erythrocytes. *J. Biol. Chem.* **235**: 1796–1802.

95. R. L. Post, A. K. Sen, and A. S. Rosenthal, 1965. A phosphorylated intermediate in adenosine triphosphate-dependent sodium and potassium transport across kidney membranes. *J. Biol. Chem.* **240**: 1437–1445.

96. J. H. Quastel, 1965. Molecular transport at cell membranes. *Proc. Roy. Soc. London B* **163**: 169–196.

97. S. I. Rapoport, 1965. Ionic accumulation by water flow through a membrane. *Acta Physiol. Scand.* **64**: 361–371.

98. P. M. Ray, 1960. On the theory of osmotic water movement. *Plant Physiol.* **35**: 783–795.

99. E. Robbins and A. Mauro, 1960. Experimental study of the independence of diffusion and hydrodynamic permeability coefficients in collodion membranes. *J. Gen. Physiol.* **43**: 523–532.

100. T. Rosenberg and W. Wilbrandt, 1955. The kinetics of membrane transport involving chemical reactions. *Exptl. Cell Res.* **9**: 49–67.

101. T. Rosenberg and W. Wilbrandt, 1957. Uphill transport induced by counterflow. *J. Gen. Physiol.* **41**: 289–296.

102. L. Rothfield and A. Finkelstein, 1968. Membrane biochemistry. *Ann. Rev. Biochem.* **37**: 463–496.

103. A. Rothstein, 1968. Membrane phenomena. *Ann. Rev. Physiol.* **30**: 15–72.

104. C. A. ROTUNNO, V. KOWALEWSKI, and M. CEREIJIDO, 1967. Nuclear spin resonance evidence for complexing of sodium in frog skin. *Biochim. Biophys. Acta* **135**: 170–173.

105. H. J. SCHATZMANN, 1953. Herzglykoside als Hemmstoffe für den aktiven Kalium- und Natriumtransport durch die Erythrocytenmembran. *Helv. Physiol. Pharmacol. Acta* **11**: 346–354.

106. H. J. SCHATZMANN, 1965. The role of Na^+ and K^+ in the ouabain-inhibition of the Na^+ and K^+-activated membrane adenosine triphosphatase. *Biochim. Biophys. Acta* **94**: 89–96.

107. B. T. SCHEER, 1958. Active transport: definitions and criteria. *Bull. Math. Biophys.* **20**: 231–244.

108. E. SCHOFFENIELS, 1967. *Cellular Aspects of Membrane Permeability*. Oxford: Pergamon Press, 266 pp.

109. S. G. SCHULTZ and P. F. CURRAN, 1970. Coupled transport of sodium and organic solutes. *Physiol. Rev.* **50**: 637–718.

110. R. I. SHA'AFI and W. R. LIEB, 1967. Cation movements in the high sodium erythrocytes. *J. Gen. Physiol.* **50**: 1751–1764.

111. R. I. SHA'AFI, G. T. RICH, V. W. SIDEL, W. BOSSERT, and A. K. SOLOMON, 1967. The effect of the unstirred layer on human red cell water permeability. *J. Gen. Physiol.* **50**: 1377–1399.

112. J. C. SKOU, 1960. Further investigations on a $Mg^{++} + Na^+$-activated adenosine-triphosphatase, possibly related to the active, linked transport of Na^+ and K^+ across the nerve membrane. *Biochim. Biophys. Acta* **42**: 6–23.

113. J. C. SKOU, 1965. Enzymatic basis for active transport of Na^+ and K^+ across cell membrane. *Physiol. Rev.* **45**: 596–617.

114. J. C. SKOU, 1969. "The role of membrane ATPase in the active transport of ions," pp. 455–482, in: D. C. Tosteson (ed.), *The Molecular Basis of Membrane Function*. Englewood Cliffs, N.J.: Prentice-Hall, Inc.

115. F. M. SNELL and T. K. CHOWDHURY, 1965. Contralateral effects of sodium and potassium on the electrical potential in frog skin and toad bladder. *Nature* **207**: 45.

116. A. K. SOLOMON, 1960. Red cell membrane structure and ion transport. *J. Gen. Physiol. Suppl. 1* **43**: 1–15.

117. A. J. STAVERMAN, 1952. Non-equilibrium thermodynamics of membrane processes. *Trans. Faraday Soc.* **48**: 176–185.

118. W. D. STEIN, 1967. *The Movement of Molecules across Cell Membranes*. New York: Academic Press, 369 pp.

119. H. B. STEINBACH, 1940. Electrolyte balance of animal cells. *Cold Spring Harbor Symp. Quant. Biol.* **8**: 242–254.

120. F. B. STRAUB, 1953. Über die Akkumulation der Kaliumionen durch menschliche Blutkörperchen. *Acta Physiol. Hung.* **4**: 235–240.

121. T. TEORELL, 1953. Transport processes and electrical phenomena in ionic membranes. *Prog. Biophys.* **3**: 305–369.

122. M. G. THOVERT, 1910. Diffusion et théorie cinétique des solutions. *C. R. Acad. Sci. Paris* **150**: 270–272.

123. D. C. TOSTESON, 1963. Active transport, genetics, and cellular evolution. *Fed. Proc.* **22**: 19–26.

124. D. C. TOSTESON, 1966. Some properties of the plasma membrane of high potassium and low potassium sheep red cells. *Ann. N.Y. Acad. Sci.* **137**: 577–590.

125. D. C. TOSTESON (ed.), 1969. *The Molecular Basis of Membrane Function.* Englewood Cliffs, N.J.: Prentice-Hall, Inc., 598 pp.

126. D. C. TOSTESON and J. F. HOFFMAN, 1960. Regulation of cell volume by active cation transport in high and low potassium sheep red cells. *J. Gen. Physiol.* **44**: 169–194.

127. A. S. TROSHIN, 1966. *Problems of Cell Permeability.* Oxford: Pergamon Press, 549 pp.

128. H. H. ÜSSING, 1949. The distinction by means of tracers between active transport and diffusion. *Acta Physiol. Scand.* **19**: 43–56.

129. H. H. ÜSSING, 1949. The active ion transport through the isolated frog skin in the light of tracer studies. *Acta Physiol. Scand.* **17**: 1–37.

130. H. H. ÜSSING, 1960. The frog skin potential. *J. Gen. Physiol. Suppl. 1* **43**: 135–147.

131. H. H. ÜSSING, 1966. Anomalous transport of electrolytes and sucrose through the isolated frog skin induced by hypertonicity of the outside bathing solution. *Ann. N. Y. Acad. Sci.* **137**: 543–555.

132. H. H. ÜSSING, 1969. "Summary address," pp. 577–588, in: D. C. Tosteson (ed.), *The Molecular Basis for Membrane Function.* Englewood Cliffs, N.J.: Prentice-Hall, Inc.

133. H. H. ÜSSING and K. ZERAHN, 1951. Active transport of sodium as the source of electric current in the short-circuited isolated frog skin. *Acta Physiol. Scand.* **23**: 110–127.

134. R. WHITTAM, 1962. The asymmetric stimulation of a membrane adenosine triphosphatase in relation to active cation transport. *Biochem. J.* **84**: 110–118.

135 R. WHITTAM, 1967. "The molecular mechanism of active transport," pp. 313–325, in: G. C. Quarton, T. Melnechuk, and F. O. Schmitt (eds.), *The Neurosciences—A Study Program.* New York: Rockefeller Press.

136. R. WHITTAM and M. E. AGER, 1965. The connection between active cation transport and metabolism in erythrocytes. *Biochem. J.* **97**: 214–227.

137. W. WILBRANDT and T. ROSENBERG, 1961. The concept of carrier transport and its corollaries in pharmacology. *Pharmacol. Rev.* **13**: 110–183.

MEMBRANE BEHAVIOR
IN EXCITABLE CELLS

chapter four

Of the many important problems that must be solved by organisms to remain alive, one is coding, transmitting, and integrating information to direct the sundry and complex activities that characterize nonrandom biological systems. The behavior of the plasma membrane of neurons—and other excitable cells—deserves extensive study since it is clear that not only are these membranes involved in cell-cell communication but they also have responsibilities in the integration and liveliness of the entire animal. While of great biological importance, the problems associated with the executive activity of nervous systems of the many-celled animals are subjects that will be ignored here.

4.1 The Classical
Nerve Cell

From today's viewpoint, it is not easy to see the basis for the arguments among the great neurohistologists of the 19th century. One group led by Gerlack and Golgi and later defended by Held saw the nervous system as one great branching network of interconnected fibers and nodes—the reticular theory. The rival group supported the concept of a nervous system made of discrete units or cells—a view first proposed by His and Forel and then

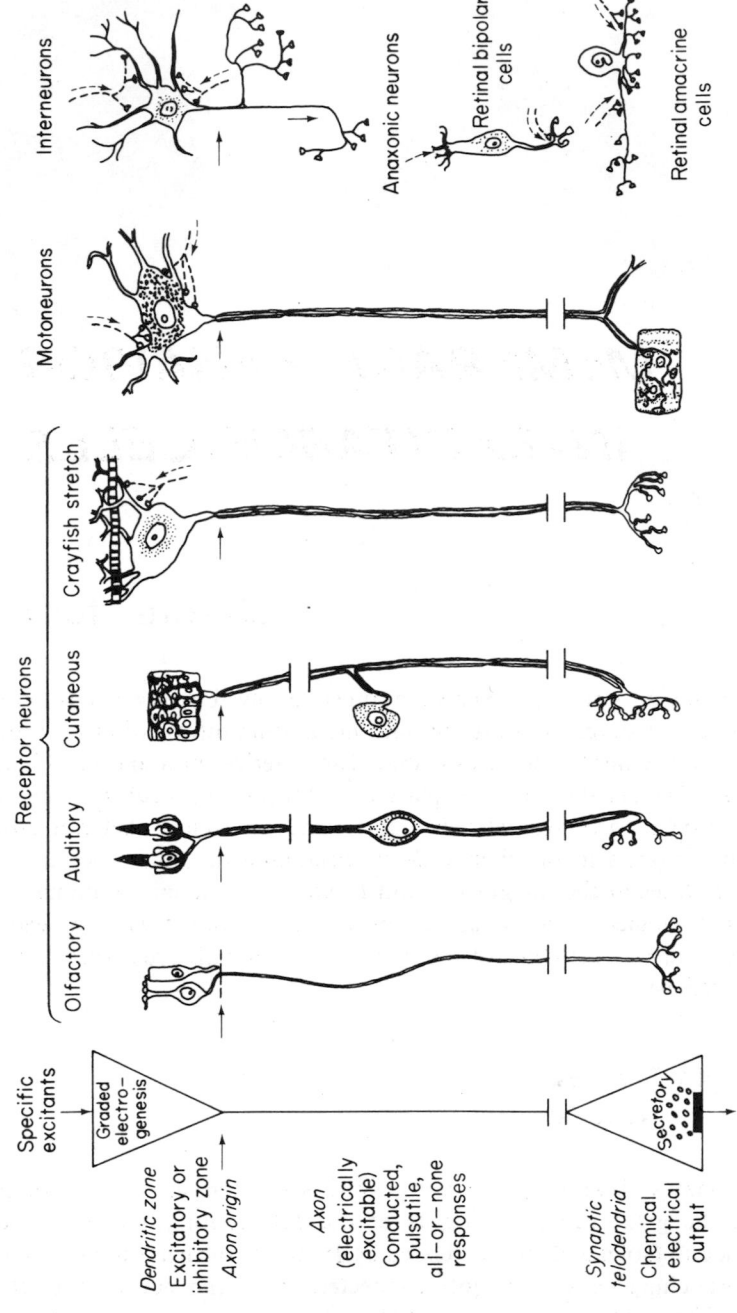

independently by Cajal. Later the name *neuron* was suggested by Waldeyer for the nerve cell and *neuron theory* for the concept of the structural independence of nerve cells. It was the major achievement of Cajal and his delicate silver-staining methods to establish the neuron theory and the functional connections effected between individual neurons by close contacts—called *synapses* by Sherrington—and not by protoplasmic continuity in a syncytial network as proposed in the reticular theory. It was not until 1934, however, when Cajal [31] published his critical summary, that the neuron theory set down all the remaining serious challenges. Curiously, the two leading antagonists shared the 1909 Nobel Prize for their work: Cajal, for uncovering evidence for independent neurons, and Golgi, for developing the method used by Cajal.

The structural units that comprise nervous systems are the nerve cells, but there are an equally abundant number of glial cells, of several kinds, that nearly fill the interneuronal space. The typical nerve cell has three main parts: the dendrites, which receive input from other cells; the axon, which transmits information over great distances; and the soma, which houses the nucleus and most of the other metabolic structures of the cell (Fig. 35). The most commonly used example of a neuron is a motoneuron from the spinal cord of a mammal even though it is not the most abundant structural arrangement found in nervous tissue throughout the animal kingdom. The soma of this cell is typically 70 μ across and thus just beyond the range of unaided vision. The dendrites or branching processes radiate from the soma for long distances, as much as 1 mm, before breaking up into small terminal twigs. Expanded axonal terminals (synaptic knobs) of other neurons make their contact with this cell on the surface of the soma and the basal region of the dendrites. The distribution of contacts is progressively sparser at the more peripheral ends of the dendrites. No part of a neuron has been found to be immune to synaptic contacts but the density of synapses is greatest on dendrites, less on the soma, and rare elsewhere. The axon arises from a hillock on the soma and narrows before it becomes encased in a myelin sheath, a wrapping of connective tissue formed by Schwann cells, some 50–100 μ away from the soma. The axon hillock and nonmedullated portion of the axon are morphologically and physiologically distinct from other portions of the motoneuron and the term *initial segment of the axon* or simply *initial segment* is used. The axon is normally covered by a heavy myelin sheath.

Figure 35. *Diagrams of a sample of neurons:* to illustrate structural and functional similarities, receptor and effector neurons are aligned at the point where impulses normally originate. Compare the features of these neurons with those found in the generalized neuron (left) proposed by Grundfest. The location of the perikaryon—the nucleated portion of the cytoplasm—is not related to impulse propagation. Neurons without a spike-conducting region still possess dendritic and telodendritic characteristics.

Postganglionic sympathetic fibers, sensory fibers of visceral nerves, and all invertebrate axons are considered unmyelinated and have only a thin wrap of Schwann cells. Peripheral axons of vertebrates are also encased in a neurilemma, a nucleated sheath of connective tissue. These sheaths are interrupted at about 1-mm intervals exposing bare axon, the nodes of Ranvier. Nodes are unique to vertebrates and greatly speed the propagation of impulses by permitting an impulse to jump from node to node instead of passing smoothly along the length of fiber as in muscle cells or invertebrate axons. Generally, myelinated axons conduct impulses much more quickly than others of equal diameter. Nerve cells display greater diversity in form than generally recognized [22, 23] but other neurons of widely differing types have been sufficiently studied to provide assurance that the mammalian motoneuron yields valid information about nerve cells in general.

Since excitable phenomena are almost exclusively cell surface phenomena, the structural and functional variation found among excitable cells suggests per se that membrane properties must vary among the different types of cells and also among the different parts of a single cell while, at the same time, showing the basic features common to all cells. Interest in the study of excitable cells is increased by the possibility of finding a dramatic elaboration of a cellular property, which may also explain one of the specialized features of such cells.

4.2 The Classical View
of a Nerve Impulse

The integrative action of a population of nerve cells depends on the nature of the influence of one cell upon another, as demonstrated by the magnificent contributions of Sherrington [232]. Differences in basic properties are amazingly minor even between cells of widely differing function; small variations in specific properties, not differences in fundamental mechanisms, are usually sufficient to explain differences in phenomena expressed. For example, an obvious feature of nerve cells is impulse propagation. While differences in opinion appear in describing the mechanism for impulse propagation, the same basic plan appears to be used by small invertebrate axons which conduct impulses as slow as 10 mm/sec and by large myelinated mammalian motoneurons which conduct as fast as 100 m/sec. Many phenomena accompany the passage of a nerve impulse along an axon such as changes in heat production [1, 117], volume [118], opacity [38], and metabolic events [42], but the electrical events [154] are much more easily recorded and measured. Each nerve impulse is accompanied by a brief pulse of electric current flowing across the plasma membrane through the axoplasm and through the body fluids surrounding the nerve. The physical properties of the axon membrane and the metabolic activities of the axon maintain a constant imbalance in the distribution of certain ions across the membrane and a resulting steady potential difference of about 60–90 mV; axoplasm is more negative than the extracellular fluids. The invasion of the axon membrane by a nerve impulse is accompanied by a brief but dramatic change in the electrical potential across the membrane; the electric currents associated with the nerve impulse are carried by a progression of ionic rearrangements across the membrane and the electrical signs constitute the "action potential," which is the waveform that appears on the recording instrument. As the wave of activity travels along the axon, each point reached along the outer surface becomes electrically negative to the inactive regions on either side of the active region. The measurement and analysis of this membrane potential and its fluctuations have been one of the major tasks of neurophysiology.

The easiest way to observe this electrical activity is to place two electrodes (e.g., metal wires) on an isolated nerve, and when the nerve is stimulated,

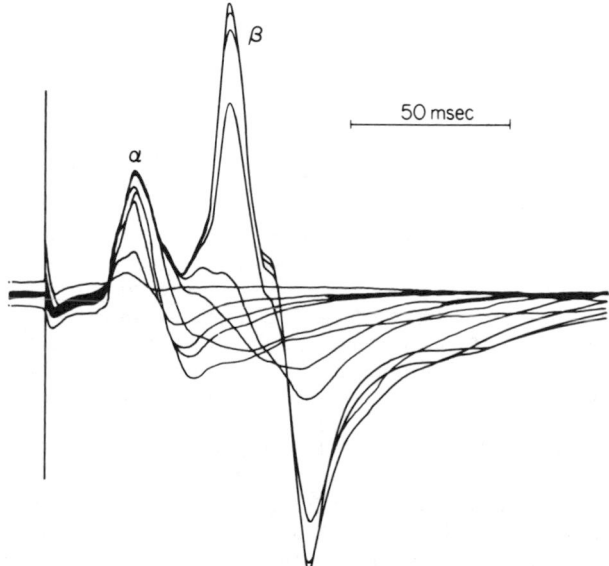

Figure 36. *Diphasic recording of action potentials:* these superimposed traces are a series of responses of the cerebro-visceral connective nerve of the surf clam *Spisula solidissima* to increasingly more intense stimulation. According to velocities of impulse conduction, this nerve contains at least two populations of axons; the α fibers have lower thresholds than the β fibers. The diphasic nature of action potentials recorded by a pair of electrodes placed 5 mm apart on a conducting region is especially clear in the maximal responses of the β fibers.

any adequate recording system will register a sudden potential difference between the electrodes at the external surface of the nerve as the wave of activity passes by. When the evoked activity reaches the region under the first electrode, it becomes negative to the region under the second recording electrode. As the activity proceeds past the second electrode, it will in turn be driven negative with respect to the first but the current through the instrument will appear reversed since the two electrodes are connected to opposite inputs of the recording instrument. The record, therefore, is a negative wave followed by its inverted mirror image and is conventionally recorded as an upward deflection followed by a downward deflection (Fig. 36). (Active tissues usually become negative at their external surface relative to inactive regions and electrophysiologists like their records to deflect upward so the polarity conventions used are, at times, opposite to those used in the physical sciences. In this situation, negativity is indicated upward.) In rapidly conducting axons, the length of axon that is active at any one instant, the wavelength, may be as much as 5 or 6 cm and when the electrodes are closer than that distance apart, the doubly recorded wave of activity becomes partly superimposed and the algebraic sum is recorded. Because the electrical signs of a nerve impulse are effectively recorded twice,

Figure 37. *Local response curves for the cerebro-visceral nerve of the surf clam Spisula solidissima*: the amount of residual excitation evoked by a subthreshold stimulus to a nerve (S indicates the stimulus artifact) can be tested by a second— also subthreshold—stimulus applied at time intervals later (stimulus artifacts of the second pulses are indicated by numbers 1–8 corresponding with the subsequent numbered nerve responses). If the residual excitation produced by the first stimulus and that of the second stimulus sum to exceed threshold for that axon, a propagating spike is evoked. Since this clam nerve contains a large number of axons with a range of thresholds (see Fig. 36), the recorded spike height, given as a fraction of the maximal response, is a measure of the relative number of fibers excited. When only a small amount of excitation remains from the first stimulus to sum with that produced by the second, only the fibers of lowest threshold can be excited sufficiently to generate a spike [(b) 1, 2]. But, with greater summation, even fibers of high thresholds can be excited to fire [(b) 7, 8]. Plotting the relative spike height of responses, such as illustrated by (b), as a function of the delay between two subthreshold stimuli (a) exposes the residual excitation produced by the first pulse as a function of time. When only the membrane capacitance is charged (see Fig. 41) and in the absence of any active responses by the axon membranes, the curve labeled "exponential decay curve" should describe the results of this experiment, and it does for stimuli less than half-threshold in magnitude. For larger subthreshold stimuli, the axon membranes show active responses; two examples are plotted as "experimental curves" (a). The difference between the experimental curves and the exponential decay curve represents the active local responses of the axons to the applied (first) stimulus. These data are similar to those obtained by Katz in 1937 although by a different experimental approach.

once under each electrode, the waveform is called the *diphasic action potential*. The action potential can be recorded as a smooth wave without the complications of the passage of the nerve impulse under the second electrode by simply locating the second electrode on some inactive region, crushing or otherwise destroying axon tissue under the second electrode; the waveform

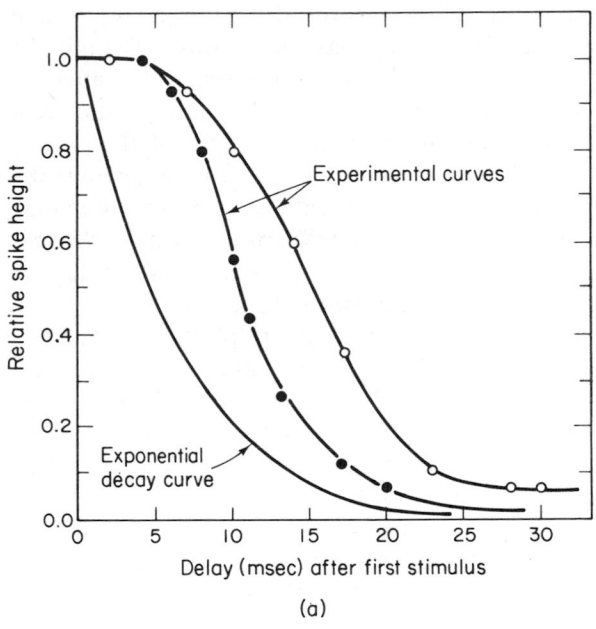

(a)

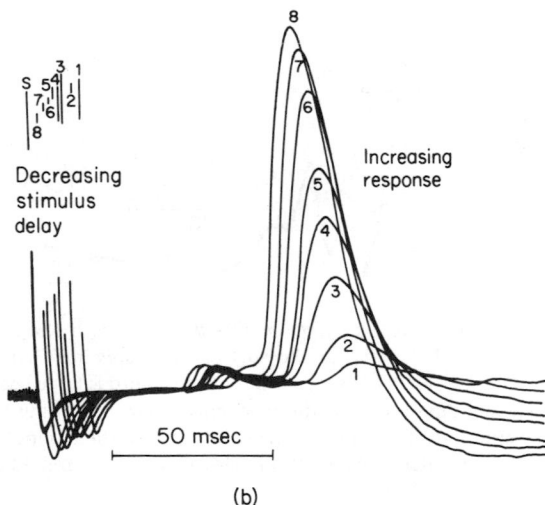

(b)

recorded is called the *monophasic action potential*. When the recording electrodes are very close together, the shape of the diphasic action potential approaches the mathematical first derivative (dV/dt) of the monophasic action potential (V). The sharp peak in the record, between 0.5 and 1 msec in duration, is descriptively called the *spike*.

The electrical signs of other events occurring in an excited membrane can also be distinguished in the records. Local responses of axon membrane appear at the base of the rising phase of the spike. The initial portion of the local responses is due to current flowing through the capacitative and resistive elements of axon membrane and the later portion represents active responses of membrane in the immediate area of the recording electrode. These local graded "prepotentials" (Fig. 37) are analogous (but not homologous) to sensory generator potentials, receptor potentials, pacemaker potentials, end plate potentials, or synaptic potentials that can be recorded from other regions or other cells. A "negative afterpotential" can be distinguished as a delay late in the return of the spike toward the base line and is followed by a smaller but longer lasting "positive afterpotential," an overshoot below the base line. These afterpotentials are named for their polarities as recorded with external electrodes. With external electrodes, it is difficult to see the smaller components of an action potential. A nerve can contain thousands of axons, and even axons of similar characteristics do not carry impulses at identical velocities. Hence, the individual impulses pass beneath the recording electrodes out of phase with each other and the sum of the action potentials is a compound waveform (Fig. 38).

The energy for impulse propagation is derived from the several condi-

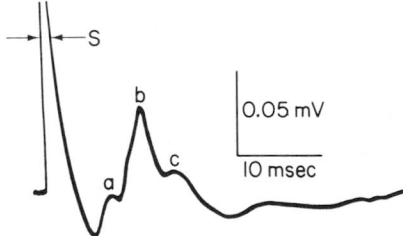

Figure 38. *Compound action potential:* this oscilloscope tracing of the response of the anterior pallial nerve of the surf clam *Spisula solidissima* to maximal excitation (stimulus artifact, S, at the left of the record is large because the ample seawater surrounding the preparation in this *in vivo* recording situation short-circuits the stimulus). In this nerve, there are at least three populations of axons (a, b, and c) that differ in conduction velocities (82, 58, and 42 cm/sec, respectively).

tions, e.g., ionic distribution across the axon membrane, of the fiber over which it travels. Impulses, once initiated, will course the length of an axon without decrement in either velocity or magnitude. Whatever type stimulus is used, the events surrounding excitation are such that electric current is forced through the membrane. Once the current density through an area of membrane is sufficiently intense, that area will switch from a resting to an active state. Current derived from excited regions of membrane usually exceed threshold, by several times, in immediately adjacent quiet regions. These regions become excited and, in turn, excite others. The system is self-stimulating and propagation proceeds at a speed dependent on the physical properties of the axon and its insulating sheaths. Once excited, it takes time for a segment of membrane to return to the unexcited state, which it must do before it can be excited again. This refractoriness limits the impulse frequency in an axon; impulses in a single axon cannot follow each other closely enough to fuse action potentials. Individual axons in a multifibered nerve possess different thresholds and refractoriness. Axons that recover quickly generally have lower thresholds. The electrical activity of such a nerve is seen by external electrodes as the sum of the action potentials generated by the individual axons in that nerve so that the height of the record is directly proportional to the number of axons carrying impulses beneath the electrodes at that specific instant. At early times after the passage of a first volley of impulses, only the few axons of highest excitability and quickest recovery will be reexcited by a second stimulus, and the summed activity recorded is small (Fig. 39). Longer delays between the two stimuli will allow more axons to be excited by the second stimulus and the record is correspondingly larger. In this way, excitability, thresholds, and refractoriness can be measured in a population of axons.

The older electrophysiologists described a number of other phenomena associated with excitable cells [27]. During the past three decades, the introduction of electrode systems that can record from just one cell at a time permitted work on the mechanisms responsible for these phenomena and specific membrane properties became described in molecular terms. The description above of electrically excitable all-or-nothing responses characteristic of axons and other long-fibered excitable systems illustrates the degree to which membranes can alter some of these properties to changes in their environments. Other excitable cells or parts of cells behave differently than do axons. Part of the value in studying excitable cells to membranologists is also due to the ease in following their activity with electrical recording equipment of increasing sophistication associated with the growth and development of the electronic industry. The remainder of this chapter is devoted primarily to these membranes: the measurement of their properties and the analysis of their mechanisms of operation.

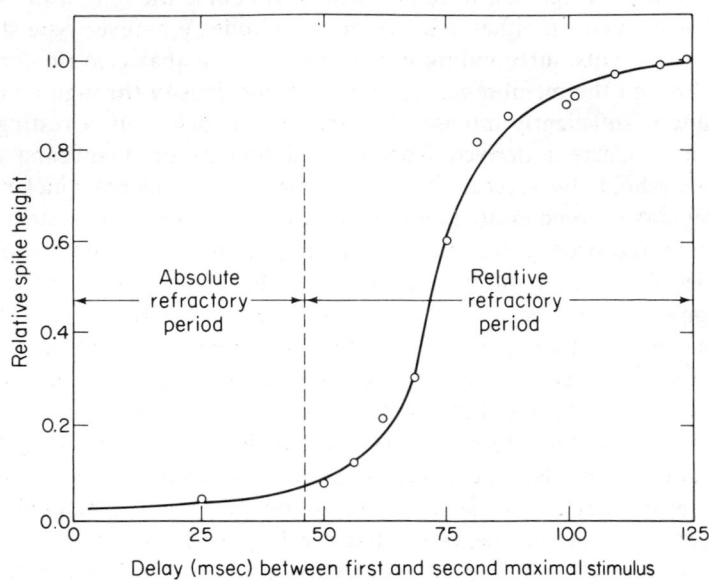

(a)

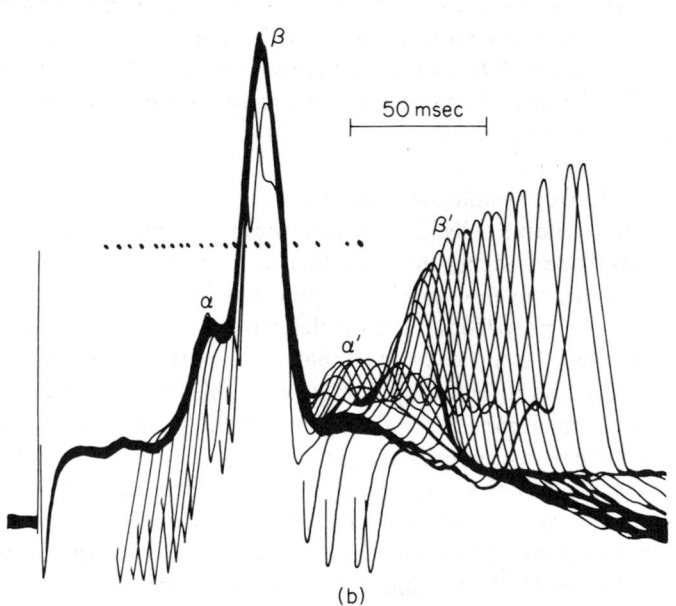

(b)

4.3 A Simple View of the Nature of Bioelectrical Potentials

Biological systems are predominantly aqueous solutions of electrolytes. Electrolytes spontaneously dissociate into ions when dissolved in a suitable medium, such as water, so that there are always equal quantities of positive and negative charges regardless of concentrations. Since there is no excess of any one kind of charge in the bulk solution, no potential difference can occur between one part of the bulk solution and another; any imbalance in the concentrations of the charges must result from other factors. The application of an electrical potential to an electrolytic solution causes the ions to move and, using the convention developed by Benjamin Franklin, the direction of the flow of the positively charged particles indicates the direction of current flow. The current, carried by ions and not electrons, is accompanied by the transfer of matter and changes in concentrations.

In any physical system, an electrical potential field exists where there is a concentration of charges of similar sign. The word *potential* indicates that, in order to bring about the particular arrangement of charge accumulation, work had to be done. The work can, in theory, be recovered because the charges tend to repel one another and to disperse with force. In other words, there is a difference in electrical potential between any two points whenever work is involved to move a charge from one point to the other. Unless there is a sustained source of fresh charge or unless the equilization process is actually going on, however, potential differences will not occur in the bulk-conducting medium.

Figure 39. *Refractoriness in nerve:* (a) is a plot of data obtained from experiments similar to that illustrated in (b) by the superimposed records of action potentials from the cerebro-visceral connective nerve in the surf clam *Spisula solidissima*. The relative spike height is an estimate of the fraction of axons excited by the applied stimulating pulse. At varying intervals after the application of a maximal stimulus [stimulus artifact appears at the far left of the record shown in (b)], which elicits maximal responses from both α and β populations of axons in this nerve (see Fig. 36), a second stimulating pulse of the same magnitude of the first is applied (artifacts appear as a series of dots across the peak of the first nervous response). Only those axons that have recovered sufficiently from the previous excitation can respond to the second pulse; the lower threshold and faster recovering fibers fire. The α fibers, which propagate impulses at higher velocities and have lower thresholds (see Fig. 36), also recover faster (α'). The β fibers recover more slowly, as illustrated by the increasing number of axons (relative spike height) responding (β') as the interval between the first and second stimulus increases. (a) graphs this distribution of fibers as a function of their refractoriness. Few fibers recover any excitability before 50 msec after the first excitation (absolute refractory period) but nearly all have recovered after 125 msec (following recovery from the relative refractory period).

Consider a cell in an animal body and the spaces outside the cell. These extracellular spaces are at a uniform potential because the whole medium is an electrolytic solution. As the animal is usually electrically continuous with the ground, it is satisfactory to call this potential zero or at least to consider it as the reference potential. The inside of the cell is also at a uniform potential throughout but this is not zero. Since cellular interiors are normally found to be negatively charged with respect to their exteriors, the system can be simplified by considering that the internal potential is set by all the excess or unbalanced negative charges within the cell and all the excess positive charges outside it. It makes no difference here that the excess charges are located adjacent to the membrane boundary between the two phases. The resultant potential difference between these two origins of opposite effect defines the magnitude of the uniform potential within the cell, which can be detected with a probe, with respect to zero potential.

When an electric field is suddenly altered, the equilization process that follows is usually rapid—a few millionths of a second. Also, it is difficult to detect electrically the presence of a small secluded concentration of ions. Those that are imprisoned can only be detected if a probe is placed among them and the potential there measured relative to that of the space outside. Those that cluster around the imprisoned charges are at the same potential as the whole of the surrounding space and so cannot be detected at least until some change has occurred in the charges within the enclosure. The practical result of this is that electrophysiologists must either penetrate the cell or other secluded ion concentration and measure the potential within it or make measurements on the system while it is supplying an external electric current flow. Bioelectricity is studied both from the viewpoint of the cell, which is the source of electrical energy, and also from the viewpoint of current flow, which relates the remote electrical fields to their sources within cells.

The interiors of cells contain concentrations of potassium, sodium, chloride, protein, and other ions that differ from the concentrations of these same ions in the extracellular spaces. The system within the cell is not in complete equilibrium because, for example, potassium, a relatively small and mobile ion, tends to diffuse out of the cell at a faster rate than other ions and to leave a small but definite ion imbalance. The diffusion rates of all ions through the plasma membrane including potassium are extremely small and yet because of a differential in diffusion rates a substantial potential difference is set up across that membrane. Of course, the membrane potential cannot be blamed solely on the activities of potassium and more will be said about this later.

A common practice of electrophysiologists is to neglect the fact that the cell or membrane potential is a property of a dynamic system in which the relative velocities of different ions are involved. Most aspects of bioelectric potentials are discussed as if a static charge separation exists across a resting cell membrane. For this reason, we have little to say here regarding the postulated sodium pump that sustains the ion concentrations despite constant diffusional leakage and losses when the membrane is active. The pump carries quantities, often equal, of sodium and potassium in opposite directions (sodium outward and potassium inward) across the membrane by an enzymatic process (see Chapter 3). Concentrations are sustained but electrical potentials are generally not thought to be directly modified by its operation.

The electrical potential wave detected near excitable tissue during breakdown and recovery of normal membrane permeabilities is called an *action potential*. It is detected with maximal amplitude when one electrode is placed as near as possible to an active area and the other electrode is located in a completely inactive or remote region. It is detected with reduced amplitude when the electrodes are placed closer to each other so that they intercept smaller increments of potential difference, or *IR* drop, within the potential field set up around the active region of the cell. Since the external potential wave is generated by the current that flows through the external fluids during cellular activity, the shape of the potential wave variation with time is related to the variation of the current with time. If the area of the cell being depolarized is static, i.e., if the depolarization is not traveling, the external potential-time variation would follow exactly the external current-time variation. This is seldom the case in experiments and the potential wave detected along the outside of a cell is nearly always caused by a traveling wave of depolarization. The external potential field originates from a source of current that rises to a maximal value at some point along its path—sometime during the breakdown phase of the cellular membrane at that point. The active region of the cellular membrane, which completely encircles the cell, migrates at a constant velocity along the length of the cellular process. The external current waveform is related to the transmembrane potential waveform and, at times, it approaches the mathematical time derivative, although it is not necessarily a precise measure of the time rate of change of this potential (Fig. 40). The actual relationship between the action potential waveform as detected by external electrodes and the real cellular potentials is a complex one [184, 248].

When a part of a cell is damaged, the transmembrane resting potential in the area may be zero and permeability to ions may be very high. If the damage is not too extensive, ions may flow in and out of the cell through the damaged membrane for a considerable time. It appears that the remaining part of the cell, perhaps with raised metabolic activity, can continue to sustain ionic separations despite the loss through the injured area. A steady current flow known as the *injury current* is detected near the damaged tissue and the potential difference causing the current flow is known as the *injury potential* or the *demarcation potential*. The direction of the current is such

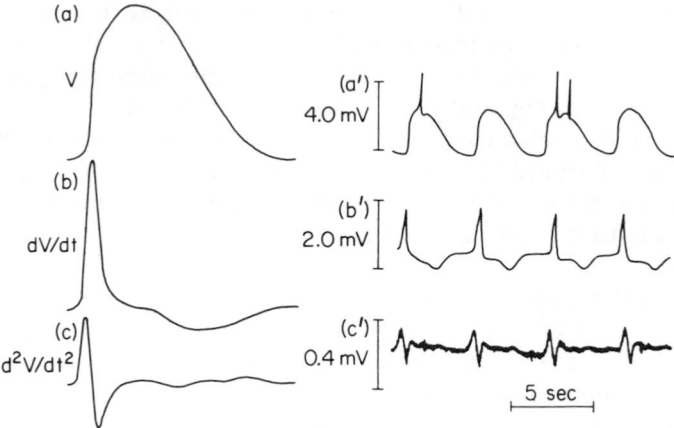

Figure 40. *Configurations of a propagated transient change in membrane potential recorded from a core conductor:* (a)–(c) are theoretical configurations and (a')–(c') are comparable recordings of electrical activity in segments of cat intestinal muscle (jejunum) using glass capillary or wick electrodes. Recording the transient response in a monophasic manner, such as with intracellular electrodes, produces a figure like (a). Graphically differentiating (a) results in a figure like (b), which resembles bipolar recordings with two external electrodes very close together. Graphically differentiating (b) results in a figure like (c), which resembles monopolar recordings with one electrode on the active tissue and the other some distance away in the volume of conducting medium. Compare these theoretical curves with actual recordings: (a') is a monopolar record obtained with a glass capillary electrode exerting a 2-g pressure on the tissue, (b') is a bipolar recording with cotton wick electrodes separated by less than 2 mm, and (c') is a monopolar recording using the same situation as in (a') but now the recording electrode is separated from the intestine by about 1 mm. The pressure electrode used in (a') yields a recording that is similar in shape to that obtained with an intracellular electrode.

that the damaged area appears to be a source of negative charges. In addition to the steady current and related external potential, the leaking ions tend to form a low resistance path directly to the interior of the still healthy part of the cell. If this part of the cell is triggered to give an action potential, electrical activity can be recorded from the injured area that closely duplicates the time course of the variation in transmembrane potential as it occurs in the healthy area. This low impedance pathway to the interior of the cell forms a sort of intracellular probe or an extension to the electrode and thus gives access to the undamaged and active region of the cell. In any injured area, many cells are likely to be contributing to the external current flow so that relatively high potentials may build up despite the relatively low resistance of the path along which current is flowing.

4.4 Electrodes for Recording
Bioelectrical Potentials

A recording or stimulating electrode must form the connection between the metallic conductors, usually copper wire, leading to amplifiers and other assorted recording apparatus and the electrolytic solutions that are surrounding, permeating, or within the biological complex [54, 81, 85]. Electric current flow within metals is in the form of the flow of electrons, whereas, within an electrolyte, it is in the form of the movement of ions. Therefore, at the junction of the electrode and the preparation, chemical reactions take place. With a small current flowing through the system, there will be only small reactions occurring and, hence, it is desirable to use high impedance recording instruments to reduce the magnitude of current flow through the electrode junction. Another reason to reduce current flow through the recording system is to reduce the amount of energy drained from the cell's normal functions and, thus, to minimize the interference into the cell's normal activities. Relatively high levels of current flow are often used to stimulate or to influence cells and, as a result, residues such as hydrogen bubbles accumulate around the stimulating electrodes that lead to the formation of an insulating layer and higher electrode resistance. Since this "polarization" of the stimulating electrodes does not form uniformly, spurious interpretations of data result. This can be reduced by rapidly changing the polarity at the electrodes and by the proper selection and manufacture of electrodes.

The ideal electrode introduces (1) no potential into the circuit and (2) no resistance into the circuit. Neither requirement is ever met. Requisite 1 is impossible since there will always be a liquid-metal junction somewhere in the circuit. In practice, both "active" and "reference" electrodes are constructed similarly in order to balance the batteries that are generated at the liquid-metal junctions, or when this is not feasible, a variable voltage source (the bucking voltage) is placed in the "indifferent" or "reference" lead and is adjusted to balance the voltage generated by the "active" lead. Some of the commonly used electrodes include Ag-AgCl wires, either singly or with connecting salt bridges; Pt-PtO wires; stainless steel needles; gold-plated gallium or indium or some of their alloys with tin; platinum-plated indium, tungsten; and various combinations of these. Some are used as large electrodes to record activity from a population of cells and some can be sharpened into fine tips (tip diameters between 1 and 100 μ) to snuggle up close to a single cell.

One of the favorite electrodes used today for recording activity from single cells is the glass micropipette first introduced by Ling and Gerard in 1949. A glass capillary tube drawn to a point less than a micron—usually close to 0.2 μ—in diameter and filled with an electrolyte solution can be inserted into a cell's interior to measure directly its transmembrane potential. These microelectrodes are filled with various solutions such as 2.7 or 3.0 M KCl, 1% or 2 M NaCl, 0.6 M K$_2$SO$_4$, saturated AgNO$_3$ or CuCl$_2$, or a dye such as saturated trypan red or procion yellow, but 3 M KCl is the most common choice. Normally the electrode is firmly attached to a micromanipulator and carefully lowered into a cell. Occasionally the preparation wiggles, as do muscle cells. To avoid or to minimize breakage of the fragile tips, a flexible coupling, e.g., fine tungsten wire or thin latex tubing, is used between the manipulator and the electrode tip. Flexible mounts are adequate since it is not always necessary to be able to see the cell being penetrated. In fact, it is often difficult to determine which cell is being monitored by a microelectrode.

The best evidence for the penetration of a microelectrode into a cell is not always a visual observation but rather the sudden appearance of the negative 40–90 mV resting potential, the reversal of polarity of the action potentials, and the increase in resistance in the recording system. Some investigators electrophoretically inject a dye into the cell at the end of the experiment and later examine the tissue histologically for the stained cell. Others will look for the damage track made by the electrode as it passes through the tissue in order to locate the cell in question.

There are a number of limitations in microelectrode procedures that can lead to errors in interpretations of the data. Diffusion potentials are not zero. Whenever the mobilities of the individual cations and and anions in a solution are not equal, the charges will separate as one ion diffuses more rapidly than the other to create a battery. The diffusion potential generated at the electrode tip generally reduces the value measured for the membrane potential especially in cells where much of the internal negative charges appear immobile. The choice of KCl as the electrolyte for filling micropipettes minimizes this source of error because the mobilities of K$^+$ and Cl$^-$ are very close, although not identical. The slight difference in mobility between K$^+$ and Cl$^-$ is exaggerated by charges which may develop on the glass along the tip of the micropipette or by protoplasmic plugs which can jam the tip of the electrode. The cell needs to make a tight seal around the electrode after it penetrates the cellular membrane in order to obtain good values for membrane potentials. Sealing around the electrode is indicated by a rise in the resistance between the electrodes after entry due to the high resistance of the membrane, steady values for the membrane resting potential and the action potential maintained after entry, and a negligible change in the resting potential after penetration of the cell by a second microelectrode. Microelectrodes have limited usefulness in small fibers because the area damaged by the penetration is large relative to the total surface area of the cell and any leakage of electrolyte from the electrode tip can change the cell's ionic composition and, thus, its behavior. Investigators must be especially careful of long experiments during which a cell can accumulate sufficient potassium from electrode leakage to become hyperpolarized to the point of complete inhibition.

Other recording techniques obtain information from single cells. The "suction" electrode and the "pressure" electrode [25] damage cells or parts of cells so that there is electrical continuity between the electrode and the interior of neighboring intact cells similar to the situation described earlier for the demarcation potential. The suction electrode works well on vertebrate heart tissue and the pressure electrode works well on vertebrate heart tissue and the pressure electrode works on mammalian intestinal muscle. These electrodes obtain records approaching those obtained by intracellular micropipettes in polarity and waveform but not in amplitude. They do not work on all kinds of tissues, however. They probably depend on the peculiar mechanical sensitivity of the cell membranes and the functional organization of the tissues in question. A third technique is the "sucrose gap." Replacing the extracellular fluid surrounding cells with a nonelectrolyte like sucrose results in a higher resistance path through the sucrose than through cell membranes and interiors. Thus, the current path through the cell interiors is more heavily traveled than the current path along the extracellular route. The recording electrodes are placed on either side of the gap and when one side is completely depolarized by an isotonic solution of KCl, full values for the transmembrane potential can be obtained [30, 236]. The technique works best with long cells that are continuous through the sucrose gap but will also work with vertebrate intestinal muscle, which has short cells but a peculiar organization: The cells, although morphologically isolated from each other, are physiologically connected by relatively low resistance paths through protoplasmic bridges or nexuses [58], which are areas where cellular membranes of neighboring cells, although still intact, are in intimate contact.

4.5 Views on the Origin
of Membrane Potentials

Over a hundred years ago biophysics became preoccupied with the mechanism by which signals are rapidly transmitted over long distances in the body. Matteuci and du Bois Reymond discovered the demarcation potential or injury potential in muscle, the electric currents generated between damaged and intact regions of muscle cells. They also discovered that nerves are not only excited by electric currents but produce in the course of their activity electrocurrents of their own. Today we know that demarcation potential, over which they had heated arguments, does not run down and hence is not a capacitative charge but must be produced continuously. Hermann suggested that the propagation of a signal along a nerve fiber or axon involves a process of a recurrent electrical stimulation from point to point. Hermann also suggested that nerve fibers possess properties similar to cables although he realized that nerve signals could not travel over any appreciable length by passive cable transmission alone. In 1902, Bernstein [19], under the influence of Nernst, developed the basis of what has since become the modern membrane theory. According to Bernstein, the mechanism of a nerve impulse is to be found in the properties of the surface membrane of the axon and especially its selective and changeable permeability, or conductance, to ions. He discovered that the potassium concentration is higher on the inside of cells than it is on the outside and he suggested that neither sodium nor anions can cross the cellular membrane. The resulting potential due to the difference in concentrations of these ions across the membrane thus resulted in the membrane potential. His theory has been modified in important details but, in essence, it still stands, and advances that have been made in this field especially during the past 20 years have confirmed the usefulness of Bernstein's basic concept.

The view that has now emerged most popular is that the electrical events in a nerve fiber are governed by the differential permeability of its surface to sodium and potassium ions and that these permeabilities themselves depend on the electric field across the surface. The interaction of these two factors lead, at a certain critical threshold level, to excitation or to a regenerative release of electrical energy from the axon membrane. The propagation of this change along the fiber in the form of a brief all-or-nothing electrical impulse is the spike or action potential. It is appropriate to question the level of confidence in the popular view by outlining the nature of the evidence that relates to the origins of the resting potential of cells. The spike will be considered later. The temperature coefficient of the membrane potential, which is 1.1 in frog muscle [176], suggests a physical process. The ideas presented to explain the membrane potential involve liquid junctions, Gibbs-Donnan equilibria, redox potentials, and diffusion potentials derived from active ion transport. This last mechanism is the one most commonly accepted as the primary one.

4.6 The Membrane Potential as a
Diffusion Potential: An Example

It is not unusual to find that the ion pumps in neuronal membranes maintain potassium ten times more concentrated in axoplasm than in the outside medium and sodium ten times more concentrated outside than inside. If all the positive charges of the cations are balanced, both inside and outside, by chloride ions, there will be no net charge to either bulk solution and no potential difference across the neuronal membrane. Therefore, to understand how that membrane can effect a sizable electrical potential across it, the permeability properties of that membrane must be defined as well as the concentration gradients across it. Permeability, in this case, is measured electrically as conductance, which can be described as the frictional resistance of the membrane to the movement of specific ions. Membranes typically have low conductances to ions but the conductance to sodium is generally much lower than is the conductance to potassium. To simplify this example, assume that the membrane conductance to anions is zero so that anions can never contribute directly to any potential difference that may finally develop across the membrane.

Since these initial conditions specify that there is no electrical driving force to act upon the ions, only the concentration differences and the conductances can have any effect during the first instant in time. That effect is the exit of potassium from the neuron at a faster rate than the entrance of sodium since the potassium conductance exceeds that of sodium and the concentration gradients are identical. At the end of the first instant, there is an excess of positive charges on the outside and a deficit of positive charges on the inside—anions are prevented from crossing the membrane in this example—and there is now a potential difference across the membrane.

Similar movements of potassium and sodium occur during the second instant but because of the potential difference that now exists across the membrane that slows the exit of cations and speeds the entrance of cations, the difference between the rates of movements of the two cations is smaller: Less potassium exits and more sodium enters during the second instant than during the first. Still, the potassium egress exceeds sodium entrance because of the important differences in membrane conductances. The result is that the potential difference continues to increase but at a decreasing rate. The membrane potential continues to build during successive instants until a steady state is reached. At this time, the potential difference developed accelerates the sodium movements and hinders the potassium movements until they are equal. The excess of positive charges on the outside and the deficit inside will have caused a membrane potential of sufficient magnitude to counterbalance the net outflux of potassium ions resulting from transmembrane concentration differences and conductance differences.

Although there are movements of ions, concentration differences between the bulk solutions are not altered since the absolute number of ions that traverse the membrane is small during the few minutes required to reach the steady state and since the sodium pumps operate continuously to maintain constant the concentration differences and generally to move sodium out as rapidly as it enters.

The rate at which a steady state is reached in this example is a function of the physical characteristics of the membrane. The relationships between the several membrane conductances have already been used to explain the final value of the resting potential developed across the membrane, but the rate of its development also depends on conductances. If the several conductances are large, a greater number of ions can flow across the membrane each unit of time and the difference in distribution of charge, and thus the membrane potential, develops rapidly; if the conductances are small, ions cross at a slow rate and the voltage changes more slowly. A second important physical characteristic is membrane capacitance. Capacitance includes the membrane properties (e.g., membrane thickness, lipid content) that cause it to behave as a condensor or capacitor. Under changing conditions, as in the example considered, not all the current is carried through the membrane by the flow of ions (Fig. 41). When an internal negative charge is not neutralized by a neighboring cation, it can drive a negative charge away from the external surface of the membrane, identical to the behavior of charges on the plates

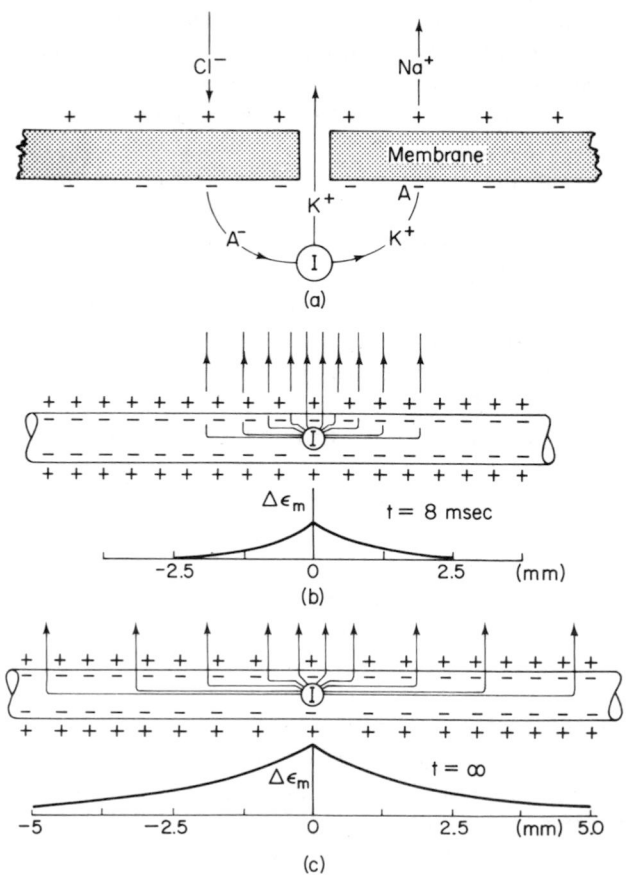

of a capacitor. Charges can move through the internal and external media (i.e., current) but the charged molecules that carry the current change at the membrane; there is no actual penetration of the membrane by the charged particles. It does not matter which particular ion carries the current; the chemical identities of charges are indistinguishable electrically. As long as the steady state conditions have not been reached, current will continue to pass through the membrane capacitor. At the steady state, the capacitor will be either fully charged or fully discharged and any continuing current flow must involve ion penetration. In our example, the capacitance will slow the development of a potential difference across the membrane caused by the migration of ions through the membrane (ohmic current) since some of the current will return by charging the membrane capacitor (capacitative current). Concerning rates, if the membrane capacitance is large, a quantity of ions crossing the membrane during a unit of time will produce only a small change in voltage and the final membrane potential will be reached more slowly. Capacitance has the opposite effect to that of conductance on the rate of achievement of the steady state membrane potential. A longer time is required to charge a larger capacity up to a final value. A shorter time is required to move a quantity of ions across a membrane of larger conductance. The ratio of membrane capacitance (C_m) to membrane conductance ($1/R_m$) is a convenient measure of this rate and is called the time constant of the membrane ($\tau_m = R_m C_m$).

In summary, the steady state membrane potential can be adequately explained by the ion pumps that maintain the concentration differences and the resting values of membrane conductances.

Figure 41. *Current flow patterns in elongated cells:* diagram (a) shows both ohmic and capacitative types of current flow. I represents a source of current (e.g., from the tip of an intracellular microelectrode) and the arrows indicate some of the paths of charge movement (current flow). The one arrow shows K^+ passing through a pore and represents ionic (ohmic) current flow through membrane resistance that can also be carried by an inflow of Cl^-. The arrow at the right shows cytoplasmic K^+ migrating to the inside of the cell membrane to neutralize A^- and release a positive charge (Na^+) from the outside of the membrane into the bathing solution. Capacitative current can effectively flow through the membrane without the actual penetration of charge. Diagram (b) illustrates a portion of the current distribution in a fiber 8 msec after the application of a pulse of current (upper) and the transmembrane potential as a function of distance along the fiber (lower). At this time, most of the current is capacitative (indicated by current lines interrupted at the membrane) and confined to the region immediately around the electrode. The spacing between $+$ and $-$ signs signifies the magnitude of the depolarization produced. Diagram (c) is similar to (b) but maps the current distribution for longer time periods. All current flow through the membrane is resistive and has spread much further along the axon. To show the current patterns more clearly, the diameter of the fiber is exaggerated by about ten times in relation to the length.

4.7 The Membrane Potential
as a Diffusion Potential:
Some Evidence

Several studies were cited in Chapter 3 to show that a nerve cell at rest has a composition quite different from the solution in which it lives. The composition of the external solution is normally well established because this solution is usually the same as blood plasma and represents a sizable fraction of the total volume of the animal examined. The composition of the internal solution is known only approximately in most cases because of the sampling problems. Indirect evidence indicates that the concentrations of sodium and chloride ions outside the cell are some ten to fourteen times larger than the concentrations inside the cell (e.g., see Table 6, p. 103). In contrast, the concentration of potassium ions inside a mammalian neuron is about thirty times greater than the concentration outside [43]. The inside of a typical nerve cell is negatively charged with respect to the outside by about 90 mV. This internal negativity drives chloride anions outward through the membrane and impedes their inward movement. In fact, a potential difference of 90 mV is just sufficient to maintain the observed disparity in the concentration of chloride ions across the cellular membrane when, at the steady state, chloride diffuses inward at the same rate as it diffuses outward. The 90-mV potential across the membrane therefore defines the equilibrium potential for chloride ions unequally distributed across the boundary. To obtain a

Figure 42. *Active and passive Na^+ and K^+ fluxes of a frog skeletal muscle fiber in the steady state:* the circle within the membrane represents the ion pump driven by metabolic energy. Each one-way flux is represented by a transmembrane band; the magnitude of each flux (given in picomols per square centimeter second at the left; $1 \text{ p}M = 10^{-12} M$) is proportional to the band's width and the direction is indicated by an arrow. The driving force on each ion—the difference between the steady membrane voltage (E_s) and the ion equilibrium potential (E_{Na} or E_K)—is shown by the slope of each band and the total height difference (shown at right). The net passive flux of an ion is proportional to the product of the driving force and the ionic conductance g_{Na} or g_K. The net fluxes are zero in the steady state. Since the passive efflux of Na is negligible, passive influx = active efflux = 1.8 pM/cm² sec. This flux results from a large driving force (-155 mV) and a small conductance ($0.07 g_K$), and is illustrated by a steep band and a small bar representing g_{Na}. Active efflux of Na must be accomplished against this electrochemical gradient. For K^+, net passive efflux (1.8 pM/cm² sec) = passive efflux (5.4) − passive influx (3.6). In contrast to Na^+, the passive flux of K^+ is the product of a small driving force (11 mV) and a large conductance. Since active fluxes of Na^+ and K^+ are equal and opposite, the net passive efflux of K^+ must also equal the net passive influx of Na^+ = 1.8 pM/cm² sec. The equal and opposite exchange diffusion fluxes of Na^+ are shown at the bottom of the diagram.

concentration of potassium that is thirty-nine times higher inside the cell than it is outside would require that the interior be some 97 mV more negative than the exterior. Since the measured membrane potential is but −90 mV, it falls short of the equilibrium potential for potassium. Evidently the thirty-ninefold concentration gradient can be achieved and maintained only if there is some auxiliary mechanism for pumping potassium cations into the cell at a rate equal to their spontaneous net outward diffusion. The pumping mechanism has a still more difficult task of pumping sodium cations out of the cell against a potential gradient of 156 mV. (156 mV is obtained by adding the 90 mV of internal negativity to the equilibrium potential for sodium, which is 66 mV of internal positivity.) If it were not for the postulated pump, the concentration of sodium across the cellular membrane would be almost the reverse of what is found (Fig. 42).

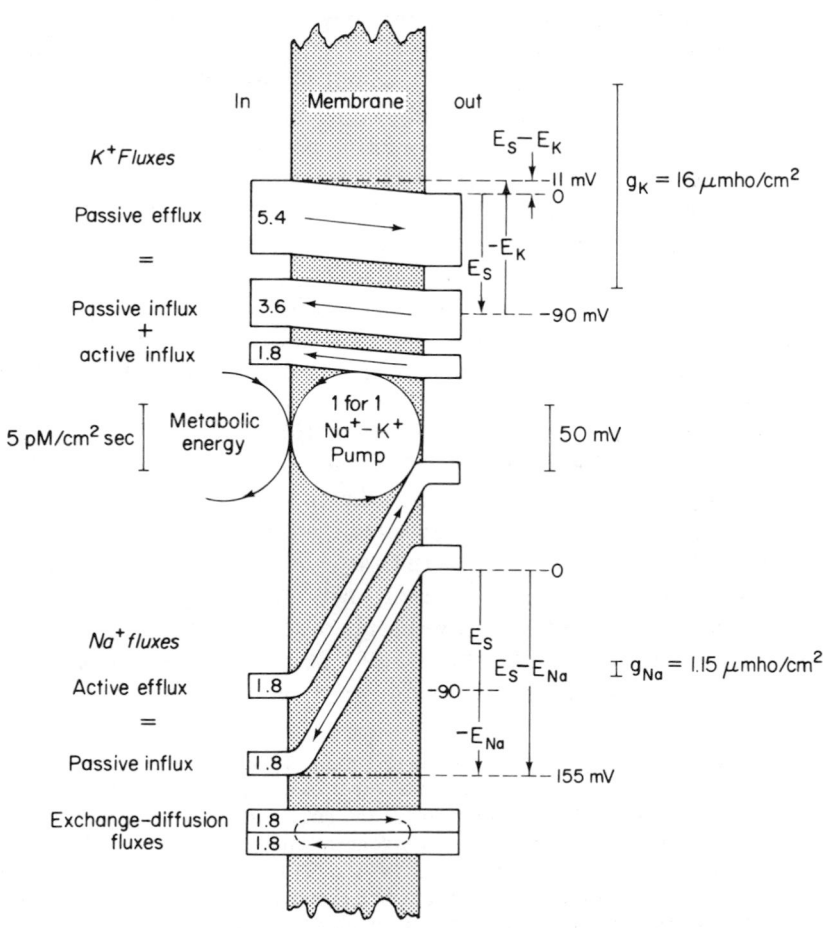

This concept of a diffusional potential and the quantity of supporting data make up the basis for the magnitude of its popularity: It is intellectually satisfying and it even has some experimental support. The outward potassium flux is balanced by positive charges, primarily sodium ions, on the outside of the membrane. Large organic molecules, functioning as anions, are contained within the cells and are not able to escape. At equilibrium or under conditions where no net work is performed, therefore, the Nernst relationship holds:

$$E_K = \frac{-RT}{nF} \ln \frac{[K_i]}{[K_o]}$$

In 1902, Bernstein essentially said that $E_K = E_m$ since K^+ is the only ion capable of being distributed across the membrane. In frog sartorius muscle, the ratio of K_i to K_o is 50:1 or the calculated E_K at 20°C from the Nernst equation is -99 mV [43] and E_m is measured as -95 mV [173, 174]. This does not present conclusive support for only a diffusional potential or a Gibbs-Donnan equilibrium mechanism since, if E_m were due to redox potentials or to some other mechanism, K^+ would be forced into the same distribution across the membrane and E_K would be the same. If the outside concentration of K^+ is raised, the membrane potential falls linearly except at low values of K_o or at high values of E_m as illustrated in Fig. 43. This deviation is commonly explained as a fall in the permeability to potassium as the membrane potential increases or as a result of the increasing influence of the sodium permeability on the potential. To satisfy Donnan equilibrium considerations, chloride should be distributed reciprocal to that for potassium and, indeed, is for most nerve fibers examined [121] but not for a number of muscles, which appear to bind large amounts of potassium in unionized complexes.

Changing the ion concentrations in the outside medium generally produced data expected from the Nernst relation, but the experiments during the 1950's involving attempts to change the concentrations of ions within muscle cells proved to be more complicated than was at first recognized by the various investigators. These experiments showed that the resting potential of muscle cells could not be altered easily by ion injection; especially noticeable was the failure to increase sharply the resting potential by raising the internal concentration of potassium [124]. Because of these discrepancies, Grundfest and his colleagues [106] proposed modifications in the basic

Figure 43. *Resting potentials of muscle cells as a function of the potassium concentration:* (a) shows the deviation of experimental points from the theoretical line given by the Nernst expression at the larger values of the ratio $[K_i]/[K_o]$. This ratio was altered by changing the concentration of potassium in the external medium. (b) is similar to (a) but shows that at low external concentrations of potassium a modified equation based on a significant permeability to sodium fits the experimental points better than a simple Nernst equation.

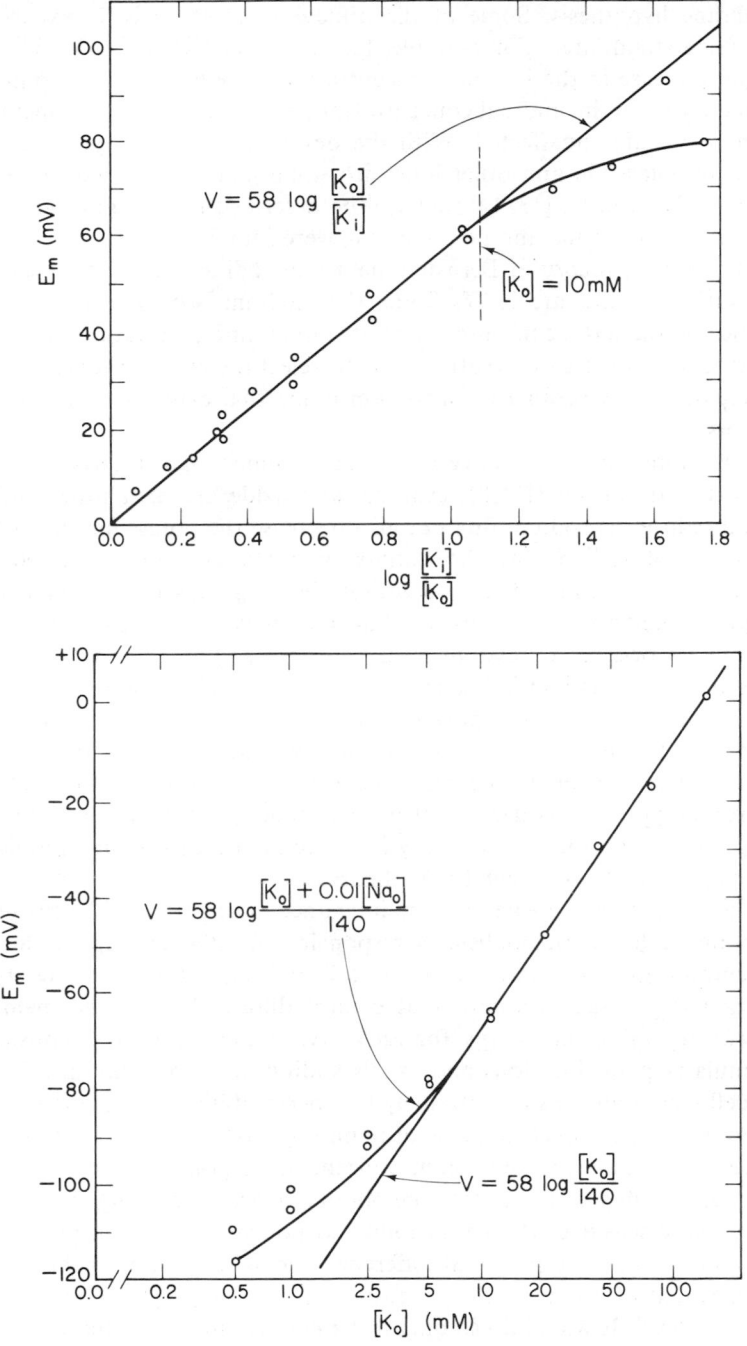

membrane hypothesis. Some of the difficulties in these early experiments were understood later. For example, Hodgkin and Keynes [133, 134] found that an increase in the internal concentration of potassium is accompanied by an increase in the internal concentration of chloride so that the membrane potential remains unaffected. With the development of the procedures for perfusing potassium and other ions into squid giant axons, like the experiments of Baker et al. [13, 14] that will be described later, significant effects of internal ions on membrane potentials were found.

The original theory of Bernstein held that sodium was impermeable but observations in the late 1930's found that sodium is only relatively impermeable and there is a continuous flux of sodium into and out of the cell. The first reaction to these observations was to reject the entire theory. But Dean [52] repaired it by postulating a sodium pump that expels sodium as fast as it enters the cell.

It is more efficient to have an exchange pump than a straight sodium pump. Dinitrophenol (DNP), cyanide, and azide are metabolic inhibitors and all reduce the sodium flux but also decrease the potassium inward flux. The influx of sodium and the outflux of potassium are unaffected [132]. Sodium can be extracted from solutions bathing cells that originally contained no sodium and no potassium but the rate is accelerated if the solution contains its original potassium level and zero sodium, hence, the sodium efflux and the potassium influx are coupled but loosely. The outward flow of sodium is due to active transport and must be associated with the mechanism for producing the potential. Grundfest [99] said that the sodium pump potential is large enough to explain the greater part of the resting potential. The resting potential is dependent on the mode of operation of the sodium pump and not on the ratios of any ion species and this is the modification that he presents for the membrane theory.

The present view of the ionic maintenance mechanism of the resting cell holds that cellular metabolism is responsible for the upkeep of the ionic concentration differences between the cell and its surroundings by (1) synthesizing large organic anions that cannot diffuse through the membrane and (2) providing the energy for an active ion-exchange mechanism that accumulates potassium ions and expels sodium ions, possibly one for one. The cell membrane has an extremely low permeability to ions in general so that even in the complete absence of pumping action it will take many hours before the sodium and potassium concentration gradients run down. The ionic permeability of the resting membrane, apart from being small, is also differentially selective; the permeability to potassium is much greater than to sodium so that the potential difference across the resting membrane approaches that of a potassium concentration cell, although it does not quite reach this level. It would do so only if the permeability to sodium was negligible.

Thus, if sodium and chloride were entering the fiber, the system would be running down and the potential would be represented by a more complex expression than the Nernst equation to include the influence of net movements of these other ions. The expression usually employed to describe—and successfully—the potential under these conditions is the constant field equation as developed by Goldman [96] and modified by Hodgkin and Katz [130]:

$$E_m = \frac{-RT}{nF} \ln \frac{P_K[K_i^+] + P_{Na}[Na_i^+] + P_{Cl}[Cl_o^-]}{P_K[K_o^+] + P_{Na}[Na_o^+] + P_{Cl}[Cl_i^-]}$$

where P_K, P_{Na}, and P_{Cl} represent the permeability coefficients for potassium, sodium, and chloride, respectively.

According to this hypothesis, sodium, potassium, and chloride occur as free ions in the axoplasm and both the resting potential and the action potential are believed to result from the diffusion of these ions across the membrane. [121, 122]. To account for the membrane potential of axons in the resting state and its dependence on the external concentration of potassium, sodium, and chloride, it is necessary to assume that potassium ions diffuse across the membrane much more readily than either of the other ions [130]. On the other hand, at the peak of the action potential, the state of the membrane appears to change so that it is much more permeable to sodium than to potassium since the membrane potential at this time is reversed in polarity and its magnitude is determined primarily by the gradient of sodium ions across the fiber surface [130]. As an action potential propagates along an axon, there occurs at each point an inward migration of sodium ions and an outward movement of potassium ions, thereby producing small changes in the ionic composition of the axoplasm [227, 228]. Thus, according to the membrane hypothesis, the nerve fiber is continuously dissipating the ionic concentration gradients that are the source of the electrical potential. If such a picture is correct, it is necessary to assume that the neural membrane contains a special transport mechanism for maintaining these concentration gradients.

That the surface membrane presents a formidable barrier for ion diffusion is indicated by at least two lines of evidence. Tracer experiments clearly define the low permeability of the plasma membrane to charged particles (see Chapter 3). Transmembrane electrical measurements also indicate a high resistance to the flow of current (i.e., ions) across the membrane. AC impedence measurements [40] and the attenuation and distortion of small voltage-step signals applied across the cell surface [184] both indicate that the interior of axons and muscle fibers behave as electrical conductor cylinders of slightly higher electrolytic resistivity than the outside fluid and that the inside is separated from the outside by a surface layer of low conductance (10^{-4}–10^{-3} ohm^{-1} cm^{-2}) and of high capacitance (about 1 μF cm^{-2}).

That the intracellular ions are able to freely diffuse about has been a controversial point. According to one point of view, potassium is combined selectively with intracellular anions in a special region of the axon, while the remaining portion of the fiber, containing but a small fraction of the cell water, is in diffusional equilibrium with the external solution. This view leads to the conclusion that the resting and action potentials result from chemical reactions occurring within the membrane [184]. The evidence that supports the conclusion of a selective binding of potassium is based on three major observations. (1) The failure of metabolic inhibitors to affect the potassium content of frog muscle at low temperatures has been interpreted to indicate that the cells need not expend metabolic energy to maintain their ionic distribution [162]. (2) In muscle cells, the intracellular quantities of sodium, chloride, lithium, and several other ions appear to be directly proportional to the external concentrations of these substances, whereas potassium appears to consist of two fractions: one relatively independent of external potassium and the other directly proportional to it. These observations have been interpreted as indicating that a muscle cell contains two compartments, one of which is in diffusional equilibrium with the external solution [86, 233]. (3) If the kinetics of isotope exchange of potassium in muscle and in red blood cells are assumed to be determined by intracellular diffusion, the calculated self-diffusion coefficient for potassium has a value only 10^{-5} times that of the self-diffusion coefficient of this ion in free solution indicating that potassium ions may be largely bound within these cells [113]. According to the other point of view, the ions are free in the intracellular solutions. Studies carried out by injecting sodium and potassium into squid axons indicate that the diffusion coefficient and the mobility of the injected substances are only slightly less than the values of these parameters in free solution [131, 133]. Additional evidence of the lack of binding is provided by the observation that the specific resistance of the axoplasm is approximately that expected if the ions were in free solution [41]. Furthermore, nearly all the ionic constituents of squid nerve have been accounted for and it has not been demonstrated that the anions have any special affinity for potassium. Moreover, there exists extensive evidence that indicates the operation of special ion-transport processes in nerve cells [141].

This view—that the sodium, potassium, and chloride exist as free ions in the intracellular medium, that special ion-transport mechanisms are located in the plasma membrane to establish and to maintain concentration gradients, and that the electrical properties of the membrane are diffusion potentials permitted by specific membrane conductances—leads directly into the explanation—often called the sodium theory—commonly given for the active membrane phenomena. As described in detail later, Hodgkin and Huxley [125–129] were able to formulate a general theory in quantitative terms from their measurements of three main parameters: (1) the membrane conductance to sodium, (2) the membrane conductance to potassium, and (3) the inactivation of the transport mechanism by which sodium is carried into the cell. All three parameters are functions of membrane potential and each includes a time constant. Their equations relating the functions and the passive electrical characteristics of the membrane yield strong numerical agreement with directly observed quantities for a variety of responses including the form of the subthreshold response, threshold, the form of the spike, refractory period, propagation velocity, and the effects of a number of alterations in the external medium. Discrepancies have been reported, however, and the theory—the sodium theory as well as its underlying diffusion potential hypothesis—cannot be cited as entirely satisfactory.

4.8 The Membrane Potential as a Redox Potential

Jahn [148] reevaluated the Lund theory [185, 187] that transmembrane potentials are primarily oxidation-reduction potentials and incorporated modern ideas of electron transport in organic compounds so as to include electron flow through the lipid substance of the membrane. There are at least several possibilities of electron movement across membranes and one of these possibilities does not assume layered enzymes in the membrane. This possibility, implicit but not really stated in the original Lund redox theory for transmembrane potentials [185, 187], was discarded by Osterhout [210] and Beutner and Lozner [20] primarily on the basis that only metals could conduct electrons. In view of the modern concepts of electron transfer in organic reactions and electronic resonance in proteins, prosthetic groups of enzymes, lipids, and all molecules containing alternating series of single and double bonds, the older criticisms now appear naive and unjustified. Moreover, discussion has been considerable concerning the possibility of electronic conduction across enzymes, the parietal cells of the stomach, chloroplast lamellae, mitochondria, muscle proteins, visual pigments, and biological processes in general.

Therefore, Jahn [148] revised and extended the theory to include these modern concepts and to outline a general theory of ion transport in which it is assumed that transmembrane redox potentials are the driving forces. Jahn's suggestions are an alternative to the ideas that the basic mechanism producing potentials is ion secretion and that the major source of membrane potentials is ion diffusion. Diffusion potentials, which can exist even in dead skin [7], may also be present but are assumed to have but a minor role and not to be the major origin of the electromotive source.

Any long lipoid molecule containing alternating single and double bonds, such as β-carotene, can bridge the lipid film of a plasma membrane and form a perfect pathway for electrons produced by an oxidizing enzyme on one side of the membrane and received by a reducing enzyme at the other side— a situation comparable to Calvin's proposal [32] for electron passage through chloroplast lamellae. Since cellular membranes consist of a lipoid film with proteins attached on both sides, there seems to be no reason why the membrane cannot serve the same function as a metallic electrode, with the measured EMF being determined largely, if not entirely, by oxidation-reduction reactions at the surface of the membranes and perhaps linked to more complex systems. Redox reactions can be coupled across the membrane in such a way that electrons released by an ATP-coupled oxidation on one side of the membrane can be transmitted through the membrane to produce reduction on the other side. The ATP may couple either directly or through other reactions, or even substances other than ATP might serve as a source of electrons. It must be assumed that the second side also has a reversible oxidation-reduction system to be reduced by these electrons and to be eventually restored by an oxidative process or the replenishment of oxidant from external sources. If the electron pressure on the first side of the membrane is higher than on the second side, the EMF developed is positive to the second side. This type of system permits the direct measurement of redox potentials somewhat as originally assumed by Lund, who suggested that the measured potentials were the difference between the oxidation-reduction potentials at two such membranes. Thus, membrane potentials may primarily be redox potentials resulting from the presence of oxidation-reduction enzymes on the two sides of the plasma membrane connected by a conjugated lipoid bridge (e.g., carotene, astacene, retinene, vitamin A, crocetin) and activated by hydrolysis of a highly resonant phosphoric acid compound that serves as a source of electrons (Fig. 44). Such a system is capable of producing the large currents found in rapidly metabolizing cells and tissues and may explains the confusing data on the effects of various factors: metabolic inhibitors, temperature, and anoxia, as well as the effect of imposed currents on growth, development, and ion transport.

In order to explain the permeability properties of biological membranes, four types of pores are described by Jahn:

Pore type I: These are small diameter pores whose charge may be positive, negative, or neutral but which are filled or lined with a continuous transmembrane water lattice which permits hydrogen ion transport through it.

Pore type II: These are larger diameter pores whose positively charged walls are of the proper diameter for chloride molecules but which are lined with a discontinuous water lattice to reduce hydrogen ion transport but to permit chloride ion transport.

Pore type III: These are larger diameter pores, as in pore type II, but lined with a continuous water lattice to permit transport of both hydrogen

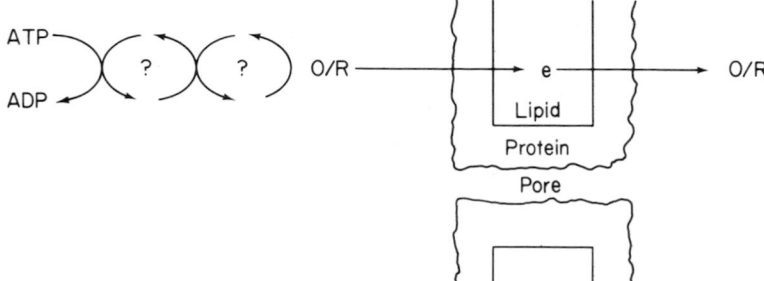

Figure 44. *Jahn's view of the origin of a redox potential across a biological membrane:* this diagrammatic cross section of a membrane illustrates the migration of electrons across conjugated lipoid substances because of electron pressure to produce the EMF.

and chloride ions. The oxidation-reduction differential between the two sides is the source of the transport energy and is also the primary source of the commonly measured potential.

Pore type IV: These are pores with negatively charged walls that have the proper diameter for the passage of sodium ions but which are lined with a discontinuous hydrated water lattice.

The apparent rapid transport of hydrogen and hydroxyl ions is almost automatic if an additional assumption is made: Proton charge can transfer through the continuous hydrogen-bonded water lattice lining the hydrated pores of small diameter [76] given the energy provided by the potential maintained by metabolism (Fig. 45). Water transport then results from electroosmosis through the pores of somewhat larger diameter [231]. For example, if the EMF of the stomach wall is assumed to be an oxidation-reduction potential, then the hydrogen ion can be actively moved by proton-charge transport through pores of type I. The chloride ion is transferred passively across the stomach wall to maintain ionic charge equilibrium through pores of type II, hence, the negative sign of the potential is explained without assuming active transport of chloride ions [218]. It is also possible that the hydrogen ions are produced in the same oxidation-reduction reaction that produces the electrons:

$$AH_2 \longrightarrow A + 2H^+ + 2e^-$$

so they are immediately available for charge transport. As the oxidant is later reduced, more hydrogen ions will be needed but these can be obtained

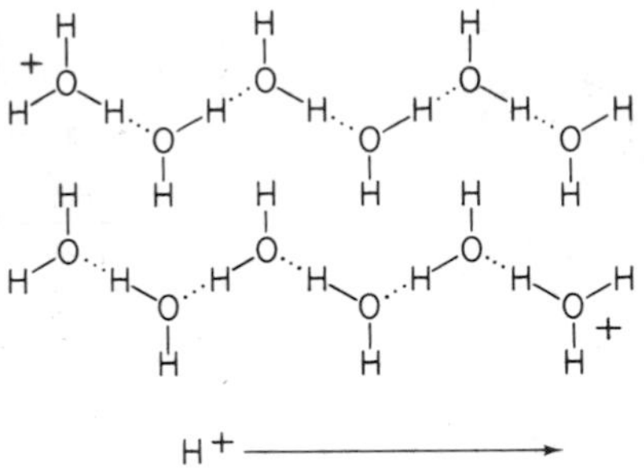

Figure 45. *Migration of H^+ ions by proton charge transfer:* a hydronium ion initially appears at the left end of the top series of water molecules—a continuously hydrogen-bonded water lattice. Solid lines indicate valency bonds and dotted lines indicate the hydrogen bonds. According to the theory of Eigen and DeMaeyer, a transfer of charges and a shift of bonds results in an apparent migration of the H^+ ion to the other end of the chain (right end of bottom series).

from carbonic acid as generally assumed for the secreted hydrogen ion [214]. The direct secretion of hydrogen ions formed from carbonic acid is also in accord with Jahn's thesis.

For transport of only chloride or any other negative ion, the water lattice in the positively charged pores of the proper size for chloride ions should be discontinuous, as in type II pores. As hydrogen ions accumulate at the end of type I pores, they would cause a reverse polarity on a micro scale through type II pores and tend to cause a migration of the negative ions through the pores by electrostatic forces. Also there is no reason why a single pore with a continuous lattice like type III pores should not function for both hydrogen ion transport and for chloride ion transport provided no other cation is transported.

The EMF across the stomach wall need not be greatly reduced by transport. The oxidation-reduction potential, which is the driving force, adds algebraically with the potentials across the membrane to produce the measured potential. Therefore, the transfer of proton charges should reduce the total measured potential but this reduction should soon be counteracted by chloride ion transport, perhaps through the same pores. There should be a positive correlation between the measured rate of HCl transport and the magnitude of the net potential as well as redox potential.

Discontinuities in the water lattice block proton charge transfer through negatively charged pores of exactly the proper size and shape for sodium, such as in type IV pores. The discontinuities need be only about 2 Å—Eigen and DeMaeyer [76] estimate proton shifts greater than 1 Å—and these can be maintained by some type of hydration on the protein lining of the pore that would still permit passage of a sodium ion and its primary water of hydration, the innermost layer of three molecules, through the pore. Mullins [195] proposed a similar requirement of a definite size of pore for potassium passage.

The effects of potassium and other ions in reducing the potential measured across membranes may be an ion effect on the outer surface rather than a diffusion potential as commonly supposed. Equations for redox potentials indicate that any direct inactivation of either the oxidant or reductant, possibly through some stereo change, would alter the EMF by some 58 mV per tenfold change in the deactivating agent and, hence, is electrically indistinguishable from a diffusion potential.

There need not be a one-to-one relationship between electrons transported and the number of ions transported. In the stomach, the negative oxidation-reduction potential causes proton transfer and the resulting excess of protons attracts chloride ions electrostatically through the same or nearby pores of a proper size for chloride ion transport. Since proton charge transport is rapid and the transfer of chloride ion is much slower, chloride ion transfer is the rate limiting step. The rate of this process depends ultimately on the rate of the oxidation-reduction reaction. The primary need for energy is to overcome the friction or kinetic energy of the chloride ion, hence, the total number of HCl molecules secreted does not necessarily have a one-to-one relationship with the number of electrons. This principle can also be applied to other ions except to hydrogen or to hydroxyl ions that are transferred as a proton charge defect.

The direction of the active transport needs to depend not only on the sign of the potential but also on the geometry and charge of the pores. Assuming two main functional types of pores, one with the inner surface positively charged and the other negatively charged, the situation in which ions will pass in one direction or in the other direction will depend on the diameter of the pore. For each type of ion, a pore of some given specific size will give a maximum rate of transport. Thus, an ion of either sodium or chloride tends to move faster through its own type of pore than the movement of the other ion. The faster ion causes the other to move in the same direction to maintain electrostatic neutrality. The faster ion will go in a direction dependent on the sign of its charge; the other ion, in following, will cause the ultimate direction of transport to depend not on the sign of the potential but on the ease with which either ion can pass through its respective pore.

In summary, Jahn proposes that most of the EMF measured by salt bridges can well be (1) oxidation-reduction potentials, (2) usually fueled directly or indirectly by ATP, and (3) expressed through oxidation-reduction enzymes on the two membrane surfaces connected electrotonically by means of conjugated transmembrane lipoid substances. Jahn's present theory still lacks, like other theories [70, 89], a detailed explanation for many phenomena, especially sodium transport and potassium effects.

4.9 The Membrane Potential
as a Fixed-Charge Effect

Many of the characteristic properties associated with large protein molecules are due to their interaction with small ions. Ling [172], in a monograph with an amazing title, developed at great length a quantitative theory concerning this interaction to explain diverse properties of living systems ranging from ion potentials to developmental embryology. Ling's chief assumptions are (1) charged particles in protein solutions do not act independently but in association and (2) a small change in the number or kind of ions attached to one portion of a protein molecule can, by induction, trigger larger changes in the state of the entire protein structure. The first assumption is generally accepted and Ling supports it with solid arguments. The second assumption has less experimental support. It is difficult to demonstrate that the effects of exchanging one sodium ion for one potassium ion are transmitted through the entire protein structure. Ling attempts to explain all the specificity observed in biological phenomena by using only the electrostatic properties of protein and ions, but we are concerned here mainly with what he has to say about bioelectric phenomena.

In the living state, the system of proteins, ions, and water must exist in close association and suffer abundant interactions. The theory of protein behavior in dilute salt solutions developed by Linderström-Lang [169, 170] deals with long-range attributes, such as net charge and dielectric constant, but not short-range interactions. Thus, association between proteins and interacting particles is the critical difference between the behavior of proteins in dilute salt solutions as treated by the Linderström-Lang theory and the behavior of proteins, salt, and water in biological systems as described in Ling's present theory. Ling also points out that the Linderström-Lang theory is contraindicated by the observations that specific ions are important, heavy water behaves differently than ordinary water, and both net charge and molecular weights are insufficient to describe the activity of molecules.

The primary function of the highly polarizable resonating chain (a protein is a linked polypeptide) is to provide a vehicle for ready transmission of an inductive effect from one functional group to another; hence, specificity of protein is in the side groups and not in the chain. By an inductive effect, Ling means the mechanism that brings about the difference between weak acids, such as CH_3COOH, and strong acids, such as Cl_3CCOOH. (The chloride is more electronegative than the hydrogens and decreases the negative charge of the distant carboxyl group to reduce its ability to hold protons.) Primarily, this involves the distribution of electrons, hence, polar groups are dominant. Short-range forces are operative: Water, ions, and proteins are in close association in the living state with charged ions associated with fixed charges located on the protein. The C-value mentioned by Ling [172] is a measure of electron distribution. A weak acid has a high C-value of approximately -2 Å. A strong acid has a low C-value of approximately -4 Å. The C-values are determined in part by polar groups of amino acid residues located next to the charged group in question.

The indirect or F-process mentioned by Ling is the sum of all the inductive effects and the direct electrostatic effects transmitted through space. Keep in mind that the "freezing" of water molecules surrounding associated ion pairs may produce a dielectric constant of water close to one in biological systems (and not eighty-one as is the case in free solution), which gives rise to high electrostatic energies. The charge is altered when the counterion is adsorbed onto the fixed site. The C-value of neighboring charges is altered and their affinities are altered and, hence, so on down the chain. The F-effect works at a distance and is consequently called an indirect F-process. The cardinal sites are small in number but of strategic importance in that they can, through indirect F-processes, control and modulate large numbers of sites, for example, the effects produced by ATP, hormones, and drugs.

The association-induction hypothesis of Ling involves nearly complete association of living protoplasm in a fixed-charge system and nearly complete ionic association. Unlike free solutions, the Debye-Hückel limiting law does not apply to protoplasm.

Ling has criticized several basic aspects of the generally accepted membrane theory. The original proponents—Ostwald, McDonald, and Bernstein— said that the semipermeability of cellular membranes excludes negatively charged anions while it permits only some cations, notably potassium, to pass. Many kinds of cells—nerve [50, 144], muscle [43, 135, 175], eggs [186], etc.—display a membrane potential proportional to the logarithm of the external potassium concentration. The first difficulty with the membrane theory occurred when Heppel [114] demonstrated that membranes are permeable to sodium as well as to potassium. Many investigators have since confirmed this and an expression for sodium is included in the Goldman equation [96, 130]. The second difficulty with the membrane theory is the poor correlation between the membrane potential and the concentration of potassium inside the cell as predicted by the theory. Although there are difficulties in performing the experiments, several investigators have injected potassium chloride, sodium chloride, and sodium glutamate into muscle fibers and have measured no change in the membrane potential [77, 106, 230]. A muscle soaked in sucrose loses 80% of the internal potassium without any drop in membrane potential [165]. The third difficulty mentioned by

Ling involves the lack of a correspondence between the constants listed in the Goldman equation and the results of actual measurements. The Goldman equation fits the data only when the permeability constants of potassium, sodium, and chloride are in the ratio of 1.0:0.04:0.45, respectively [121], but various other ions can substitute for chloride ion and no lasting change occurs, hence, chloride plays only a minor role. Eliminating chloride from the equation and reevaluating the constants to find an equation to fit the data results in a 40:1 ratio between the permeability constants of potassium to sodium [176]. Results of actual measurements show a ratio of less than 10:1 [172]. A fourth difficulty again involves a disagreement between experiment and theory. P_K, P_{Na}, and P_{Cl} are essentially diffusion coefficients and, like all rate processes, contain the term $\exp(\epsilon_i/RT)$. ϵ_i is the activation energy of the ith ion and the expression

$$g = \frac{P_K[K_i] + P_{Na}[Na_i]}{P_K[K_o] + P_{Na}[Na_o]}$$

is temperature dependent unless $\epsilon_{Na} = \epsilon_K$. The experimental data [172] show this not to be true. If g is temperature dependent, the membrane potential will not be proportional to the absolute temperature, but the experimental data [176] show that it is.

This evidence shows that the measured cellular potential does not depend directly on the gross internal concentration of potassium of sodium nor does it depend on surface permeabilities of these ions. If theory and experiment do not agree, something must be wrong with one or the other (or sometimes both). Ling rejects the membrane theory. Instead, he suggests that the membrane potential of living cells corresponds to the potential at cellular surfaces similar to a surface potential between solid-liquid interphases [171]. This is based on the concept that the entire cell is a fixed-charge system and, as such, constitutes a phase separate from the plasma or Ringer's solution.

The features of Ling's fixed-charge hypothesis are (1) the ionic selectivity is due to differential adsorption of various cations onto fixed anionic charges and, since this arises under equilibrium conditions, a metabolic pump is unnecessary; (2) Ling's equations exclude the term for chloride and replace the P_K and P_{Na} in Goldman's expression with equilibrium constants of adsorption of the ions onto the surface ionic sites of the fixed-charge system; (3) the magnitude of the potentials are related to the concentration of the surface fixed anionic charges and the nature of these charges but not to the bulk phase of potassium concentration. In Ling's model, the alternative metastable states are such that in one—the resting state—the fixed anions have a C-value that prefers potassium exclusively. During activity, the C-value is such that they prefer sodium exclusively. Flexibility of the C-value in both the resting and the active state in various types of cells can account for the many varying observations and must be anticipated.

The main experimental evidence summarized by Ling to support his hypothesis includes: (1) In model systems used to demonstrate membrane potentials as given by the Nernst equation, like the pH electrode, some investigators feel that the potential measured is the algebraic sum of two surface potentials contrary to the conventional view that only H^+ is permeable and potassium, etc., are not. (2) In a second model system, a collodion membrane separating two solutions of potassium, the potentials recorded agree with the membrane theory [189] but if glass is coated with collodion, the data obtained are much like that with collodion alone [171]. Also, collodion develops electrode behavior after acquiring fixed carboxyl groups as a result of oxidation [234, 235]. (3) By varying the external potassium concentration, an instantaneous change in the surface equilibrium is expected and a new stable membrane potential should result. The fixed-charge theory predicts that the time course of the membrane potential is independent of the change in the internal potassium concentration. The membrane theory holds that the time course of the membrane potential should parallel the change in potassium concentration inside the cell. Experimental data support the fixed-charge hypothesis and an example is illustrated in Fig. 46. (4) Additional support comes from the work of Tasaki and Teorell. For example, they found that the time constant for the loss of radioactive chloride ions injected into cells is an order of magnitude greater than for intercellularly injected monovalent cations [244]. This suggests that the squid axon membrane, which they used, has a fixed negative charge similar to a cation exchange membrane.

Keynes [161], a strong supporter of membrane theory, severely criticized Ling's explanations of bioelectric phenomena on the basis of fixed charges. Keynes accused Ling of either ignoring or misstating experimental evidence for which the existence of a surface membrane provides a simple and satisfactory explanation. Keynes pointed to the strong experimental support for the membrane theory, as refined by Hodgkin and Huxley, provided by the work of Baker, Hodgkin, and Shaw [13, 14]. They rolled the axoplasm out of the end of a squid giant axon with a squeegee and, with a syringe, refilled the axon with a test solution. The axon behaved normally although more than 90% of its contents was exchanged, however, the reduced internal resistance due to loss of axoplasm did lower the height of the spike and the conduction velocity. The conduction velocity also correlated with the changing fiber diameter caused by shrinking and refilling the axon. Axons remained excitable in perfusates of isotonic potassium sulfate, potassium methyl sulfate, potassium ethane sulfonate, potassium chloride, and potassium isethionate. Axons filled with potassium isethionate of potassium sulfate carried between 3 and 5×10^5 impulses, which is the about the same maximal number of impulses that can be carried by intact axons. An axon filled with K_2SO_4 had the same temperature dependence as a normal one. Dropping

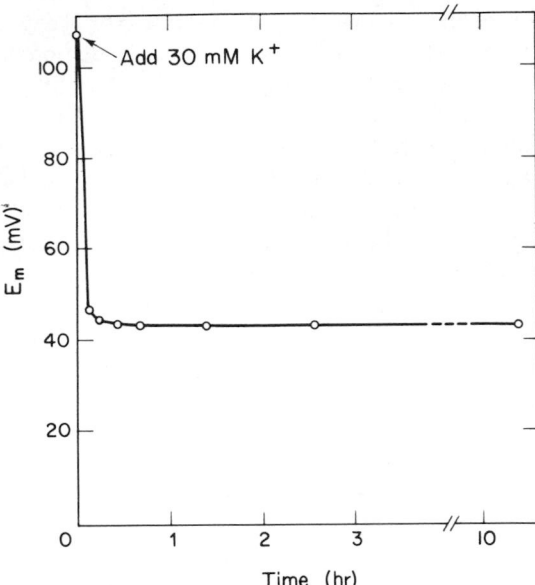

Figure 46. *Time course of the effect of high external potassium on the resting potential of frog sartorius muscle cells at room temperature:* at zero time, the Ringer's solution bathing the muscle fibers was changed from one containing 2.5 mM K$^+$ to one containing 30 mM K$^+$; the concentration of other ions was kept constant. The resting potential of individual cells was measured with an intracellular microelectrode. Within the time required to make the first measurement, the resting potential had already reached a new equilibrium in contrast to the 6–10 hr required for equilibration of the intracellular potassium to its new concentration.

the internal concentration of potassium to zero reduced the membrane potential to zero. Increasing the internal concentration of potassium raised the membrane potential. If the potassium inside was replaced by sodium and the outside sodium by potassium, the potential across the axon membrane reversed and became +55 mV inside. The curves do not fit a potassium electrode well, however. Increasing K$_i$ from 150 to 600 mM increased the membrane potential by only 5–10 mV, whereas a good potassium electrode would show an increase of 33 mV. Apparently, potassium permeability decreases with increased membrane potential.

Another set of experiments that contraindicates Ling's views was provided by Adrian [2] who, in contrast to other workers of the 1950's, showed a clear relationship between the internal concentration of potassium and the membrane potential in frog muscle. Adrian maintained the external concentration of potassium at 75 mM and altered the concentration of sucrose to retain a uniform osmotic concentration gradient. As a result, he was able to produce changes in the concentration of potassium inside. At physiological external concentrations of potassium, the membrane potential is not independent of the internal concentration, which can be altered by changes in the osmotic pressure.

Ling concluded that the low internal concentration of sodium cannot depend on the operation of a metabolic pump because at 0°C a pump would make impossible demands upon metabolism. Ling compared the rate of splitting of energy-rich phosphate bonds in a poisoned muscle with the energy requirement for sodium extrusion. Keynes [161] doubts Ling's energy numbers since he used only tracer experiments and short loading times and he ignored possible errors due to exchange diffusion with extracellular sodium, which does not operate at the expense of cellular metabolism. Moreover, Ling showed no washing out curves to substantiate his method for measuring the exchange rates and the rates he obtained are higher than those observed by many other investigators. The evidence presently available about the movements of labeled ions in muscle at 0°C is unsatisfactory and Keynes does not have the confidence in the existing data to agree with Ling's conclusions.

Keynes [161] summarized the main weaknesses of Ling's theory: (1) Ling failed to explain the precise difference between proteins in an intact cell and those in a ruptured cell or in a cell extract that results in potassium retention in one case and an almost complete lack of selectivity in the other. (2) Ling did not account for low internal values for the concentration of chloride. (3) Ling's hypothesis leads to vague and qualitative explanations of diverse physiological phenomena rather than to quantitative predictions that can be experimentally tested.

To balance the presentation, it should be noted that Keynes is committed to defending the view that ion movements are a diffusional process in excitation phenomena, the idea so effectively developed by Hodgkin and Huxley. But even this view is difficult to reconcile with a variety of experiments and has been repeatedly challenged. The work of Nachmansohn and Tasaki will be described later.

4.10 Gibbs-Donnan
Equilibrium Potentials

A system consisting of two compartments separated by a membrane that is permeable to some but not to all charged particles present will, at equilibrium, have unequal concentrations of the permeable ions and an electrical potential difference across the membrane. The common situation in biological systems is the presence of large organic anions imprisoned within cells. If all the permeable salts are considered to be NaCl, then, at equilibrium and at conditions when no work is being performed, the concentration of sodium inside the cell will exceed that outside and vice versa for the chloride ions. Gibbs theoretically and Donnan experimentally showed that the products of the diffusible ions on both sides of the membrane are equal at equilibrium; in this case,

$$[Na_i][Cl_i] = [Na_o][Cl_o].$$

$[Na_i]$ is clearly greater than $[Cl_i]$ since some of the Na_i is associated with the organic anions and the remainder with the Cl_i. On the outside of the cell, $[Na_o] = [Cl_o]$, preserving the electroneutrality in the bulk extracellular solution. Hence,

$$[Na_i] + [Cl_i] > [Na_o] + [Cl_o] = 2[Cl_o].$$

If water is permitted to permeate the membrane, osmotic pressure differences can result as water enters the cell. As sodium diffuses out of the cell, down its concentration gradient, it carries positive charge with it to leave the interior now more negative and to cause the exterior to become more positive. The development of this potential across the membrane impedes further migration of sodium. A similar situation holds for chloride (Fig. 47). At equilibrium then, a balance will occur among the net migration of diffusible ions across the membrane, the electrical potential difference generated, and the osmotic forces. In such an equilibrium, the direction of the potential change depends on the charge of the nonpenetrating ion. A positive potential will develop on the side of the membrane with a nonpermeating positive ion and a negative potential will develop on the side of the nonpermeating negative ion. The magnitude of the potential depends on the ratio of the concentrations of the permeable and nonpermeable ions.

While the steady membrane potential of a cell, especially an excitable one, is more complex than a simple Gibbs-Donnan equilibrium potential and probably incorporates a variety of types of EMF's, such as those mentioned in previous sections, phase boundary or equilibrium potentials that arise from Donnan effects are not excluded as participants [249, 250]. Donnan relations demand fixed ratios, however, between all permeant ions:

$$\frac{C_i^+}{C_o^+} = \frac{C_o^-}{C_i^-}$$

which is not the case for sodium as well as other ionic species. Sodium is almost impermeable under most conditions, although tracer experiments show that sodium can enter and leave a cell. Chloride ion concentrations are higher than what is expected from a potassium Donnan ratio [160] and, in some preparations, the membrane is virtually impermeable to chloride

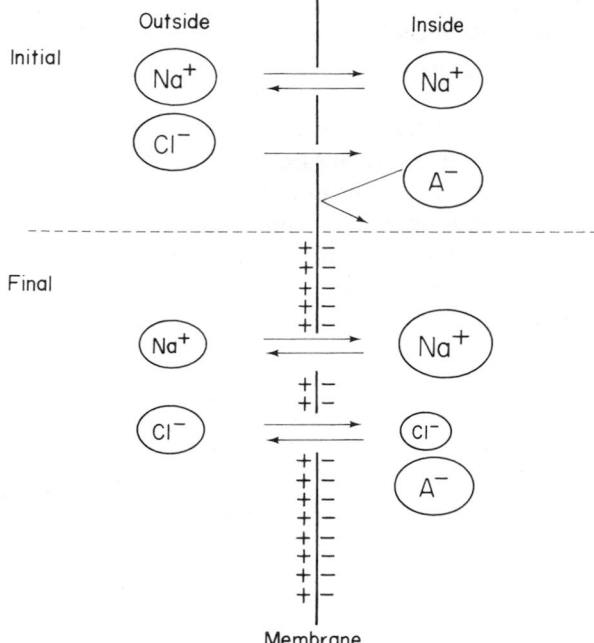

Figure 47. *Development of a membrane potential due to a nonpermeating ion:* initially, equal concentrations of NaCl and NaA are placed on each side of a membrane. If Na⁺ and Cl⁻ can pass through the membrane but A⁻ cannot, the initial net flux of Cl⁻ will cause the inside to become negative with respect to the outside. The membrane potential developed will in turn cause Na⁺ to redistribute passively according to the Nernst relation and will progressively reduce the net Cl⁻ flux. A steady state (Gibbs-Donnan equilibrium) will be reached when the electrical potential gradients exactly counteract the net diffusional gradients including the osmotic gradients.

ions. Some of the discrepancies result from ion pumps that can eliminate or introduce ions apart from diffusional mechanisms of electrochemical gradients. A pump that separates charges is called an *electrogenic pump* since it is the direct cause of the membrane potential. A pump can also cause two simultaneous active fluxes without net current flow and is called a *nonelectrogenic pump*. The membrane potential resulting from the latter pump is established by the passive redistribution of the pumped ions. In both cases, continuous active transport is required to sustain the membrane potential. The usual view of the functioning of membrane pumps is a one-to-one exchange of sodium for potassium [132]. If this one-to-one coupling is lacking, the sodium pump is an electrogenic one and electrogenic pump activity is indicated by many recent experimental findings [87, 88].

4.11 The Ionic Basis for Action Potentials

While it is obvious that beliefs concerning the mechanism responsible for generating action potentials are biased by beliefs concerning origins of membrane potentials in general, the descriptions here are based upon potentials that result from diffusion of ions. Figure 48 tabulates the variety of types of responses of nerve cells. Since axonology has dominated during the first half of this century, the mechanisms for spike production are appropriate for a starting point.

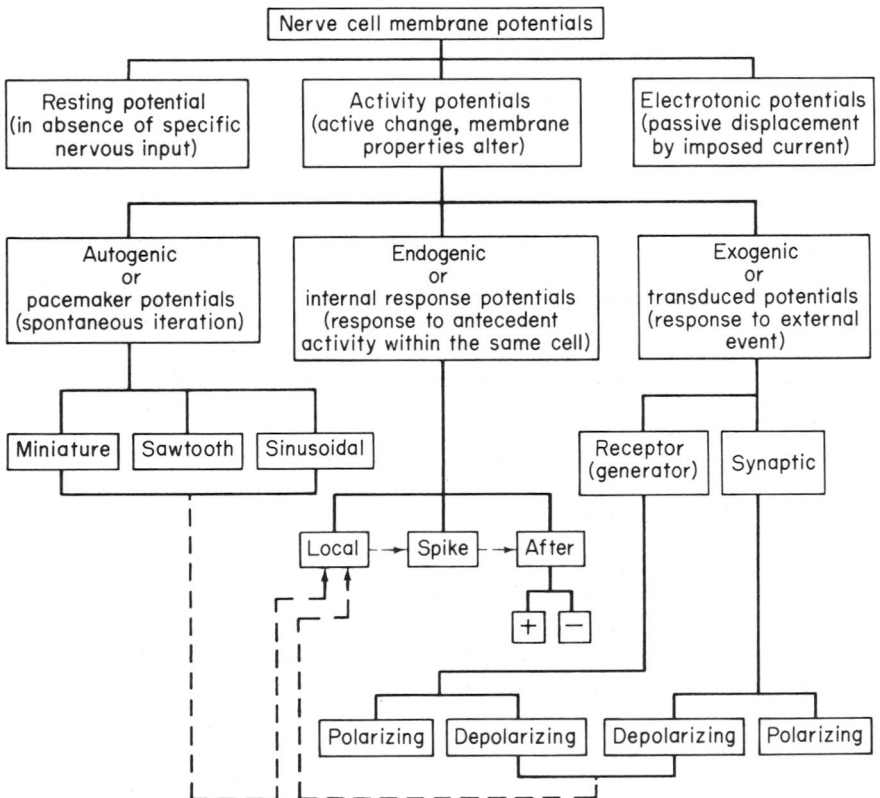

Figure 48. *Chart of the types of potentials found across membranes of nerve cells especially during activity:* no single cell displays every type. Arrows indicate that a sufficient level of one potential may cause the initiation of the other. Polarizing (or hyperpolarizing) and depolarizing potentials are indicated by + and − signs, respectively, i.e., afterpotentials that increase (+) or decrease (−) the membrane potential.

The classical studies of nerve impulse transmission in squid giant axons that have been summarized by Hodgkin [122, 123], Huxley [143], Cole [39], and Katz [154] demonstrated that the propagation of the impulse coincides with abrupt changes in the permeability of the axon membrane or, more specifically, a sequence of alterations in the sodium and potassium conductances.

Before becoming more quantitative, it is useful to return to the oversimplified example described earlier. The initial conditions are that potassium is ten times more concentrated inside the axon than outside, sodium is ten times more concentrated outside than inside, the potassium conductance of the neuron's membrane is considerably greater than the sodium conductance, and the membrane potential is at the steady state level determined by the concentration and conductance differences. If, as when a nerve impulse is triggered in some way, the sodium conductance suddenly becomes much larger than the potassium conductance, the diffusional flow of sodium across the membrane also becomes larger than that of potassium. This results in a net flux of positive charges from outside to inside. Moreover, as the neuron is initially positive on the outside relative to the inside, the membrane potential also causes positive charges (sodium and potassium) to migrate inward. As this occurs, the initial charge difference across the membrane diminishes— the axoplasm becomes less negative with respect to the outside medium— and eventually reverses sign. As the axoplasm becomes positive, the membrane potential will begin to drive positive charges out of the neuron counteracting the inward flow caused by concentration and conductance differences until a new steady state potential is achieved across the membrane. The neuron is now positive on the inside rather than negative. So, under resting conditions when potassium conductance exceeds that of sodium, the axoplasm is negative with respect to the outside, during active conditions when the sodium conductance exceeds that for potassium, the membrane potential reverses or the inside is positive with respect to the outside. Maximal values for the potentials are obtained when the conductance to one of the ions is zero (the membrane is impermeant to that ion). For situations where different concentration gradients occur from the 10:1 gradient mentioned here, the larger the concentration gradient, the larger is the ionic flow due to diffusion and, therefore, the larger will be the potential difference across the membrane that is required to counterbalance net ionic movements. For example, if sodium conductance is zero, the maximal negative membrane potential that develops depends on the magnitude of the potassium concentration gradient across the membrane. That potential is known as the potassium equilibrium potential and is described by the Nernst equation. Similarly, exchanging the word sodium for potassium in the sentence above will result in the positive sodium equilibrium potential.

To give an oversimplified summary, the first event in a nerve impulse is a rapid cathodal (outward) current that triggers a rapid increase in the sodium conductance. The increased sodium conductance quickly surpasses the potassium conductance. As sodium flows into the axon faster than potassium exits, the electrical polarity of the membrane falls and eventually reverses its sign. The swing in membrane potential, away from a potassium equilibrium potential and toward a sodium equilibrium potential, will in turn cause an increase in the potassium conductance. At the peak of the action potential, the sodium conductance begins to fall to its initial low level, while potassium conductance continues to rise. The inward flux of sodium becomes less than the efflux of potassium and the membrane potential is returned toward its initial level. As the resting potential is approached, potassium conductance falls to its normal resting level and the axon again becomes quiescent.

The work of especially Hodgkin and Huxley exposed some of the general characteristics of these conductance changes in axons: (1) The greater the depolarization of the fiber, the greater is the conductance of the membrane to both sodium and potassium. The changes in conductances are quick but not instantaneous. (2) The sodium conductance increases more rapidly than does the potassium conductance. This observation is one that is unique to membranes like those of axons, which are capable of producing large spike potentials that overshoot zero. (3) When a neuron is depolarized, the sodium conductance will at first increase and then decrease even if the neuron is maintained in a depolarized condition, a phenomenon known as sodium inactivation. The first two generalizations are responsible for the explosive nature of axons. Once a depolarization initiates increased sodium conductance above that for potassium, the increased flux of sodium ions will cause further depolarization of the neuronal membrane and thus cause further rapid increases in sodium conductance—a positive feedback mechanism. The third characteristic plus the increased potassium conductance are responsible for the return of the action potential to the resting level.

Although the remainder of this chapter is devoted to much elaboration and variation in these basic phenomena, no explanations will be found for the changes in membrane conductances in response to alterations in membrane potential. That work remains for the future.

The voltage-clamp experiments and subsequent analysis, especially of Hodgkin and Huxley, using classical cable theory [248] have placed the hypothesis involving diffusion potentials (called at times the sodium theory or the ionic theory) on quantitative terms. Data obtained from the voltage-clamp methods [39, 193] show that, during a spike, an initial surge of inward current is carried by sodium ions and a subsequent outward current is carried by potassium ions. The following quantitative description of a nerve impulse is derived from work summarized by Hodgkin [122, 123] and Huxley [142, 143].

The transient sodium-permeability change during depolarization can be described as the simultaneous operation of two independent processes—one process that rapidly increases the membrane permeability for sodium from its low value at rest and another process that, less rapidly, inactivates the permeability change. The inactivation process deserves attention since it has a direct role in determining the excitability of nerve. If the parameter h is used to distinguish the fraction of the maximum sodium permeability that can be activated by a small depolarization, the quantity $1 - h$ is the direct measure of inactivation. *Note:* The definition of inactivation used here, as originally described by Hodgkin and Huxley, is a more restrictive one than the definition used by Grundfest, which will appear later. The analysis of the spike and the equivalent circuit that describes the electrical behavior of axon membrane as envisaged by Huxley and Hodgkin are diagrammed in Fig. 49.

The sodium conductance (g_{Na}, conductance = 1/resistance) and the potassium conductance (g_K) vary with time and with membrane potential; the other components including the permeability to all other ions (leak conductance, $\bar{g}_l$) are assumed to remain constant (although this is not strictly true). The total membrane current can be divided into a capacity current and an ionic current:

$$I = C_m \frac{dV}{dt} + I_i$$

where I is the total current density, inward current is positive; I_i is the ionic current density, inward current positive; V is the displacement of the membrane potential from its resting value, depolarization is negative; C_m is the membrane capacity per unit area, assumed to be constant; and t is time. The ionic current can be divided into components carried by sodium ions (I_{Na}), potassium ions (I_K), and other ions (I_l) so that

$$I_i = I_{Na} + I_K + I_l.$$

The ionic permeabilities of the membrane can be satisfactorily expressed in terms of ionic conductances (g_{Na}, g_K, and $\bar{g}_l$) or

$$I_{Na} = g_{Na}(E_m - E_{Na}), \qquad I_K = g_K(E_m - E_K), \qquad I_l = \bar{g}_l(E_m - E_l),$$

where E_m, E_{Na}, and E_K are the membrane potential and the equilibrium potentials for sodium and potassium, respectively. E_l is the potential at which the "leakage current" due to chloride and other ions is zero. For practical application, it is convenient to write these equations in the form

$$I_{Na} = g_{Na}(V - V_{Na}), \qquad I_K = g_K(V - V_K), \qquad I_l = \bar{g}_l(V - V_l),$$

where $V = E_m - E_R$, $V_{Na} = E_{Na} - E_R$, $V_K = E_K - E_R$, $V_l = E_l - E_R$, and E_R is the absolute value of the resting potential. V, V_{Na}, V_K, and V_l can be measured directly as displacements from the resting potential.

From careful but empirical curve fitting of experimental data, the potassium conductance is described as

$$g_K = \bar{g}_K n^4. \qquad \text{Also } \frac{dn}{dt} = \alpha_n(1 - n) - \beta_n n,$$

where $\bar{g}_K$ is a constant with dimensions of conductance per square centimeter; α_n and β_n are rate constants which vary with voltage but not with time and which have dimensions of time^{-1}; and n is a dimensionless variable that can vary between 0 and 1. To give these expressions a physical basis, assume that potassium ions can only cross the membrane when four similar particles occupy a certain region of the membrane, e.g., inside a pore in the membrane; n represents the proportion of molecules in position and $(1 - n)$ represents the proportion that is somewhere else, e.g., outside the membrane. α_n determines the rate of transfer from outside to inside, while β_n determines the transfer in the opposite direction. If the particle has a negative charge, α_n should increase and β_n should decrease when the membrane is depolarized.

There are at least two general methods of describing the transient changes

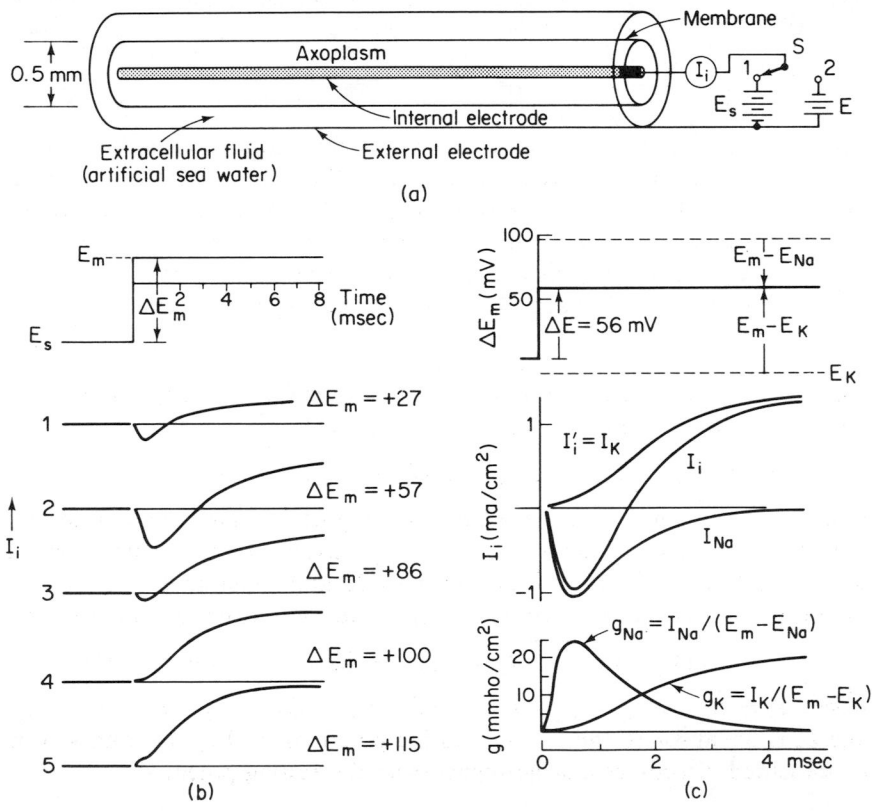

(a)

(b)

(c)

in sodium conductance that correspond roughly to the proposed types of general mechanisms for sodium inactivation. Sodium movement depends on the distribution of charged particles which do not act as carriers in the usual sense but which allow sodium to pass through the membrane when they occupy particular sites in the membrane. One proposal suggests that the activating particles undergo a chemical change after moving out of the position occupied when the membrane potential is high and the other suggests that relatively slow movements of another particle block the flow of sodium ions when they reach a certain position in the membrane. The alternative chosen by Hodgkin and Huxley to describe the sodium conductance is that determined by two variables, each of which follows first-order equations, instead of the alternative determined by a second-order differential equation. It was simpler to apply to the experimental result. So

$$g_{Na} = m^3 h \bar{g}_{Na}, \qquad \frac{dm}{dt} = \alpha_m(1 - m) - \beta_m m, \qquad \frac{dh}{dt} = \alpha_n(1 - h) - \beta_n h,$$

where $\bar{g}_{Na}$ is a constant and α and β are functions of V but not of t. Hodgkin and Huxley suggested a physical basis for these equations by assuming that the sodium conductance is proportional to the number of sites on the inside of the membrane which are occupied simultaneously by three activating molecules and which are not blocked by an inactivating molecule. m then represents the proportion of activating molecules on the inside of the membrane

Figure 49. *Voltage clamping in squid giant axons:* (a) diagrams the basic experimental arrangement that holds the transmembrane voltage (E_m) constant over a considerable length of membrane by connecting the internal and external media to a battery through long electrodes. E_m can be suddenly changed from E_s ($I_i = 0$) to any other value chosen by flipping the switch (S) to position (2). The total current (I_i) through the membrane is measured as a function of time by an ammeter (cathode-ray oscilloscope). (b) shows representative data obtained by Hodgkin and Huxley of transmembrane current flow as a function of time (top scale) after a sudden change in E_m. Curves 1 to 5 show the membrane current that flows after the membrane is depolarized by increasing amounts (given at the right in millivolts). In curve 4, the depolarization was near E_{Na}; in curve 5, E_m was greater than E_{Na}. Thus, for all but the largest depolarizations, the early component of current flow is in a direction opposite to that expected from the change in E_m and the late current flow is in the same direction. (c) plots the components of the total membrane current and capacitance. The top curve gives E_m as a function of time (scale at bottom); E_{Na} and E_K are indicated by dashed lines. The middle curve gives the total ionic current (I_i) broken into its two components I_{Na} and I_K. (I_{Cl} is neglected since it is small and constant.) The separation was made by reducing [Na_o^+] to the value at which a depolarization of 56 mV equal E_{Na}. Since $I_{Na} = 0$ under these conditions, the total ionic current (I_i') is equaled to I_K as labeled. The bottom curve gives g_{Na} and g_K as functions of time for the step change in E_m shown in the top curve. The conductances are the same shape as the current curves because they are calculated by dividing the ionic current by the effective voltage driving ion (indicated in top curves).

and $(1 - m)$, the proportion on the outside. h is the proportion of the inactivating molecules that are on the outside of the membrane and $(1 - h)$, the inside. α_m or β_m and α_n or β_n represent the transfer-rate constants in the two directions.

In summary, the empirical Hodgkin-Huxley equation becomes

$$I = C_m \frac{dV}{dt} + \bar{g}_K n^4 (V - V_K) + \bar{g}_{Na} m^3 h (V - V_{Na}) + \bar{g}_l (V - V_l).$$

The treatment here is obviously incomplete and is intended in part to illustrate the degree of success that can be obtained by mingling sound experimental technique and careful data analysis with insight and patience. This work has been repeatedly tested both theoretically and experimentally and while modifications have been suggested, it has been essentially confirmed. One caution has appeared: This work was based on data obtained from squid giant axons; other preparations, especially nonaxon material, behave differently. Also, the suggestions concerning the physical basis for some of the constants have not yet been adequately tested and some of the alternatives are yet to be suggested.

Despite dramatic success with the work on axons, it is a mistake to conclude that such responses are general or even typical of the response capabilities of excitable membranes. Other parts of neurons behave differently from axons and differences are also found in other long-distance signal-transmitting systems such as muscle fibers. Clearly, it is time to examine the responses of other kinds of excitable membranes.

4.12 Properties of
Junctional Membranes

At the start of the century, the Spanish anatomist Santiago Ramon y Cajal [31] showed that the nervous system is built of individual nerve cells. Physiologists have since then been concerned with the mechanisms by which these individual elements influence each other. The electrical impulse which travels along the axon ceases abruptly when it arrives at the termination of the fibers which make contact with another cell—at the synapse. Beyond the synapse, the impulse must be generated anew. Partly resulting from the emphasis on axons by older neurophysiologists—studies led by the greats Lucas, Gasser, and Erlanger—many prominent physiologists held that transmission at synapses was predominantly, if not exclusively, an electrical phenomenon [64, 65, 83, 98]. Today the abundance of evidence shows that transmission is effected by the release of specific chemical substances that trigger the generation of a new impulse in the second cell [67, 188]. In fact, the first strong evidence suggesting that a chemical substance acts across the synapse was provided long ago by Sir Henry Dale and Otto Loewi, who together laid the foundation for the concept of chemical transmission across junctions.

Without documenting the evidence, the widely accepted concept of chemical transmission can be formulated as follows [84, 88]: Although the transmitter may be concentrated in nerve endings, it or its immediate precursor occurs everywhere in the neuron; the transmitter is stored in a physiologically inactive bound form. Either the axon and nerve endings contain the enzyme system capable of synthesizing the transmitter or the nerve cell body does and the transmitter or its precursor then migrates down the axon to the terminals. The transmitter is released in nearly constant amounts by the nerve ending upon the arrival of a nerve impulse and is free to diffuse within the synaptic gap. Subsequently, it reacts with specific receptor molecules located in the subsynaptic membrane giving rise to the postsynaptic response and, finally, it is inactivated by an enzyme system located in the synaptic region. Enzymatic inactivation is not absolutely essential to the theory since diffusion alone may effectively remove the transmitter from its original point of action. Also some of the transmitter may be taken up by the presynaptic ending, as found in some central adrenergic synapses [166] or by the postsynaptic membrane, as described in some sympathetic fibers [147]. Only the small subsynaptic patches of the postsynaptic membrane are specificially sensitive to the transmitter agent. In many systems, however, for example, some invertebrate and vertebrate muscle, the entire surface of the postsynaptic cell is sensitive to the transmitting agent or, after denervation of vertebrate skeletal muscle, the sensitive areas expand over the whole surface of the postsynaptic cell [190, 191].

The electron micrographs have revealed structural details of synapses that fit nicely into the picture of chemical transmission [55, 258]. The nerve endings impinging upon another nerve cell are enlarged into knobs, the synaptic knobs or boutons of Couteaux. Enclosed in these synaptic knobs are many vesicles or tiny sacs that are thought to contain transmitter substance. There are also mitochondria located near the membrane of the postsynaptic cell and in the presynaptic knob. Between the presynaptic and postsynaptic cells is a remarkably uniform space of about 200 Å, which is termed the *synaptic cleft*. Deviations from the 200 Å synaptic cleft are clefts at the motor end plate (about 1200 Å wide) and at electrically conducting synapses (less than 20 Å wide). The membrane bounding the cleft shows structural variations from the trilaminar construction of other plasma membrane and many of the synaptic vesicles are concentrated adjacent to this cleft. Contrary to the suggestions of many of the earlier workers, there appear to be no specific structural correlations with the functioning of a synapse as excitatory or inhibitory, which is really a postsynaptic phenomenon. After excitation the number of vesicles appears to remain the same but their contents diminish, suggesting that they are concerned with the release of the transmitter substance [153]; the arrival of a nerve impulse at the synaptic ending is supposed to cause the release of transmitter into the cleft from vesicles adjacent to the synaptic membrane. This hypothesis is supported by the discovery that the transmitter is released in packets of a 20–60,000 molecules per packet—a quantum of transmitter—that elicit miniature junctional potentials out of the subsynaptic membrane [53, 155].

The introduction of intracellular recording techniques in the early 1950's permitted significant progress in the study of synaptic and neuromuscular transmission. These early investigations include work on frog neuromuscular junctions, mammalian spinal cord motoneurons, motoneurons in the electric lobe of *Torpedo*, motoneurons of toad spinal cord, crustacean neuromuscular junctions, stellate ganglion of *Loligo*, mammalian sympathetic ganglia, and giant ganglion cells of *Aplysia*. The more recent work is summarized in McLennan [188] and Eccles [67].

In view of the interest in the subject and the number of workers in the field, there are remarkably few junctions that have been carefully and completely described [84, 255]. The major ones are the neuromuscular junction of vertebrates, a few synapses in the spinal cord and in autonomic ganglia of vertebrates, and a few isolated junctions among invertebrate animals. The scarcity of information is caused by the large number of individual experiments that must be performed before a completely satisfactory identification of the transmitter substance of that junction can be made. These experiments include demonstrating that (1) the substance is released into a perfusion medium during stimulation of the presynaptic neurons; (2) either excitatory or inhibitory action on the postsynaptic membrane by nerve action can be imitated by artificially applying the substance to the junction preferably with micro-injection apparatus so that the material is applied only on the synaptic surface of the postsynaptic cell and only in physiological quantities; (3) the enzymes necessary for the biosynthesis of the compound occur in the presynaptic nerve fibers; (4) an adequate means is available to destroy the substance or that the mechanism is reversible; and (5) the potentiation or inhibition of nerve action is the same as that of the test transmitter by the same potentiating or blocking agents and at similar concentrations; the pharmacology of nerve action and test substance needs to be alike. All of these experimental approaches are indirect so that unless all aspects—biophysical, biochemical, and pharmacological—are considered together, the mechanism for transmission at a particular junction must remain available for further questioning.

As an example of behavior at junctional membranes, consider the work of Eccles [66] on motoneurons in the spinal cord of cats. Fibers coming from special stretch receptors located in a muscle were stimulated. These fibers form a part of a typical reflex arc such as those responsible for the patellar reflex or knee jerk. The impulses generated travel toward the spinal cord almost synchronously as a volley. About 0.5 msec after the detection of the arrival of a volley in the spinal cord, there is a wavelike change in voltage inside a motoneuron located within the spinal cord that has just received the volley (Fig. 50). The normal membrane potential becomes progressively less negative as more of the fibers impinging on the cell are stimulated to fire. This observed depolarization is in fact a simple summation of the depolarizations produced at each individual synapse. When the depolarization of the motoneuron reaches a critical point, a spike suddenly appears on the oscilloscope tracing indicating that the cell has generated a nerve impulse. The critical depolarization necessary to regularly elicit a spike is usually between 10 and 20 mV below the normal resting potential. The effect of a further increase in the magnitude of the synaptic stimulus is to shorten the time needed for the motoneuron to reach the firing threshold. Such depolarizing potentials produced by the subsynaptic membrane have been called excitatory postsynaptic potentials or, more succinctly, EPSP's.

The resting potential of a cell can be either increased or decreased by applying background current through one barrel of a double-barreled intracellular microelectrode [47]. When the membrane potential is made more negative, the EPSP rises more steeply to a higher peak; when the potential is made less negative, the EPSP rises more slowly to a lower peak. By actually reversing the membrane potential—the interior of the cell positive to the exterior—the excitatory synapse gives rise to an EPSP that is inverted. Finally, by artificially adjusting the membrane potential of the cell, one can establish that there is no flow of ions and, therefore, no EPSP when the voltage drop across the membrane is zero (Fig. 51). Electrophoretically injecting a number of different ions into a cell does not essentially change the EPSP. Such experiments were initially interpreted as indicating that during excitatory synaptic activity the postsynaptic membrane becomes permeable to all species of ions [48, 49, 80] but, as first shown for frog end-plate junction [238, 239], the present thinking is that all known cases are due to a selective sodium and potassium activation. During synaptic activity, the ions are permitted momentarily to flow freely down their electrochemical gradients so that sodium enters the cell and, to a lesser degree, potassium leaves. It is this net flow of positive ions that creates the EPSP. The flow of negative ions, such as chloride, is apparently not involved. The accumulation of evidence also shows that the subsynaptic membrane is not excitable by electrical means [101, 103]; neither the presynaptic action currents nor artificially applied pulses elicit more than passive responses.

It is currently accepted that the molecules of transmitter substance, once released from presynaptic vesicles by the arrival of a nerve impulse, diffuse across the synaptic cleft to become attached to specific receptor sites on the

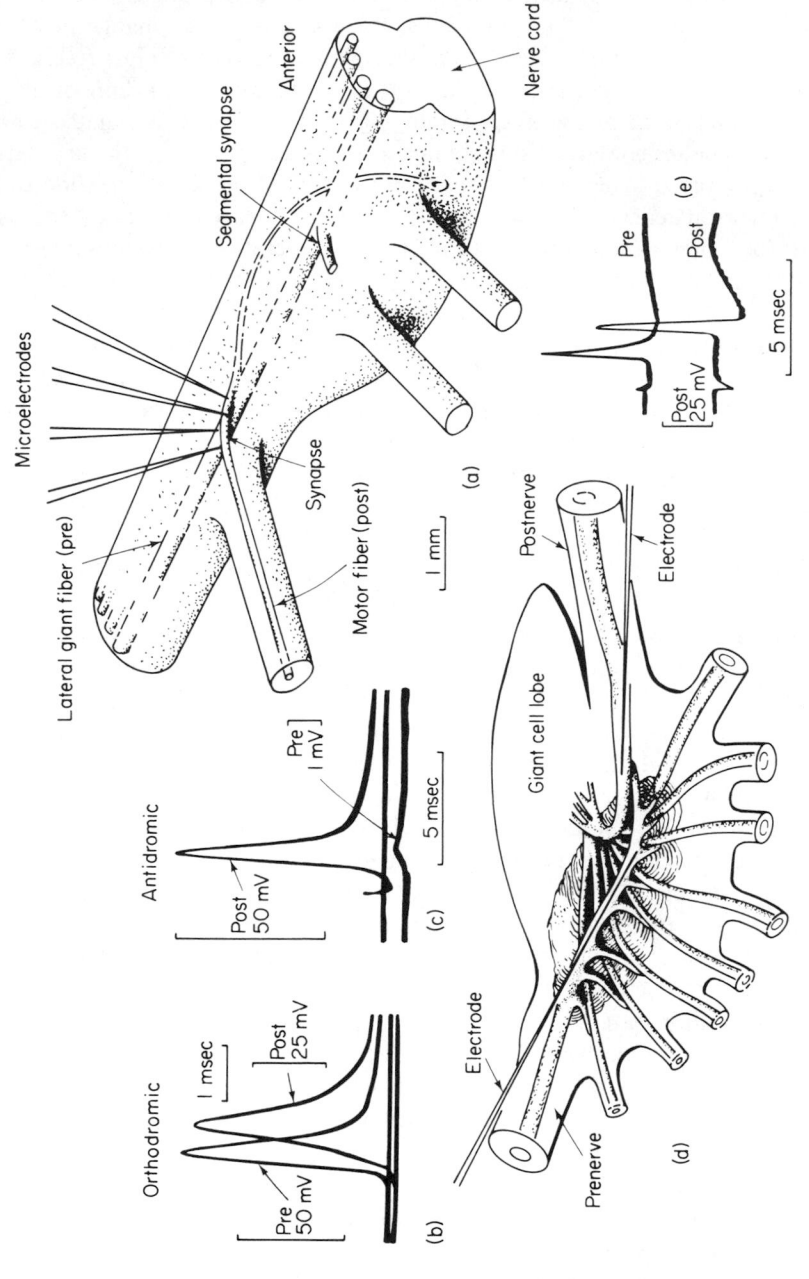

Microelectrodes

Segmental synapse

Anterior

Nerve cord

Lateral giant fiber (pre)

Synapse

Motor fiber (post)

1 mm

(a)

Pre

Post

Post
25 mV

5 msec

(e)

Antidromic

Pre
1 mV

Post
50 mV

5 msec

(c)

Orthodromic

1 msec

Post
25 mV

Pre
50 mV

(b)

Giant cell lobe

Postnerve

Electrode

Electrode

Prenerve

(d)

subsynaptic membrane of the adjacent nerve cell. Presumably, the receptor sites are associated with fine channels in the membrane that become opened by the action of the transmitter, thus converting the synaptic membrane from a strong ionic barrier to an ion-permeable one. With the channels now opened, sodium and potassium ions flow through the membrane thousands of times more readily than normally to produce the intense ionic current flux that depolarizes the cellular membrane and produces the EPSP [67]. In many synapses, the current flows strongly for about a millisecond before the transmitter substance is eliminated from the synaptic cleft either by diffusion out of the synaptic regions or as a result of enzymatic destruction. This later process is best known to occur when the transmitter is acetylcholine and the hydrolytic enzyme is acetylcholinesterase [21].

The substantiation of this general picture for synaptic transmission requires the solution to many fundamental problems. The specific transmitter substance is known for only a few synapses. Practically nothing is known about the mechanism by which a presynaptic impulse causes transmitter to be injected into the synaptic cleft. Nor is there any information on how the synaptic vesicles not immediately adjacent to the synaptic cleft move up to replace empty vesicles. It is conjectured that the vesicles contain the enzyme systems needed to recharge themselves and that the process must be swift and efficient. While some recycling of transmitter is known to occur, the total amount of transmitter within synaptic vesicles is thought to supply only a few minutes of synaptic activity at normal operating rates. There are also some knotty problems to be solved on the other side of the synaptic cleft, for example, the nature of the receptor site and the mechanism for opening the ionic channels in the membrane. In fact, despite the general acceptance of chemical transmission, the evidence available greatly depends on indirect pharmacological data and is still per se inconclusive. Although such experimental findings—that the action current of the presynaptic nerve

Figure 50. *Electrical and chemical synaptic transmission:* (a) diagrams the experimental arrangement for recording the electrical activity on either side of the electrically transmitting synapse between the lateral giant axon and the large segmental motor fiber that innervates abdominal flexor muscles. The evidence that led Furshpan and Potter to conclude that this synapse operates by electrical rather than by chemical transmission is the negligible latency between the presynaptic and postsynaptic responses (b) and the near failure of a response in the presynaptic fiber to an antidromic spike in the postsynaptic fiber (c), which demonstrates rectification in the synaptic membranes. In contrast, the more common case of chemical transmission is illustrated by synapses in the stellate ganglion of the squid. In the preparation diagramed in (d), one electrode is located in the internuncial giant fiber (left) and another in the giant motor fiber (right). The considerable latency between presynaptic and postsynaptic spikes (e) is indicative of chemical transmission. Of course, much additional evidence is required before either chemical or electrical transmission can be concluded.

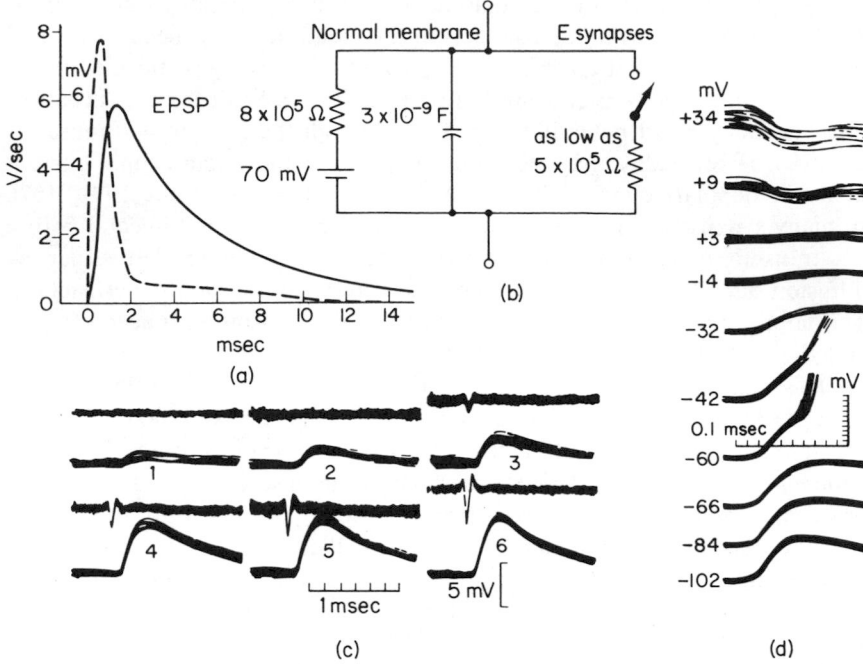

Figure 51. *Excitatory postsynaptic potentials:* (a) plots the time course of a typical EPSP recorded intracellularly from a motoneuron in cat spinal cord generated by the postsynaptic current (broken line) and (b) is an electrical circuit analog of postsynaptic membrane showing the short-circuiting action of excitatory synapses (right side) on normal membrane. (c) shows several super-imposed traces of monosynaptic EPSP's (depolarizations) recorded intracellularly (lower records 1–6) from a median gastrocnemius motoneuron of cat that are evoked by afferent volleys in the median gastrocnemius nerve of progressively increasing size as indicated by the upper extracellular recordings 1–6 from the dorsal root. (d) shows a series of EPSP's set up in a cat biceps-semitendonosus motoneuron at various levels of membrane potential as indicated. Each record is formed by the superposition of about twenty faint traces. The membrane potential was shifted from its resting value of −66 mV by steady currents passed through the other barrel of a double microelectrode

does not elicit the postsynaptic potentials of motoneurons [65], electro-plaques [4], or squid giant axon synapses [111], and that the muscle spike does not initiate the end plate potential [80]—are compelling evidence for chemical transmission, it is still indirect evidence. But extensive studies now show a lack of electrical excitability in receptor membrane in general [104] so if synaptic membrane is electrically inexcitable, chemical transmission is obligatory.

4.13 Normal Spike Initiation
in a Motoneuron

Work in the 1950's started investigators to seriously ponder the significance of different behavior patterns in membrane from different locations within the same cell. One such demonstration was the spike recorded from a motoneuron that is, on careful analysis, a complex waveform consisting of several components. If a motoneuron is excited antidromically, that is, if the axon is stimulated so that the impulse generated travels toward the cell body opposite to the direction of normal impulse flow, three components can be identified in the spike recorded intracellularly at the soma [45, 46]. The first portion of the recorded potential is all-or-nothing in character, has a short refractory period, follows high frequencies, and in general resembles a spike passing through the medullated region of axon, hence its name, *M* spike. The second portion of the record is the *IS* spike (Fig. 52). The *IS* spike

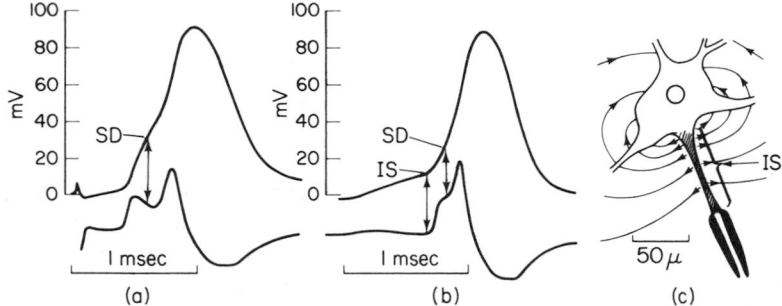

Figure 52. *Spike origin at the axon hillock region of a motoneuron:* antidromic (a) and monosynaptic (b) stimulation of a motoneuron evoke these intracellularly recorded spike potentials (upper traces) and their electrically differentiated records (lower traces). (c) diagrams the lines of current flow that occur as a synaptically induced depolarization of the soma-dendritic membrane (*SD*) spreads electrotonically to the membrane of the initial segment (*IS*)—axon hillock and unmyelinated initial portion of axon. A comparison of records (a) and (b) shows that, following synaptic activation, the spike originates in the *IS* membrane and then invades the soma of the cell at the same time that it begins its journey down the axon.

is larger (30–40 mV) than the *M* spike indicating that it occurs in an area of membrane closer to the soma of the cell where the electrode is located. Blockage sometimes occurs at the *IS* site because of the increase in surface area and the higher threshold of soma-dendritic membrane. The *IS* refers to the "initial segment" of the axon, which is a composite of the axon hillock and the unmedullated region of the axon. Other experiments show that the *IS* spike and EPSP's from synaptic activity are additive. The third component of the potential recorded during antidromic stimulation is the *SD*, or the soma-dendritic, spike. The *SD* spike appears full-sized (80–100 mV) so it must be generated in that part of the motoneuron that is impaled by the intracellular microelectrode or in the membrane of the cell's soma and adjacent dendrites. This conclusion is supported by observations of destruction of preformed EPSP's by *SD* spikes and the considerable diminution of any EPSP set up by a later excitatory presynaptic volley. This indicates that the full-sized spike has invaded those regions of membrane on which synaptic knobs are concentrated, which have been shown to be the soma and adjacent dendritic regions [28]. Cells stimulated synaptically and by directly applied current show a notch on the rising phase of the recorded potential at the same position as during antidromic stimulation, the point of origin of the *SD* spike out of the *IS* spike. Both events occur at about 30 mV depolarization in the cell; hence, synaptic and direct stimulation also produce an *IS* spike that precedes the *SD* spike (Fig. 52). In all three cases, it is envisaged [9] that the invasion of the soma-dendritic membrane is preceded by a spike in the initial segment and the threshold of 6–15 mV that has been measured for both synaptic and direct stimulation is therefore the threshold for generating an impulse (*IS* spike) in the initial segment. Approximately three times that depolarization (or 20–40 mV) is required to generate a spike in soma-dendritic membrane [66]. Normally this current results from that produced by the *IS* spike. This explains the much higher threshold for antidromic invasion of the soma-dendritic membrane (30 mV) and that (10 mV) for synaptic and direct stimulation. To explain the high threshold for soma-dendritic membrane, it has been suggested that the large number of synaptic knobs, which may cover as much as 50% of the total soma membrane, and glial cells [265] may insulate much of the soma-dendritic membrane from the current flow [66]. An alternative suggestion is an intrinsic difference in membrane [103], which has been supported by cytological evidence that distinguishes between the soma and the initial segment [35, 37]. The latter is also supported by differences in the afterpotentials: The afterhyperpolarization is large after the *SD* spike but is undetectable after the *IS* spike [47].

4.14 Inhibitory Synapses

Another type of synapse has been identified that is able to inhibit the firing of a nerve cell despite the simultaneous reception of excitatory impulses. While there may be subtle structural differences that are not yet clear, inhibitory synapses look much like excitatory synapses in the electron microscope [67].

As a sequel to their experiments with excitatory synapses described earlier, Curtis and Eccles [51] stimulated nerve fibers coming from an antagonistic set of muscles to drive a volley upon their impaled motoneuron. The effect was not EPSP's but inhibitory action upon the motoneuron. Microelectrode recordings of the activity of single motoneurons and other nerve cells show that the inhibitory postsynaptic potential (IPSP) is similar to a mirror image of the ESPS although it has a slightly faster decay. Moreover, inhibitory synapses, like excitatory synapses, have cumulative effects. Generally, the chief difference is that the IPSP makes the cell's interior more negative to the exterior than it normally is. By hyperpolarizing the cell, inhibitory synapses oppose the action of excitatory synapses. A strong volley of inhibitory impulses can drive the cell's membrane potential from -70 mV to -75 to -80 mV and, therefore, further from the threshold point required for spike generation. The nerve cell responds to the algebraic sum of the internal voltages produced by excitatory and inhibitory synapses. By altering the internal membrane potential by a flow of an electric current through one barrel of a double-barreled microelectrode, the effect of such changes on the inhibitory postsynaptic potential can be measured through the other barrel [48]. When the internal potential is made less negative, the IPSP is deepened (Fig. 53). Conversely, when the potential is made more negative, the IPSP diminishes and finally reverses when the cellular potential is driven below -80 mV. Many workers have watched the IPSP's of mammalian moto-neurons and of other central synapses converted to depolarizing responses by the diffusion of chloride ions out of an intracellular KCl-filled micro-electrode. The inversion of the IPSP is expected if the inhibitory transmitter increases the permeability of the postsynaptic membrane to chloride ions, which are then permitted to flow down their electrochemical gradient and so depolarize the membrane [67].

The conclusion is that inhibitory synapses share with excitatory synapses the ability to change the ionic permeability of the synaptic membrane. The difference is that the inhibitory synapse enables ions to flow freely down an electrochemical gradient that has an equilibrium point at -80 mV rather than at 0 mV for excitatory synaptic action. The effect can be achieved by the outward flow of positively charged ions, such as potassium, or the inward flow of negatively charged ions, such as chloride, or by a combination of negative and positive ionic flows such that the interior reaches equilibrium near -80 mV.

To discover the permeability changes associated with the inhibitory potential, Eccles and his colleagues [67] altered the concentration of ions normally found in motoneurons and introduced a variety of other ions not normally present. They impaled a nerve on a microelectrode filled with salt

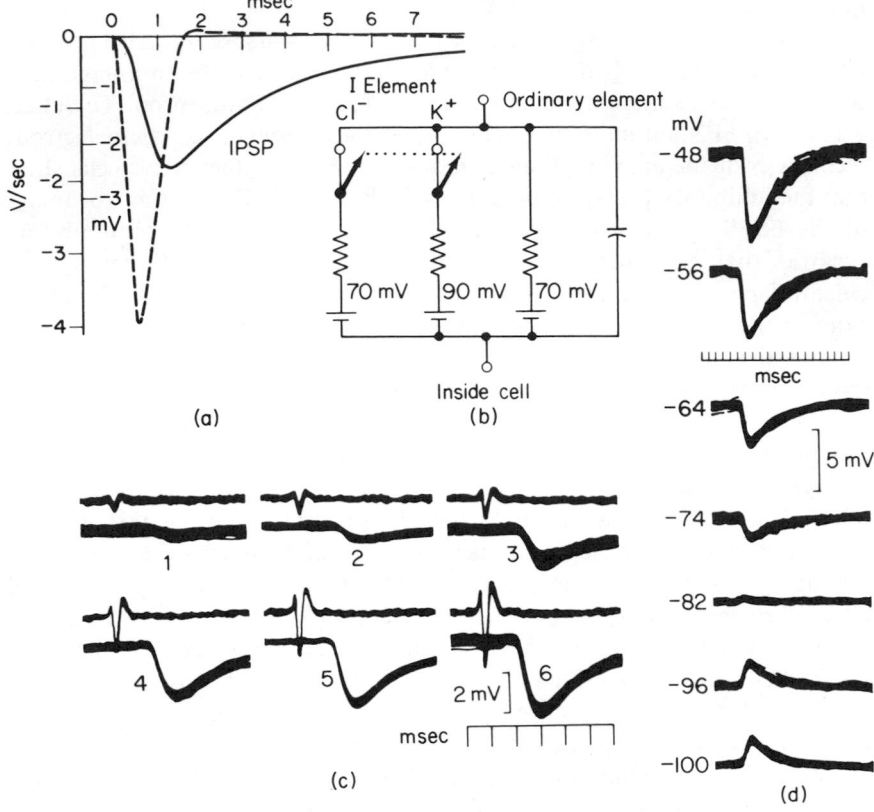

Figure 53. *Inhibitory postsynaptic potentials:* (a) plots the time course of a typical IPSP as recorded intracellularly from a motoneuron in cat spinal cord that is generated by the postsynaptic current (broken line). (b) is the equivalent electrical circuit showing capacity, resistance, and battery of the soma membrane and, to the left, the additional components representing the subsynaptic areas that are activated to produce the IPSP; maximal activation is indicated by closing the switches. (c) presents typical intracellular records of IPSP's (hyperpolarizations) of progressively increasing size together with the extracellularly recorded dorsal root potentials (upper traces), which show that an increase in number of excited afferent fibers from the opposing muscle increase the inhibition of the impaled motoneuron. (d) gives a series of IPSP's recorded intracellularly from a biceps-semitendonosus motoneuron. The records, formed by the superposition of about forty faint traces, show the IPSP's set up by a quadriceps afferent volley. By means of a steady background current through one barrel of a double-barreled microelectrode, the membrane potential has been preset at the voltages indicated on each record. The original resting potential of this cell was −74 mV.

solutions containing the ion to the injected. The actual injection was achieved by passing a brief pulse of current through the micropipette (iontophoretic injection). Increasing the internal concentration of chloride threefold reverses the IPSP into a depolarizing current resembling an EPSP. There is no such reversal when the cell is heavily injected with sulfate ions, also negatively charged. More than thirty different anions have been electrophoretically injected into motoneurons and either the injection is without effect on the IPSP or it is fully effective in converting the IPSP into a depolarizing potential; the anions are unambiguously segregated into two categories. Including the contribution of the adhering sphere of water molecules to the diameters of hydrated ions, the ones that are permitted to permeate the cellular membrane under the influence of the inhibitory transmitter are all smaller than the nonpenetrating ions. The only exceptions are the small HS^- ion, which rapidly binds to protein molecules [146], and the larger HCO_2^- ion, which really has an elipsoidal shape [8]. Apart from formate, all the permeating molecules have hydrated diameters less than 2.9 Å or not greater than 1.14 times the diameter of a potassium ion. The same permeability effects including the exceptional behavior of the formate ion have been found in fishes, toads, and snails, as well as in mammals, suggesting that the ionic mechanism responsible for synaptic inhibition is the same throughout the animal kingdom. The significance of these and other studies is that they strongly suggest that the inhibitory transmitter substance converts the inhibitory subsynaptic membrane into a sieve-like structure with accurately standardized pores that allow small hydrated ions (e.g., potassium and chloride) to pass but block all larger ones (e.g., sodium, which is somewhat larger than any of the permeating anions including the anomalous formate) [67].

It is not possible to properly test the effectiveness of potassium ions by injecting excess amounts into a cell since any excess is immediately diluted by an osmotic inflow of water. The intracellular potassium concentration is about thirty times greater than the extracellular concentration, which means that the equilibrium potential for potassium is near -90 mV. If that membrane became more porous to potassium, the resulting outflux of potassium would drive the membrane potential even more negative than it is in the resting state. This is just what happens during synaptic inhibition [68, 69]. Similar reasoning prohibits the participation of sodium since a rapid sodium influx would depolarize and more than compensate for a potassium outflow. In fact the fundamental difference between synaptic excitation and synaptic inhibition appears to be that the membrane freely passes sodium during an EPSP but largely excludes sodium during an IPSP. The synaptic mechanisms more often proposed include performed channels of at least two different sizes in the membrane. In some way the excitatory transmitter selectively unplugs the larger channels to permit the rapid inflow of sodium ions. Potassium ions are permitted to leave as well to reduce the large potential change expected by a sodium influx. The inhibitory transmitter unplugs the smaller channels that are large enough to permit potassium and chloride ions to move through but not sodium [78].

To explain certain types of inhibition, other features must be added to this hypothesis for inhibitory synaptic transmission. In the simple hypothesis, chloride and potassium ions can flow freely through pores of all inhibitory synapses, as found in crustacean stretch receptors [71, 108], where the transmitter substance has been reasonably well identified as gamma amino butyric acid (GABA) [72, 167], and in the giant ganglion cell of *Aplysia* [245]. It has been shown, however, that the inhibition of heart rate by the vagus nerve is due almost exclusively to potassium ion outflow [251, 252]. On the other hand, in the muscles of crustaceans [24] and in nerve cells in snail brains [156], synaptic inhibition is largely due to the flow of chloride ions. As an explanation for the selectivity of the permeability changes brought about by the action of the inhibitory transmitter on the subsynaptic membrane, fixed charges are suggested to reside along the walls of the channels that are opened by transmitter action. If such charges were negative, they would repel anions and prevent their passage and thus show a preference for small cations as against small anions. If the charges were positive, they would similarly prevent the passage of cations but allow anions to move. The absence of a net charge along the wall of the pore would permit approximately equal cation and anion operation. Of course there is no confirmation for such details in this suggestion but examples for each pattern of ion movement have been found, as already noted.

4.15 Receptor Membranes

Impulses coursing through sensory nerves provide the only channel for transferring information from the sense organs and, hence, the various environments to the central nervous system where perception can begin. A fundamental aspect of receptor operation is the capture of energy and the initiation of impulses in attached or neighboring nerve fibers. Whatever the form of energy received from the environment and whatever the primary process of energy reception, the action of the receptor mechanism soon manifests itself, or at least is detected, as an electrical signal. If the sensing cell is a specialized cell separate from an afferent neuron, the signal is called a *receptor potential*; if the sensing unit is a modified dendritic terminal, the signal is known as a *generator potential*. An early clear example was provided by Bernhard Katz [151, 152] when he stretched a muscle spindle in a frog muscle: The transducer mechanism is associated with a small localized electric current which parallels in intensity the degree of stretching and which, when it reaches a certain magnitude, triggers the firing of impulses in the afferent neurons leading from the spindle to the central nervous system. The measuring and processing of information are other stories that will not be told here. We are only interested in the membrane phenomena associated with the operation of a receptor.

Studies of receptor fine structure have demonstrated complex arrangements. Axon endings that are exposed beyond the end of Schwann cell sheaths are never left naked but are covered by membranes of other types of cells. The result is that it is difficult to properly assign receptor guilt to neural membrane separated from that of other cells and the local intercellular space, which is quite likely to possess a different electrical potential and a different ionic composition from that of the general body fluids [216]. Some of the mechanoreceptors are less complicated structurally, however, and their study has proved useful. One type is the Pacinian corpuscle—a large receptor, measuring almost 1 mm in length and 0.6 mm in thickness, quite easily seen in the mesentaries of cats [217]. Under a microscope, a Pacinian corpuscle looks like an onion consisting of many concentric layers or lamellae intimately enclosing a nerve ending [215]. To stimulate a Pacinian corpuscle, Loewenstein [177, 178] used a piezoelectric crystal similar to the kind used in old phonograph pickups. Normally, such crystals convert mechanical into electrical energy but they can do the reverse. The deflection of the crystal increases linearly with the applied voltage and a glass rod attached to transmit the deflection to a corpuscle provides a readily controlled and measurable stimulus. Loewenstein and Rathkamp [182] took a direct approach to locate the transducer site in a Pacinian corpuscle. They peeled off pieces of its structure, stopping at times to stimulate, until they reached a preparation of nerve ending surrounded only by a thin layer of the innermost core. That prep was as good a transducer as the intact corpuscle, quite capable of responding to a poke with generator current that triggered the firing of impulses in the outgoing nerve fiber. They were not able to dissect the inner core away from the nerve ending without damage to the nerve but they could puncture it with a fine glass needle. Now the nerve ending was the only intact structure and it still produced normal transducer action. The last experiment was the construction of an endingless core by cutting the nerve fiber of a corpuscle in a living animal and allowing 2 or 3 days for that nerve ending to degenerate. When this preparation was tested later, it failed to produce a generator current in response to mechanical stimulation. The implication of these experiments is that the nerve-ending membrane is indeed the transducer site.

The generator current produced by the nerve ending in the mechanoreceptor does not itself propagate along the nerve fiber but, instead, it triggers that nerve fiber to generate its own impulses [97]. The generator current and the impulse originate at different places in the corpuscle. The site of impulse initiation was discovered by selectively blocking activity at different locations along the fiber [61, 182]. The myelin sheath that covers the axon extends well into the corpuscle and is interrupted at 0.25-mm intervals by the nodes of Ranvier. In a dissected preparation, several nodes and the nerve ending can be seen—the first node lies within the corpuscle—and their activity blocked by applying pressure with a fine wisp of glass. The nerve fiber continues to fire impulses in response to mechanical stimulation of the nerve ending and the resulting generator current until the first node—the one within the corpuscle—is blocked. Clearly, the first node is the point where nerve impulses begin; impulses do not begin at the nerve ending itself where transduction takes place. Under resting conditions, a potential exists across the receptor membrane at the nerve ending similar to the resting potentials of other excitable cells. Normally there is no appreciable ion current leakage through it. Mechanical deformation of this membrane produces an increase in conductance to allow ions to move down their electrochemical gradients and to cause that membrane potential to drop. The transfer of charges across the receptor membrane constitutes the generator current and the resulting depolarization is the generator potential, which is what is measured on the oscilloscope. Part of the generator current flows through the first node of Ranvier where it triggers the axonal membrane mechanisms responsible for initiation of a nerve impulse. Apparently the generator current must be of at least a minimal intensity (it must reach a threshold) and it must reach this intensity at a minimal rate (to avoid accommodation) in order to have effect [177, 178].

To ascertain how the receptor membrane produces a generator current that measures the strength of the mechanical stimulus, Loewenstein [177, 178] poked only small patches of membrane, about 30 μ in diameter, and he found that the resulting generator current decreases exponentially with increasing distance from the stimulated site. Excitation of the receptor membrane is confined to the mechanically distorted regions and although current leaks to neighboring areas, they remain unexcited; the generator current flows passively or electrotonically. By stimulating two spots on the membrane about 0.5 mm apart simultaneously, the two independently produced generator currents sum to produce a larger generator current. The summation of two or more currents, each generated at separate active sites in the membrane, thus explains how the intensity of the generator current can be proportional to the strength of the stimulus. Loewenstein then applied a series of mechanical stimuli of progressively increasing intensity to a part of the nerve ending and scanned the membrane with a microelectrode. He found that as the stimulus strength increased to deform progressively more of the receptor membrane, the excitation spread over a correspondingly greater area. This suggests that the receptor membrane may contain a great number of tiny active sites that show conventional all-or-nothing responses to mechanical stimulation and yet sum to give rise to a continuously variable generator current. Theoretically, this model can account for the entire input-output relationship of the mechanoreceptor and it is an attractive model because of its simplicity. The model is at best only tentative, however, since there still may be an intensity factor at work. A membrane model that fits the experimental data as well is one that operates on the basis of spatial summation of the activity of many active sites that each generate a variable current. There is presently no way to distinguish between each of these two possibilities.

The finding that current flow increases with the area of membrane deformed by the stimulus suggests that the excitation might be a statistical process. In other words, the distortion of a given area of receptor membrane excites some fraction of the total number of available sites in that area. Each site does not remain excited and neighboring sites can substitute. The result is that the number of excited sites fluctuates and the sum of their activity produces a generator current that fluctuates statistically. The fluctuations proved, upon measurement, to be large and the randomness of these fluctuation increases with the strength of the applied stimulus, as would be predicted by the spatial summation model [45, 181]. The membrane may now be pictured as having a number of tiny holes that in the resting state are too small for certain ions to penetrate. Mechanical deformation of a given area opens or enlarges a statistically determined number of the pores to permit ions to pass and to carry the generator current. As the stimulus strength increases, the number of holes that open also increases and a correspondingly larger quantity of ions penetrates the membrane.

The electron micrographs of the Pacinian corpuscle suggest that the number of ions available for transfer must be limited. The lamellae of the receptor core, which appear to be formidable barriers for ion diffusion, are tightly wrapped around the nerve ending and leave little fluid space between the receptor membrane and the first lamella. Loewenstein and Cohen [179, 180] repeatedly stimulated a corpuscle to deplete it of ions and found a considerable reduction in responsiveness. For example, a stimulus that produces a generator current of 100 units in the fully rested receptor produces a current of but 10 units after the receptor has received 5000 stimuli at a rate of 500 stimuli per second and the receptor produces no response after 7000 stimuli. Sato [222] suggested that the transfer of charges across the membrane depends on the interplay of two competing processes: the depletion (or inactivation) and the restoration (or reactivation) of some unknown factor or unidentified chemical precursor.

Sodium is essential for production of the full-sized generator potentials and sodium ions transport most of the charge across the membrane of the receptor during mechanical stimulation. Bathing a receptor in a sodium-free solution reduces both the amplitude and the rate of rise of generator potentials and the subsequent impulse activity stops [60]. In addition, generator

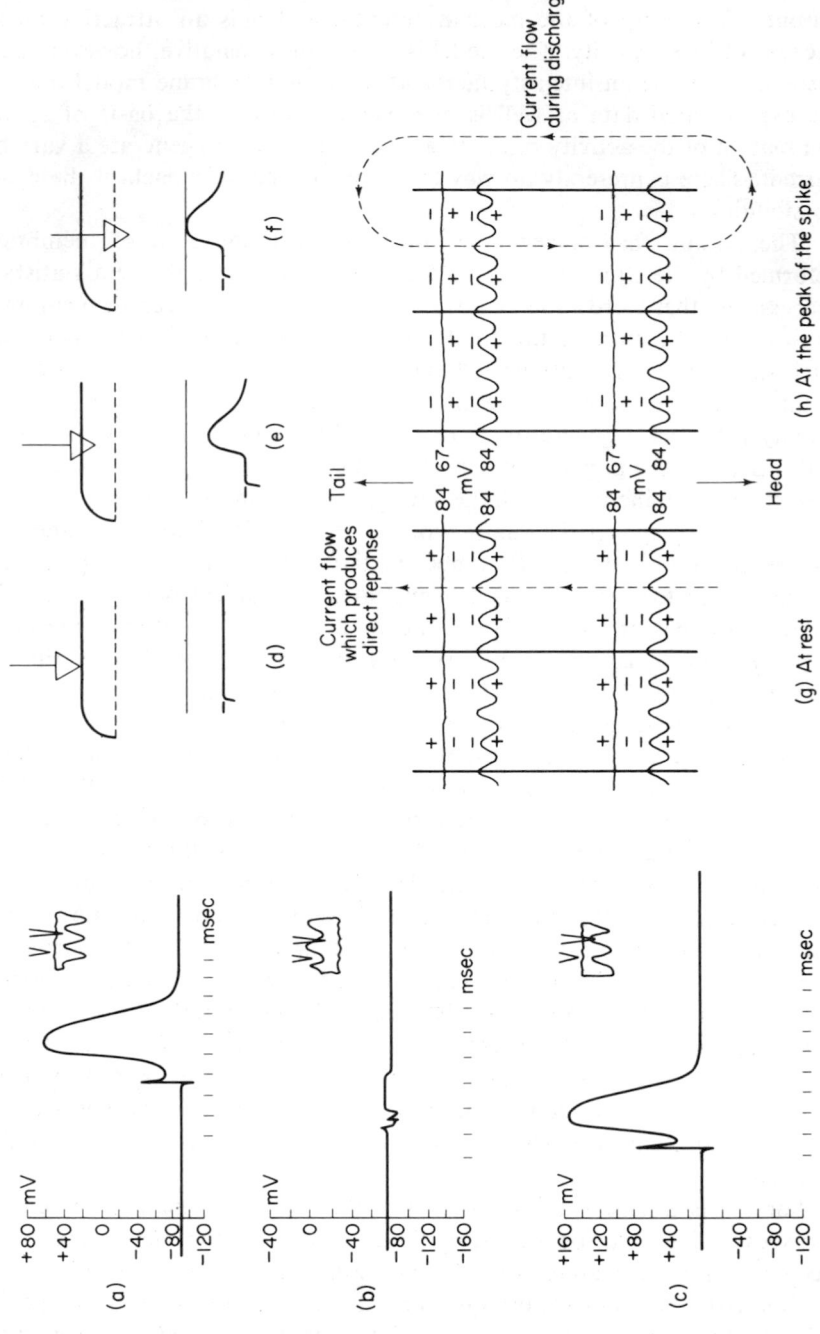

(a)

(b)

(c)

(d)

(e)

(f)

Current flow which produces direct reponse

Current flow during discharge

Tail

Head

84 67 mV 84 84

(g) At rest

(h) At the peak of the spike

potentials are depressed by antidromic impulses but less than by impulses initiated mechanically through the receptor [59]. Originally the receptor membrane was thought to be electrically inexcitable but some electrically excitable elements appear to be intermingled with the receptive elements in the Pacinian corpuscle membrane [140, 223]. Activity of the electrically excitable components can be selectively blocked by tetrodotoxin without affecting the electrically inexcitable components—the receptive ones [183, 213].

4.16 Electric Organs

The first bioelectric potentials observed by man were discharges from electric fish. The Egyptians knew of the electric catfish *Malapterurus* and the Romans used the ray *Torpedo*, an elasmobranch. Galen recommended "electric therapy" as did Indians in Guiana and 18th-century Europeans. South American explorers brought *Electrophorus*, a freshwater eel, back to Europe. Galvani, Davy, Faraday, Volta, Walsh, and Cavendish are among those who used electric fish in their laboratories and as models for their experiments on electricity, but it was du Bois Reymond who conclusively demonstrated the bioelectric nature of the discharges.

Generally, an electric organ consists of columns of connective tissue compartments each containing a flattened plate of syncytial tissue called an *electroplax*. An electroplax is normally innervated on only the electronegative—usually the posterior—surface, while the other surface is frequently papilliform and not innervated. Electroplaxes are derived embryologically similar to muscle: Some are modified muscle fibers and others are modified end plates. Electrically, the electroplaxes are arranged in series and the

Figure 54. *Sample responses of fish electric organs:* (a)–(c) are records from eel electroplax and (d)–(f) are from *Torpedo* electroplax during electrical stimulation. A full resting potential and overshooting spike are recorded when one of the recording electrodes pierces the innervated surface [(a), note insert]—only the resting potential is recorded across the uninnervated surface (b). When the recording electrodes are located across the electroplax, near both innervated and uninnervated surfaces but outside the cell, no resting potential is recorded as expected but a full spike is (c). The response of a *Torpedo* electroplax to stimulation is not recorded by an electrode located near the innervated surface [solid line, (d)]. When the electrode penetrates the electroplax, a nonovershooting spike is recorded (e). As this electrode protrudes through the uninnervated surface to again be extracellular, the action potential is still recorded (f). Comparing (d) and (f) provides evidence for the electrically inexcitable properties of the innervated surface of *Torpedo* electroplax. (g) and (h) diagram the additive discharges of eel electroplax. In (g) there is no net potential across the two electroplax but, at the peak of the spike (h), the potentials are in series and add, and the head of the eel becomes positive with respect to its tail.

columns are in parallel [100]. *Electrophorus*, a freshwater eel, can deliver pulses that measure 600 volts in open circuit and 1 amp in a short circuit; maximum power is about 100 watts. Of its four electric organs, its main one has seventy columns of 6000 plates each. The marine ray *Torpedo* can deliver a jolt at 30–60 volts but at several amperes, and it is capable of over 100 watts of power; one large specimen generated 6 kw [17, 18].

When a microelectrode is inserted into an electroplax of *Electrophorus*, a resting potential of about 90 mV (inside negative) is measured across both the innervated and the noninnervated surfaces (Fig. 54). When the electroplax discharges, the innervated (posterior) surface reverses in polarity by some 60 mV but the noninnervated surface remains unchanged. The two surfaces added in series produce a total cell potential of 150 mV [163]. Direct electrical stimulation of the innervated surface initiates a local prepotential out of which rises a full propagating action potential. Indirect stimulation through the nerve produces a graded junctional (postsynaptic) potential which is followed by an action potential which overshoots zero [100]. The junctional potential is analogous to an end-plate potential in that it can be elicited during the refractoriness that results from direct stimulation, it is fatigued upon rapid repetition, it is reduced by *d*-tubocurarine and decamethonium, and it shows facilitation (maximum occurs between 10–40 msec and persists for about 200 msec) that is increased by anticholinesterases [3, 82]. The innervated face of eel electroplax is electrically excitable after nerve degeneration. The upshot of such data is that the electroplaxes in *Electrophorus* are similar to striated muscle fibers: The innervated surfaces contain junctional regions, much like motor end-plates, intermingled with spike-producing membrane [100].

Electric organs have evolved independently, however, in several groups of unrelated fish in both fresh water and seawater. There are important differences among some of these specialized structures that have led to improved insight into the mechanisms for electrogenesis such as discussed later. A different response, for example, is found in the electric catfish *Malapterurus* found in Africa. Its electroplaxes have excitable membrane covering both innervated and noninnervated surfaces but the amplitudes of their respective responses differ so that, upon excitation, the noninnervated face becomes negative to the innervated face. The resting potential of *Torpedo* electroplax is between 40 and 50 mV. Its action potential is only 55–60 mV so there is little overshoot across the innervated surface. This electric organ can be excited only through its nerve and not by direct stimulation; it is electrically inexcitable. The total response is reduced by *d*-tubocurarine or atropine [29]. When denervated, the *Torpedo* electroplax can be stimulated by acetylcholine but under no condition is the electroplax membrane electrically excitable; *Torpedo* electroplaxes behave as motor end-plates. The high voltage of these electric organs results from a reversal, usually, of the polarization on only one side of a plate so that the voltages across the two surfaces add in series. Biologically, the discharges are emitted periodically for stunning prey or continuously for locating objects in surrounding waters or establishing domination over territory.

4.17 The Case for Membrane
Heterogeneity

No one really believes that the properties of well-known preparations are the only ones extant although most believe them to be typical of the responses to be found in nature. Grundfest [103, 104] has organized an enormous quantity of comparative data on excitable membranes to illustrate the variety of membrane properties and capabilities available within a single cell. Grundfest's thesis is that excitable membranes are composed of heterogeneous molecular structures. The weight of evidence is on his side.

Even from the few examples mentioned above, it is obvious that subsynaptic membrane has fundamentally different properties from spike-generating membrane from the same cell. There are also important differences in the mechanisms of operation and in the means required to influence the variety of electrogenic membrane found in neurons, muscle fibers, receptors, gland cells, electroplaxes, and neurosecretory cells. Parts of neurons and many muscle cells are noted for their dramatic electrical excitability and ability to produce self-propagating spikes. Glands and receptors as well as subsynaptic regions possess electrically inexcitable membrane but still electrogenic membrane in that they may generate depolarizing or hyperpolarizing potentials in response to adequate stimulation. These effects in turn play upon neighboring electrically excitable membrane to produce their physiological actions. With the few exceptions of electrical transmission, the electrical inexcitability of subsynaptic membrane by itself demands chemical transmission. Similarly, synaptic drugs exert their actions by either activating or inactivating components in the electrogenic processes of postsynaptic membrane. The nature of a particular postsynaptic membrane and its kind of electrogenesis determines its sensitivities to drugs. Conversely, drug actions can be used to assess the different components or properties of junctional membrane that may cover part of a given cell; the relations between the different units of the junctional membrane determine the net synaptic electrogenesis and its transfer to a conductile mechanism or electrically excitable spike-generating membrane. The kinetics of the electrogenic phenomena of the latter and its action in transferring excitation either to a subsequent postsynaptic element or to a contractile or secretory mechanism lead to the overt effects that are usually observed in the overall responses of a given system or of the whole organism.

Sufficient evidence to build a case for the electrical inexcitability of postjunctional membrane has been available for a number of years. The exceptions are limited to the occasional electrically transmitting synapses found originally in crayfish [92, 93] and leech [109] ganglia and on goldfish Mauthner cells [91, 94]. Essentially, the data summarized by Grundfest in 1957 show that, in many different junctional preparations, transmission across junctions involves short but measurable delays, the current derived from the prejunctional spike that flows across the postjunctional membrane is insufficient to excite, junctions show no response to strong shocks but do respond to minute quantities of a transmitter substance like acetylcholine, postsynaptic potentials cannot be elicited by antidromic stimulation, spikes in postjunctional cells are often initiated at sites far from the junctional regions, and the characteristics of postjunctional potentials are different from spike potentials; for example, they generally last longer.

Bioelectric phenomena result from the properties of plasma membranes. Indeed, many investigators have worked with squid axons that have been deprived of most of their intracellular contents. The mechanisms for bioelectrogenesis are believed to be mediated through reactive changes in membrane permeabilities for specific ions to characteristic adequate stimuli. The pumping mechanisms manufacture the intracellular ionic composition and the changes in permeabilities both represent fundamental functions of cell membrane, the latter important for excitable cells. Grundfest [104] describes two basic processes in all examples of excitation: an activation process that manifests itself as an increase in membrane permeability for one or several ions (an increase in membrane conductance or a decrease in membrane resistance) and an inactivation process that is a decrease in membrane permeability for ions (a decrease in conductance or an increase in resistance). An electrically excitable membrane possesses reactive elements that respond to electrical stimuli—either by an activation process or by an inactivation process. An electrically inexcitable membrane is one that, of course, does not respond to electrical stimuli but responds only to specific nonelectrical stimuli (chemical, mechanical, photic, or thermal). The response to the alterations in membrane permeability is a redistribution of ions and the consequent flow of current. The membrane potential can be driven in a positive direction (depolarizing electrogenesis) by an influx of cations, an outflux of anions, or both, or the membrane potential can be driven in a negative direction (hyperpolarizing electrogenesis) by an efflux of cations, an influx of anions, or both. Some reactions of excitable membranes result in little or no change in membrane potential evidenced by a change in membrane resistance or conductance. The potential does not change because ions are not permitted to flow, gradients to drive them do not exist, or the behavior of one ion is counterbalanced by the behavior of another. Presumably, each stimulus-response capability is indicative of a molecular organization specifically different from that of other capabilities and, so, the demonstration of different reactive elements within the same cellular membrane requires heterogeneity of membrane structure. Now is the time to point out that there are no data on membrane fine structure of sufficient detail to correlate with—or even oppose—this suggestion.

Another hypothesis that has been questioned is that the carefully measured perturbations of the steady resting potential are taken as indicative of the ion movements consequent to alterations in membrane permeabilities. The problem with the resting potential is that it probably incorporates other types of EMF, such as phase boundary potentials, which may arise from Gibbs-Donnan potentials at the interfaces between the two media and their separating membrane [249, 250], and contributions from other situations discussed earlier. Moreover, sodium ions in most cells and chloride ions in some (in squid axons but not lobster axons) do not fit Donnan ratios—the membranes are nearly impermeable to them—so ion pumps driven by metabolic energy are postulated to reconcile the discrepancies. If a one-to-one coupling of oppositely directed ion flux is lacking in an ion pump, the pump itself is electrogenic and examples of such have been found [87, 88]. Therefore, alternative hypotheses have been suggested [241]. But the ionic theory of electrogenesis—initiated by Bernstein, developed by Hodgkin and Huxley, and supported by quantities of data—can, with modifications, survive vigorous attack and is still used successfully to explain the diverse phenomena. The accumulation of comparative data over the years, while perhaps not eliminating alternative interpretations, at least supports the basic postulate that changes in membrane permeability for specific ions are the cause of bioelectrogenesis.

Although much space has already been devoted to spike electrogenesis, it should be restated here in Grundfest's terms. The spike is the result of two activation and one inactivation processes. Initially, sodium activation outstrips potassium activation so that the net influx of cations leads to depolarization. The channels permeable to sodium, opened by activation, become subsequently closed by sodium inactivation. The original conditions are restored by the swift decline in the depolarizing sodium influx and the continuing operation of the repolarizing potassium efflux. The latter also contributes to the hyperpolarizing afterpotentials.

To generalize details of the ionic theory from squid axon data alone to all excitable cells is, of course, a mistake—one made occasionally by proponents and antagonists alike. While not violating the basic premises, complexities in membrane properties not envisaged in the Hodgkin-Huxley theory have appeared in experiments on squid axons with ions introduced by micro-injection or perfusion. There is no reason to suppose that similar complexities do not appear in membranes from other sources. A quick survey will indicate some of the variations among action potentials that have been recorded and will make clear that the kinetics of electrogenic processes must differ in different cells or even in different regions of the same cell. Generalizations of the theory must allow for such qualitative differences. Indeed, as Nobel [209] and co-workers demonstrated on heart muscle, variations in the numerical values of the Hodgkin-Huxley parameters can account for a wide variety of different types of electrical responses.

For reference, the squid giant axon presents a spike that overshoots zero and returns to fall below the resting level in a hyperpolarizing afterpotential. In contrast, spikes of the septate giant axons of crayfish [254] and earthworms [149] terminate in depolarizing afterpotentials and axons in the skate spinal cord show no afterpotential at all [104]. Supramedullary neurons of the puffer fish have spikes with delayed falling phases [201, 202] homologous to the plateau in the muscle spikes of vertebrate cardiac fibers. Other cells including squid giant axon can be made to produce spikes by experimentally altering membrane conductances during spike recovery [12, 206]. Axon membrane behaves differently from soma and dendritic membrane. No spikes have been recorded within the soma and dendrites of crustacean cardiac ganglion cells [112] and goldfish Mauthner cells [91, 94]. Differences in threshold among these regions have been noted in other cells including motoneurons of mammals [90] and toads [9], central neurons of puffer fish [18] and molluscs [246], and crustacean stretch receptors [73].

The responses recorded from electric organs of different electric fish show great diversity in form [17, 103–105]. In some organs, the spikes recorded across one surface of an electroplax differ in amplitude, duration, or both from spikes recorded across the other surface. Some fish generate spikes only across the innervated surface and other fish show only graded junctional responses: The uninnervated surface can carry propagating spikes, graded responses, or even be unresponsive depending on the particular electroplax preparation studied. Some spikes are all-or-nothing responses but do not overshoot.

Spikes are essential for propagation without decrement but often the conduction problem is solved by a diffuse innervation of the tissue rather than by special transmitting properties built into the effector system, which are then permitted more varied responsiveness. Many invertebrate muscle fibers normally show graded potentials in place of all-or-nothing spikes [136]. The graded responses can, at times, be converted into spikes, for example, by stimulating "fast" instead of "slow" motor axons in grasshoppers [257] or by treating crayfish muscle with procaine [211, 212], lobster muscle with 5-hydroxytryptamine [256], larval *Tenebrio* muscle with tetraethylammonium [15], and some arthropod and molluscan muscle by replacing sodium in the extracellular medium by calcium or magnesium or by addition of other alkaline earth cations or onium ions to the bathing medium [107, 256]. Conversely, many examples may be cited to illustrate the conversion of spike responses into graded responses including the squid giant axon itself [39]. The eel electroplax has a low safety factor and occasionally graded responses emerge spontaneously [4].

The point is that there is much more variation and flexibility in responsiveness of excitable membranes than can possibly be imagined by reading elementary textbooks. The parameters that can be varied include the two processes (activation, inactivation), the specific ion permeabilities (usually sodium, potassium, and chloride but occasionally calcium, magnesium, and others), the time course (rate of change, duration, phasing if more than one ion flux is involved), the magnitude (intensity, area), the sensitivity (nature of stimulus: electrical, nonelectrical), and the initial electrochemical gradients across the membrane. Nature has abundant opportunities to select different combinations from among these parameters, and has used them.

The apparent ease by which all-or-nothing spikes and graded responses can be exchanged is strong evidence that the activation process that permits a depolarizing influx of cations (e.g., Na, Ca, or Mg in different cells) can be modified independently of the activation process that permits the repolarizing efflux of cations (usually K). For example, urethane reduces Na activation in *Onchidium* neurons resulting in graded potentials and TEA (tetraethylammonium) produces prolonged spikes in these same neurons by reducing K activation [110]. Similar results can be obtained by hastening the onset of K activation to produce a graded response or by delaying it to produce a prolonged spike. Hence, a graded response may be produced by a decrease in the influx relative to the efflux or by an earlier onset of the efflux. A spike can be generated from a graded response or a spike can be prolonged by a decrease or a delay in the efflux or by an earlier onset of the efflux. A spike can be generated from a graded response or a spike can be prolonged by a decrease or a delay in the efflux or an increase or prolongation of the influx [104].

There will be no attempt here to develop a comprehensive review of the complex effects of either pharmacological agents or ion substitutions on biological membranes. It is useful to point out, however, that the terms *labilization* and *stabilization* found in the literature [227, 228] are often synonyms for Grundfest's more descriptive terms *activation* and *inactivation*, respectively. It is also useful to note that the drugs used to modify responses may be specific for a single membrane component or may be unspecific; the specificity may change from one preparation to another. Procaine, for example, depresses both Na activation and K activation in squid axons [229, 247] but affects only Na activation in eel electroplax and primarily K activation in some crustacean muscle fibers ·[104]. Na activation but not K activation is blocked by urethane in *Onchidium* neurons [110]. TEA blocks K activation in *Onchidium* neurons and decreases membrane conductance, K activation, and probably also Na inactivation in lobster muscle fibers

Electrogenic membrane behavior

	Electrogenically inert	Electrically inexcitable		Electrically excitable			
Class	Electrogenically inert	Electrically inexcitable		Electrically excitable			**Potentials**
Type		Repolarizing	Depolarizing	Depolarizing		Repolarizing	
Transducer action		P_K and/or P_{Cl}	$P_{Na} + P_K$ (others?)	$P_{Na} + P_K$ (others?)		P_K or P_{Cl} (others?)	E_{Na} ca. +50 mV
Electrogenesis		IPSP's	Generator potentials / EPSP's	Over-shoot / Spike / Under-shoot / Graded response		Rectification	Reference zero / E_K or E_{Cl} −50 mV / Resting potential −100 mV / E_{Cl} or E_K
Examples	Unresponsive cells	Inhibitory synapses Receptor cells Glands	Excitatory synapses Receptor cells Primary sensory neurons Glands	All-or-none conductile membrane: axons muscle fibers neurons	Graded responsive membrane: arthropod muscle fibers	Frog slow muscle fibers	Rajid electroplaxes
Electrical analog components	C_m, E_{Cl}, E_{Na}, E_K	E_{Cl}, E_K	E_{Na}, E_K	E_K, E_{Na}		E_K, E_{Cl}	

[256]. Tetrodotoxin has been found to be a potent and specific inhibitor of Na activation in a variety of preparations including lobster axons [208], squid axons [203], frog muscle fibers [207], and eel electroplax [204].

The result of the many types of electrogenic activities found in nature is the formulation of more extensive, and hopefully more complete, descriptive diagrams (Fig. 55). Equal emphasis must be given to the electrically in-excitable as well as to the electrically excitable membranes although ion involvements and electrical behavior of the two general classes of excitable membrane are roughly parallel. A key point is that each of the ionic mechanisms can occur independently of the others. Such independence is strong evidence for separate pathways for each ionic movement and, therefore, separate pores in the molecular structure of the membrane; the plasma membrane must be heterogeneous—some of the channels are selective for actions; others are for anions. Figure 55 shows the changes in membrane potential that result from activation processes; inactivation processes are represented by increased resistance in the various reactive ionic components diagramed. Inactivation, in Grundfest's view (contrast with Hodgkin-Huxley definition on p. 171), closes membrane channels that are normally open at rest and keeps closed channels that are normally closed so that they cannot be opened by adequate stimuli. Figure 56 more clearly diagrams the different processes although the various ionic species involved are no longer evident. Some excitable membranes are known to react with permeability changes to ions other than those diagramed. For example, experimental conditions have been found that allow spikes to be generated without the use of sodium: Mg in *Tenebrio* muscle [15] Ca in the giant muscle fibers of barnacles [107] and also in squid axons [243], various other cations in lobster and crayfish muscle [79, 256], insect muscle [257], and frog neurons [164].

The usual explanation given for a membrane's ability to discriminate among cations and anions is the presence of negative and positive charges on the walls of the channels. There is, of course, no direct evidence for the existence of either the channels or their charges (see Chapter 2). The transverse tubular system of crayfish muscle fibers [95] and inhibitory synapses

Figure 55. *A diagrammatic chart of electrogenic responses to activation* (*increased conductance of different ions*): not all possibilities are shown, as examples can be found to show spike electrogenesis without potassium activation in the depolarizing electrically excitable membrane and also to show independent modification of sodium and potassium activation processes. The electrical analog components associated with these activities are diagramed below. As a dielectric, the membrane has a capacity C_m. Nonselective ion-permeable membrane is not shown. Electrogenically inert selectively permeable sites are shown as batteries with fixed internal resistances. The electrogenically reactive components are batteries that change their internal resistance in response to appropriate stimuli (increase = inactivation; decrease = activation). Batteries for other ions (e.g., Ca, Mg) and current paths due to electrogenic pumps have also been omitted.

[67] possess membrane components selectively permeable to chloride ions although the channels also accept a large variety of anions of about the same ionic radius [8, 146]. Permeabilities to cations are more selective and Na is readily distinguished from K. Most investigators feel that the K pores are too small to accept Na but it may also be that an ion with a single hydration shell must fit the channel closely; it cannot be either too big or too small. It is interesting to note that the individual channels do not appear to have a high degree of chemical specificity as illustrated by the diverse actions of acetylcholine. Acetylcholine can elicit depolarizing activation of Na and K in membrane that produces generator potentials or EPSP's but can also elicit Cl (or K) activation of inhibitory membrane; both types of responses can also be evoked by quite different chemical agents.

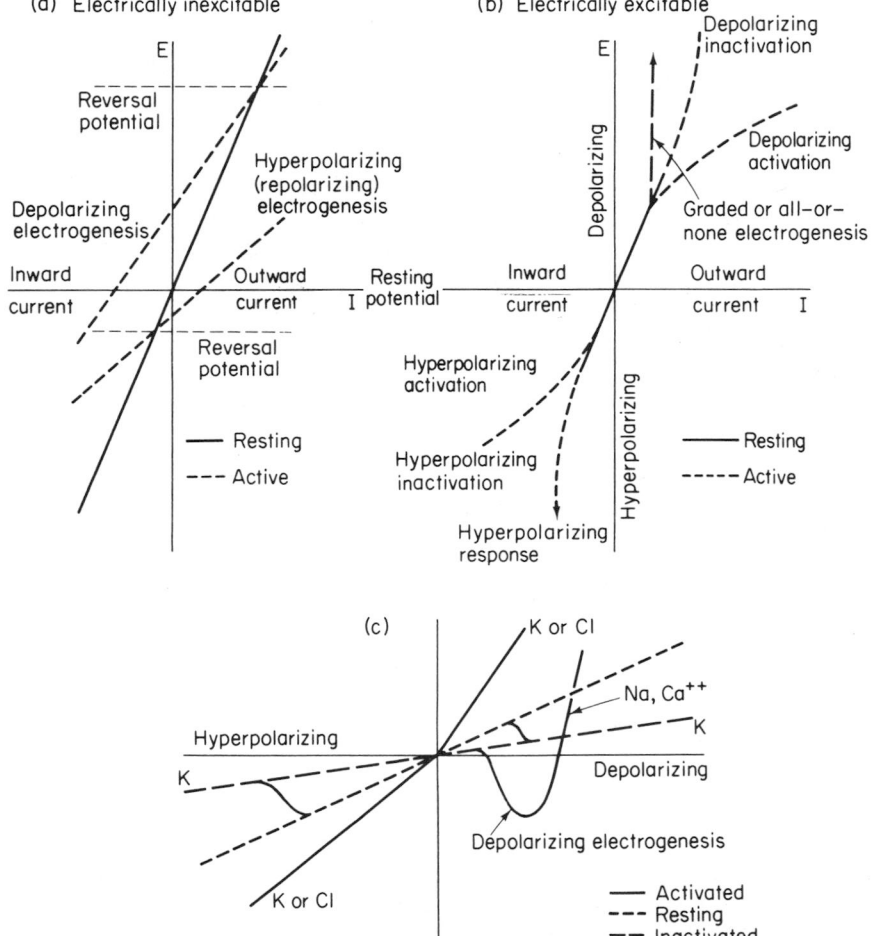

It is impossible to effectively defend any specific model for the molecular structure of excitable membrane. Presumably, membranes possess pores related to the ideas discussed in Chapters 2 and 3. The major feature of excitable membranes is that a significant number of pores can control their permeation. The size of the active pore site in both electrical excitable and inexcitable membrane may well be less than 50 Å in diameter [103] since molecules as small as procaine can block Na and K activation in squid axons [229, 247] and synaptic activity [4]; tetrodotoxin, molecular weight 319, is an especially potent inhibitor of Na activation [203, 237]. If it is true that the K channels, which must be smaller than Na channels, are clogged by the alkali metal ions (Cs and Rb), alkaline earth ions (Ba, Ca, Sr), and small organic ions (TEA), then the opening within each membrane patch must be no more than 2 Å [103]. This simple view of excitable membrane structure calls for donut-like patches—each one is a 50 Å structure ringing a 2–5 Å opening. In addition, the data presently available indicates that that the active patches are rare. Grundfest [102] calculated the area of the channels to be less than 10^{-4} of the area of free space available for ion movement in an aqueous saline medium and Moore, Narahashi, and Shaw [194] calculated less than thirteen sodium channels for each square micron of lobster axon membrane.

Figure 56. *Current-voltage relations of electrogenic membranes:* the origin of each graph is the resting potential. Electrically inexcitable membranes (a) behave as ohmic resistances—*E* changes linearly with *I*—although slopes change during activation of the membranes by appropriate stimuli. The dotted lines represent active membrane responses of (depolarizing) EPSP's and of (hyperpolarizing) IPSP's. Changes in the membrane potential alter the amplitudes of the recorded responses (given by the difference between the resting *I*-*E* line and that during activity). When *E* exceeds the reversal potential specified by the intersection of the resistance lines for active and passive membranes, the sign changes. Electrically excitable membranes (b) show nonlinear behavior characterized by changes in conductances. Most, but not all, develop graded or all-or-nothing depolarizing electrogenesis accompanied by increased membrane conductance, usually regenerative. Decreased conductance (inactivation) may be evoked by depolarizing or hyperpolarizing conductances and can be regenerative. Usually, the increased conductance for K or Cl evoked by depolarizing stimuli or for any ions by hyperpolarizing currents (rectification processes) are not regenerative. (c) presents voltage-clamp data for electrically excitable components. Increased conductance for Na or Ca causes a shift toward inside positivity; the link between the resting and activated states forms a negative slope because the voltage and current change in opposite directions. Unless the membrane potential is clamped, it undergoes a regenerative change to an all-or-nothing spike. Transitions from the resting state to the inactivated state also go through a negative slope region. The change gives rise to regenerative depolarizing and hyperpolarizing inactivation responses. The various ionic processes that cause the nonlinear relations have different time courses. Pharmacological inactivations are not shown.

4.18 A Summary of the
Rules for the Behavior of
Excitable Membranes

Some cells are specialized in their ability to respond dramatically to specific changes in their environments. They receive and detect small signals, formulate and transmit messages relevant to these signals, and execute action based on that information. The cellular properties that permit such functions reside in the plasma membrane and since an individual cell or portions thereof can perform more than one of these functions independently of the others, the molecular structure of excitable membrane is believed to be heterogeneous or a matrix of chemically different components. The unique structure of each component permits it to be selective in its responsiveness and its permeation properties.

The values given for membrane resistivity range between 10^5 and 10^{10} ohm cm, a range intermediate between that of stabilized lipid bilayers (10^{11}–10^{15} ohm cm) and the ionic solutions inside and outside cells (about 10^2 ohm cm). The difference may be evidence for aqueous transmembrane channels available for ion movement. Certainly, the number of such pores is small and the number that can control and alter their permeation properties to proper stimuli is still smaller. The capacity values for most biological membranes are close to 0.5 $\mu F/cm^2$, which is also true for artificial membranes. A few cells have capacities as high as 50 $\mu F/cm^2$ and some of these high values—but not all—can be explained on the basis of extensive folding of the cell's membrane. Conductances for the individual ion pathways are additive, which is difficult to explain of the basis of the theory that membranes behave as fixed-charge ion exchangers, particularly since cases are known where both cations and anions may be involved in carrying the current.

Some of the ionic channels do not respond in any obvious way to stimuli of any kind; they are electrogenically inert. Their relative number—reflected in the total conductance or permeability of the membrane—remains unaltered when the cell responds to stimuli. Considerable quantities and well-defined regions of a cell's plasma membrane may be electrogenically inert. The excitable regions can possess reactive elements that either increase the conductance of the membrane above that at rest (activation) or decrease the conductance below resting levels (inactivation). Some of the reactive elements respond only to specific but nonelectrical stimuli (electrically inexcitable), while other elements respond to changes in the transmembrane potential (electrically excitable). The stimuli may be regarded as causing distortions in membrane structure to effectively remove hindrances to the flow of ions. Channels that respond to electrical stimuli are expected to exhibit nonlinear behavior since changes in their configuration will also lead to changes in the transmembrane potential giving rise to opportunities for either positive or negative feedback.

The stimulus initiating regenerative depolarizing responses (spikes) is itself a depolarizing potential. In theory, conductile activity can be equally well effected by pulses that drive the membrane more negative inside. Such activity does not occur and must be regarded as a consequence of the normal resting membrane potential (negative inside) and the absence of ionic batteries across normal membranes that can drive the transmembrane potential still more strongly negative. Similarly, the electrical activity generated by an input component (e.g., generator potential, receptor potential, EPSP) must also be a depolarization if it is to exert maximal excitatory effects on the spike-generating membrane of the next link in the message chain; the ionic batteries involved must have an EMF that is positive inside (e.g., Na battery). The electrogenesis associated with the IPSP involves other batteries (e.g., K or Cl batteries) to successfully modulate the amplitude of both generator potentials and EPSP's. The electrical activity that may be associated with secretion can be of either sign as the potential is merely a concomitant of executive activity and is not itself that activity. Hence, receptor and gland cells, which do not have axons and therefore do not produce spikes, may develop any type of electrogenesis or even none at all. Some effectors are electrogenically indiscriminate also.

An ohmic (linear) current-voltage relation is exposed by subjecting the electrogenically inert and the electrically inexcitable membrane to applied currents. But the electrically excitable membranes respond to their appropriate stimulus by changing resistance and, consequently, membrane potential. The altered permeability characteristics of electrically excitable membrane caused by the electric shocks lead to nonlinearities in the current voltage relationships. This occurs both when constant currents are applied leading to changes in transmembrane voltage and when the transmembrane voltage is clamped at arbitrary values permitting the current across the membrane to change. Electrical stimuli, both depolarizing and hyperpolarizing, can elicit activation or inactivation processes and both types of processes may be regenerative. These reactions may develop independently in different ionic channels.

4.19 The Search for the
Electrogenic Protein

Recent years have seen an explosion in published theories and experiments aimed at understanding biological systems at the molecular level. This has been especially true for genetic and energetic systems but it is also now ongoing for nervous and developmental problems. Some of these ideas relate to each other, such as the RNA-memory work; some of the attacks have centered on entire organisms and complex problems (e.g., learning); others have dealt only with portions of single cells (e.g., axoplasmic flow). Investigators, at either end, have a busy future: The problems are now being defined, not solved.

Neurons do many things: They respire at high rates and synthesize RNA and protein at high rates. Yet, impulse transmission phenomena—the best known of neural behavior—are not very sensitive to respiratory poisons and RNA appears to be contained within the perikaryon. Therefore, impulse firing must be maintained and controlled primarily by proteins (including enzymes) and lipids. The current knowledge of the roles of protein in neural function has been reviewed [224, 225] but too many of the important functions are centered in speculation.

Schmitt and Davison [225] proposed an electrogenic protein (Fig. 16, p. 52) that can be induced to change configuration by an imposed electric field (or other stimuli adequate for that membrane). The protein has not been identified or characterized beyond the evidence suggesting the presence of SH groups on it [139]. Almost everyone postulates pores for the movement of ions that are so well measured by the voltage clamping devices. These channels have been characterized but never visualized and are not abundant [194]. The electrogenic protein is proposed as an ion gate. Some evidence suggests that it is located on the outer surface of the membrane, especially the work with tetrodotoxin [192, 237] but other evidence points to its presence on the axoplasmic side of the membrane, for example, the work with antibodies made against axoplasm [138] and with TEA [119]. Nachmansohn and his collaborators [196, 197] have tried to tie the ion-gating mechanism to an acetylcholine-acetylcholinesterase system although interpretations are restricted because the attack has been limited to the external surface of the membrane and has required harsh treatments (snake venom, proteases, etc.) to expose reactive sites. Antibodies against squid axoplasm can block conduction but only when applied to the internal surface of the membrane [137, 138]. The specific antigens are unknown; at least fourteen proteins have been identified in squid axoplasm and at least six different antibodies can be made but only the antibody to whole axoplasm is effective. In addition to working with small quantities of axoplasm, the electrogenic proteins guarding ion channels are neither abundant nor slow acting so an analysis of binding is equally difficult. Astute guesses and indirect experiments will have to suffice for the present in order to narrow the possibilities and to better characterize the questions that will have to be asked in the future.

4.20 Electrogenic Protein—
the Acetylcholine Story

The proposition that the electrogenic protein is an acetylcholine-acetylcholinesterase system has been researched by Nachmansohn and his collaborators [196–199]. During the past 30 years that the biochemistry of acetylcholine has been studied, the enzymes of formation and hydrolysis have been analyzed, the sequence of energy transformation has been established, and a number of chemical reactions have been correlated with physical events. In essence, Nachmansohn envisages acetylcholine to act in an intramembraneous process that is responsible for the changes in conductance that occur in conducting membrane during activity. An acetylcholine system is the specific chemical system contained in all conducting cells enabling them to generate the electric currents of propagating impulses.

The heat measurements during impulse propagation [1] suggest the chemical nature of this process. During conduction of a spike along an axon, a strong positive heat is detected (coincident with the electrical activity) followed by a negative heat during recovery. Considering that conducting membrane is only 50–100 Å thick, this positive heat is about 3 mCal/g of active tissue or about the same amount of heat generated during a muscle twitch.

According to Nachmansohn's hypothesis, acetylcholine is bound in the resting condition to a protein, perhaps a conjugated one. Any stimulus reaching the membrane releases acetylcholine, which then reacts with a receptor protein to produce a change in its configuration. This results in the release of the bound Ca^{2+} which induces further changes in the conformation of phospholipids and polyelectrolytes which are responsible for the increase in sodium conductance. The ester-receptor complex is in a dynamic equilibrium with the free ester and receptor protein. Since the free ester is rapidly hydrolyzed and inactivated by acetylcholinesterase, the receptor—and sodium conductance—return to their original condition to reestablish the ionic barrier. The turnover time for the acetylcholinesterase is 30–50 μsec [168], which permits rapid restoration of sodium conductance and allows the fibers to carry high frequencies of impulse transmission; only about 20,000–50,000 ions move in each direction during the period when conductance is high. The speed and precision of these events result from the linkage of the bound acetylcholine, the acetylcholine receptor protein, and the acetylcholinesterase into a protein assembly located within the plasma membrane. Moreover, Nachmansohn [199] does not believe that acetylcholine is a chemical mediator at synapses since he is not convinced that there is evidence to conclusively demonstrate the release of acetylcholine from a cell under physiological conditions. He believes that the action of acetylcholine is intracellular and within excitable membranes. Acetylcholine is the trigger which initiates and controls the permeability changes which permit the ion movements during electrical activity; the function of acetylcholine in excitable membranes is similar in fibers and in junctions.

The major evidence provided by Nachmansohn to support this system is that, in all fibers tested, electrical activity fails when cholinesterase activity is blocked by potent and specific inhibitors. The major detraction from the evidence is the high dose levels required to achieve a response. Nachmansohn counters that the external conditions used are irrelevant since structural barriers surround conducting membranes to limit diffusion of the compounds to reactive sites. If a nerve fiber is exposed to 1 g/ml of diisopropylphosphofluoridate, a potent anticholinesterase agent, less than 1 μg/g is found inside the axon when electrical activity fails. Electrical activity also fails in crab fibers when the enzymatic activity falls to 20% of the initial level regardless of the inhibitor used or its concentration in the bathing solution [262]. At exposed regions of frog axons, the nodes of Ranvier, a reversible block can be accomplished within 30 by 300 μg/ml eserine and within a few minutes by 30–40 μg/ml eserine, a dose comparable to that effective at neuromuscular junctions [56].

The cholinesterase enzyme has two active sites to complex with an acetylcholine molecule (Fig. 57). Choline is split out and the acetylated enzyme becomes hydrolyzed within microseconds to free the acetate and to restore the enzyme [200, 261].

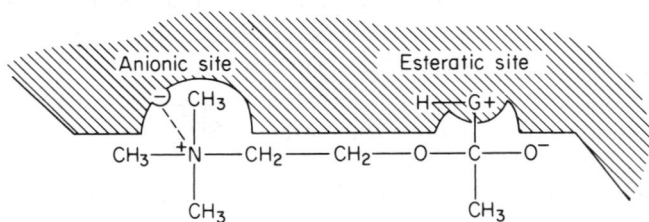

Figure 57. *Schematic representation of the interaction between the active groups in the surface of a cholinesterase enzyme molecule and a substrate acetylcholine molecule:* the negatively charged anionic site holds the positively charged quaternary nitrogen by coulomb and van der Waals forces and the esteratic site contains an acidic group (H) and a nucleophilic group (G) in its surface; the latter group forms a covalent bond with the electrophilic carbon of the carboxyl group of acetylcholine.

The organophosphorus compounds—potent insecticides and chemical warfare agents—phosphorylate the enzyme so that the hydrolysis does not occur or does so only slowly—days or weeks. The enzyme is essentially inhibited, synaptic and neuromuscular transmission is stopped, and its owner becomes moribund [259, 260]. The phosphorylated group can be displaced by a nucleophilic group [115]. Paradine-2-aldoxine methiodide (PAM) is such a compound [264] and is a powerful reactivator of inhibited enzyme [263]. Consequently, it was found to be an antidote against the organophosphorus insecticides and nerve gases, especially in combination with atropine in mice [158, 159] and can protect against 50–100 times the LD_{100} dose of organophosphates [205]. PAM is lipid insoluble and cannot penetrate the structural barriers that surround conducting membranes; but a lipid-soluble derivative, benzoylphridineoxime methiodide (BPO), is also a good reactivator of the phosphorylated enzyme [157]. Electrical activity, blocked by the organophosphate paraoxon, recovers when washed with BPO but not when washed with Ringer's solution [120]. Electrical activity can also be affected by agents reacting with the acetylcholine receptor on the membrane without affecting cholinesterase activity, for example, the blockade of activity of eel electroplax by carbonylcholine, decamethonium, and procaine [5, 6].

Chargas [36] made the first attempt to isolate the membrane's cholinergic receptor and, shortly later, Ehrenpreis [74] succeeded in extracting a curarebinding protein, 90% of which had a molecular weight near 100,000. The binding strengths of a series of different blocking agents to this protein vary but parallel the effects of the same compounds upon an isolated unicellular eel electroplax preparation [220, 226].

Local anesthetics, such as procaine, are related structurally to acetylcholine. The anesthetic potencies of such compounds parallel their binding capacities to the isolated receptor proteins [75] and their effectiveness on the electroplax preparation [220]. They appear to act as competition for the acetylcholine reactive site [116]. Since these local anesthetics block electrical activity in conducting fibers, Nachmansohn suggests that the acetylcholine receptor protein is present and essential for spike electrogenesis.

Other evidence implicating acetylcholine in impulse propagation can be presented. Curare, acetylcholine, and other lipid-insoluble quaternary nitrogen compounds do not affect conduction because they cannot penetrate the surface diffusion barriers [221], but curare blocks at nodes of Ranvier of frog sciatic nerve where the axon is more exposed [56]. Curare, acetylcholine, neostigmine, and other quaternary compounds, which were previously inactive, reversibly block electrical activity in desheathed frog neurons after treatment with detergent [253]. Acetylcholine also depolarizes desheathed cat vagal fibers [10] and crustacean somatic axons [57]. The enzymes in cobra venom lower the structural barriers sufficiently to allow curare to block conduction, as demonstrated in squid axons [219]. The venom of cottonmouth moccasin is even more effective at exposing membrane reactive groups and, after the completion of such treatment, conduction in squid axon can now be blocked by acetylcholine, curare, and a series of tertiary nitrogen derivatives that are all related in structure to acetylcholine and, in addition, are lipid soluble and show good and relatively specific binding to the receptor protein in solution. Indeed, the effective concentrations of these latter compounds are close to those required for effect at the synaptic junctions of the isolated electroplax [197].

The importance of this work, in addition to the serious attempt to identify the ion-gating protein in excitable membrane, has been the improved molecular characterization of the acetylcholine receptor in junctional membrane as a protein component of plasma membrane with multiple cooperatively interacting binding sites that involve at least a disulfide bond [150]. The molecular characterization of electrogenic protein in conductile membrane is less clear despite the theory of acetylcholine involvement championed by Nachmansohn. Work under way using compounds whose chemistry is well known—nitrobenzene derivatives—shows that these have specific reactions with the membrane constituents responsible for the changes in Na and K conductance during activity in squid and lobster axons [44].

Duncan [26, 62, 63] has presented and summarized evidence to suggest that the electrogenic protein is an ATPase system capable of responding to stimuli by an alteration in the balance between the ATP substrate and the ATPase enzyme and consequent changes in the physical conformation and charge distribution of the enzyme. Duncan [63] also associates nonspecific cholinesterase activity with the enzyme complex during excitation and a Mg-dependent ATPase with the control of passive permeability and a Na-K-Mg-dependent ATPase with the cation pump.

4.21 A Macromolecular Approach
to Excitable Membrane

An alternative explanation for the basis of excitation is given by Tasaki [240] who has interpreted work in his laboratory on perfused squid giant axons in terms of physicochemical alterations of the macromolecular constituents rather than in terms of ion gates, channels, and electrogenic proteins. Tasaki finds it difficult to use an equivalent electrical circuit model for excitation since it implies a simple homogeneous structure. Rather, he prefers to think of a complex macromolecular system of proteins and phospholipids and of an excitation process as a phase transition involving the interaction of charged macromolecules and their ionic environment. The emphasis is on fixed negative sites within critical layers of the membrane that have a cation-exchange nature.

The model presented by Tasaki (Fig. 58) for axon membrane has lipoprotein molecules bound together mainly by noncovalent bonds and subdivided into two physiologically distinct layers. The outer layer has a relatively high density of negatively charged groups, possibly carboxyl in nature, and the

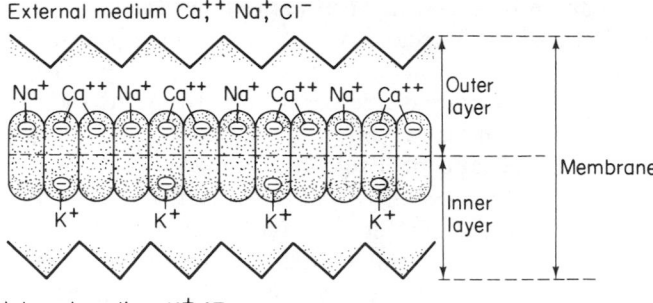

Figure 58. *Tasaki's macromolecular model for excitable membrane:* between the outer and inner protein coats lie lipoprotein molecules. The outer layer of these molecules contain sites occupied predominantly by divalent cations and the inner sites by monovalent cations. During excitation, the interior monovalent cations (e.g., K+) are driven through the membrane to displace the divalent cations (e.g., Ca2+) and, thus, lead to configurational changes in the macromolecules and the excitation phenomena.

inner layer has a low density of negative fixed charges, possibly phosphates [243]. At rest, the outer sites are occupied predominantly by divalent cations derived from the external medium and the inner sites by univalent cations derived from the axon interior.

The nature of some of the supporting evidence for this model is summarized and interpreted by Tasaki [240, 241]. The internal surface of squid giant axon membrane differs from the external surface: Several enzymes and antibodies abolish axon excitability when applied to the internal surface but not to the external surface; divalent cations are required in the bathing medium to maintain excitability but block conduction when applied internally; the inner but not the outer surface readily distinguishes among anions required to maintain excitability, and a lyotropic series can be arranged; and large differences are measured between the inward and the outward transfer rates of depolarizing cations across axon membrane. The effects of diluting both external and internal solutions with isotonic nonelectrolytes can be interpreted as cation exchange and suggest fixed negative charges on the inside of the membrane [16] and associated double-layer potential [34]. The insignificant effects of substitutions of Br, SO_4, etc., for Cl in the external medium is indicative of a large number of negatively charged groups in the outer layer of membrane that tend to exclude anions, a property common to cation exchangers. But, substitutions among anions inside have large effects suggesting an insufficient quantity of fixed negative charges in the inner layer of the membrane to exclude anions. Similarly, the smaller fluxes of anionic than cationic substances through membranes are attributed to exclusion of anions from the outer layer [244]. Divalent cations appear necessary to keep the critical layers of membrane in the resting state: Membranes are stabilized in the rest condition by excess divalent cations but their removal from the bathing solution leads to fluctuations in the membrane potential and spontaneous excitation.

The transition of membrane macromolecules from one stable configuration to a different one that is excitation in this model [240] can be produced either by lowering the ratio of divalent to univalent cations in the outer layer of membrane (experimentally by altering the ratio in the external solution) or by passing electric currents. Divalent cations in the outer layer at rest (Ca, Mg) form relatively stable complexes with the macromolecules. The small alkali metal cations (e.g., Li, Na) have weak affinities for the negative sites and little tendency to disrupt the complexes between the divalent cation and the macromolecules. The larger alkali metal cations (K, Rb, Cs) have a much stronger affinity toward the negative sites, thought to be carboxyl groups, and, consequently, a much stronger tendency to disrupt the bonds. An outward flowing electric current drives the disrupting cations (K, Rb, Cs) from the interior of the axon into the outer layer of the membrane to displace the divalent cation from the macromolecules. These disruptions lead to alterations in the configuration of the macromolecules and the subsequent physiological properties associated with excitation. In the stable resting state, the negatively charged outer sites are occupied by univalent and divalent cations but, during the stable active state, these sites are occupied primarily by univalent cations.

Instead of a voltage-dependent conductance as the basis for an action potential, Tasaki has proposed a rapidly reversible ion-exchange process that involves a transition between two stable configurations of a macromolecular complex. The underlying structural changes suggested are in essence similar to the processes of helix-random coil and oil-water transitions, which exhibit two rapidly reversible but stable states, lyotropic ion sequences, temperature effects, and physicochemical properties analogous to those observed in excitable membranes. Not all Tasaki's work is documented here [240, 241] and many of his ideas have been evolving, as additional details are uncovered, or controversial [33]. The approach generally has yet to be completely exploited to a fully mature theory for excitation and is limited to axons and squid data. In any case, although the approach is less phenomenological than that of Hodgkin and Huxley and more satisfactory to physical chemists, it has yet to receive strong support from a majority of axonologists.

4.22 Conclusions

Studies on bioelectricity have always been popular with physiologists and especially so today with the sophistication and precision of electronic technology. Certainly, the search for a better understanding of the workings of nervous systems, prominent among these studies, is one of the most important endeavors of our age; yet there are some disturbing features in the current status of our knowledge.

It is probably reasonable to conclude that the general electrical phenomena associated with the behavior of excitable cells have been described and properly assigned as functions of the plasma membrane, and the various classes of bioelectric potentials of resting and active cells tabled in Fig. 48, p. 168, illustrate this. Not at all unusual, progress has been stepwise paralleling the advancements in technology. A sample of the obvious examples of important experimental breakthroughs that have each in turn been exploited by many investigators include oscilloscopes, glass micropipette electrodes, voltage clamps, and radiotracer and perfusion techniques. Yet, the studies have emphasized axons and other long-fibered systems to detract from the importance of other excitable behavior. A survey of excitable cells—receptors, effectors, junctions, as well as conductors—shows that spikes may not be the dominant mode of behavior. Giant fibers, especially in squid, have yielded important information about conduction but their study has led to an emphasis of sodium and potassium in spike generation despite the examples that can be cited to show the involvement of other ions. Drugs are applied to reach and modify hidden sites; yet few drugs appear to have a singular action when tested upon a variety of preparations. Studies with highly modified systems (e.g., perfused squid axons) and highly complex systems (muscle fibers) have led too many workers to reject useful ideas without more adequate substitutions. We are only beginning to examine the contributions to excitable behavior from nonnervous tissues, which are present in abundance in even the more simple preparations as seen in electron micrographs. We know where some but not all diffusion barriers are located. There are real limitations in our present knowledge of morphology, molecular structure, and physical chemistry of membranes. Voltage measurements are fairly simple to make; yet there is disagreement as to the source of the potential differences inferred from these measurements [242]. The result is a confused status: too many controversies generated, too many special interests defended, and too much dependence on a few dramatic experiments and fewer tissues. Curiously, the belief in a well-understood field where experiment confirms theory is continuously sold to the neophyte in too many elementary textbooks. It is now time to be more careful in applying the conclusions derived from beautiful work on one system to all systems. The evolutionary development and phylogenetic relationships of these properties are described only by those who are willing to speculate.

It is not proper to belittle the real progress in our understanding of the behavior of excitable membranes and the fine stories that can be told about many of the aspects, however. Unfortunately, many basic problems concerning the underlying mechanisms remain essentially unsolved. That this is so cannot be blamed on any lack of attention as evidenced by the quantity of published papers, reviews, symposia, and books over the years. Nor can a lack of brilliant investigators be noted as illustrated by the sizable sample selected for Nobel awards. Moreover, the associated instrumentation has been sophisticated, techniques precise, and experiments imaginative. But the problems are now reasonably well defined and proper study underway. It should be made clear that the emphasis of this chapter and the few examples used are selected more to illustrate the basic but diverse behavior encountered than to demonstrate satisfaction with the completeness of the ideas presented. The more popular view among workers is that bioelectric potentials are diffusion potentials that result from modulation of membrane permeabilities by valves in the transmembrane ion pathways. But it is not necessary to accept this view *in toto* in order to satisfy existing evidence. Other views, such as ion-exchange effects, can be tolerated as long as some mechanism is presented that can explain the means by which the specific ions involved in each preparation can do so. Existing evidence exposed by the present phenomenological approaches identifies more the specific ions involved than the mechanisms by which they operate. Better ideas will come as more is learned about the basic molecular structure and physical chemistry of proteolipid membraneous systems.

REFERENCES CITED

1. B. C. ABBOTT, A. V. HILL, and J. V. HOWARTH, 1958. The positive and negative heat production associated with a nerve impulse. *Proc. Roy. Soc. London B* **148**: 149–187.

2. R. H. ADRIAN, 1956. The effect of internal and external potassium concentration on the membrane potential of frog muscle. *J. Physiol.* **133**: 631–658.

3. D. ALBE-FESSARD, C. CHARGAS, A. COUCEIRO, and A. FESSARD, 1951. Characteristics of responses from electrogenic tissue in *Electrophorus electricus. J. Neurophysiol.* **14**: 243–252.

4. M. ALTAMIRANO, C. W. COATES, and H. GRUNDFEST, 1955. Mechanisms of direct and neural excitability in electroplaques of electric eel. *J. Gen. Physiol.* **38**: 319–360.

5. M. ALTAMIRANO, C. W. COATES, H. GRUNDFEST, and D. NACHMANSOHN, 1955. Electrical activity in electric tissue. III. Modifications of electrical activity by acetylcholine and related compounds. *Biochim. Biophys. Acta* **16**: 449–463.

6. M. ALTAMIRANO, W. L. SCHLEYER, C. W. COATES, and D. NACHMANSOHN, 1955. Electrical activity in electric tissue. I. The difference between tertiary and quaternary nitrogen compounds in relation to their chemical and electrical activities. *Biochim. Biophys. Acta* **16**: 268–282.

7. W. R. AMBERSON, 1936. On the mechanism of the production of electromotive forces in living tissues. *Cold Spring Harbor Symp. Quant. Biol.* **4**: 53–62.

8. T. ARAKI, M. ITO, and O. OSCARSSON, 1961. Anion permeability of the synaptic and non-synaptic motoneurone membrane. *J. Physiol.* **159**: 410–435.

9. T. ARAKI and T. OTANI, 1955. Response of single motoneurones to direct stimulation in toad's spinal cord. *J. Neurophysiol.* **18**: 472–485.

10. C. J. ARMETT and J. M. RITCHIE, 1960. The action of acetylcholine on conduction in mammalian non-myelinated fibres and its prevention by an anticholinesterase. *J. Physiol.* **152**: 141–158.

11. Y. ASADA, 1963. Effects of intracellularly injected anions on the Mauthner cells of goldfish. *Jap. J. Physiol.* **13**: 583–598.

12. P. F. BAKER, A. L. HODGKIN, and H. MEVES, 1964. The effect of diluting the internal solution on the electrical properties of a perfused giant axon. *J. Physiol.* **170**: 541–561.

13. P. F. BAKER, A. L. HODGKIN, and T. I. SHAW, 1962. Replacement of the axoplasm of giant nerve fibres with artificial solutions. *J. Physiol.* **164**: 330–354.

14. P. F. BAKER, A. L. HODGKIN, and T. I. SHAW, 1962. The effects of changes in internal ionic concentration on the electrical properties of perfused giant axons. *J. Physiol.* **164**: 355–374.

15. P. BELTON and H. GRUNDFEST, 1962. Potassium activation and K-spikes in muscle fibers of the mealworm (*Tenebrio molitor*). *Am. J. Physiol.* **203**: 588–594.

16. M. R. BENNET, 1967. An analysis of the surface fixed-charge theory of the squid giant axon membrane. *Biophys. J.* **7**: 151–164.

17. M. V. L. BENNETT, 1961. Modes of operation of electric organs. *Ann. N.Y. Acad. Sci.* **94**: 459–509.

18. M. V. L. BENNETT and H. GRUNDFEST, 1959. Electrophysiology of electric organ in *Gymnotus carado. J. Gen. Physiol.* **42**: 1067–1104.

19. J. BERNSTEIN, 1902. Untersuchungen zur Thermodynamik der bioelektrischen Ströme. *Pflüger's Arch. ges. Physiol.* **92**: 521–562.

20. R. BEUTNER and J. LOZNER, 1933. Transmembrane potentials are not redox potentials. *Protoplasma* **19**: 370–415.

21. R. BIRKS and F. C. MACINTOSH, 1957. Acetylcholine metabolism at nerve endings. *Brit. Med. Bull.* **13**: 157–161.

22. D. BODIAN, 1962. The generalized vertebrate neuron. *Science* **137**: 323–326.

23. D. BODIAN, 1967. "Neurons, circuits, and neuroglia," pp. 6–24, in: G. C. Quarton, T. Melnechuk, and F. O. Schmitt (eds.), *The Neurosciences—A Study Program*. New York: Rockefeller University Press.

24. J. BOISTEL and P. FATT, 1958. Membrane permeability change during transmitter action in crustacean muscle. *J. Physiol.* **144**: 176–191.

25. A. BORTOFF, 1961. Slow potential variations of small intestine. *Am. J. Physiol.* **201**: 203–208.

26. K. BOWLER and C. J. DUNCAN, 1967. Evidence implicating a membrane ATPase in the control of passive permeability of excitable cells. *J. Cell. Physiol.* **70**: 121–126.

27. M. A. B. BRAZIER, 1968. *The Electrical Activity of the Nervous System*, 3rd ed. Baltimore: Williams & Wilkins Co., 317 pp.

28. L. G. BROCK, J. S. COOMBS, and J. C. ECCLES, 1952. The recording of potentials from motoneurones with an intracellular electrode. *J. Physiol.* **117**: 431–460.

29. L. G. BROCK and R. M. ECCLES, 1958. The membrane potentials during rest and activity of the ray electroplate. *J. Physiol.* **142**: 251–274.

30. G. BURNSTOCK and R. W. STRAUB, 1958. A method for studying the effects of ions and drugs on the resting and action potentials in smooth muscle with external electrodes. *J. Physiol.* **140**: 156–167.

31. S. R. y CAJAL, 1934. Les preuves objectives de l'unité anatomique des cellules nerveuses. *Trab. Lab. Invest. Biol. Univ. Madrid* **29**: 1–137.

32. M. CALVIN, 1959. Energy reception and transfer in photosynthesis. *Rev. Mod. Phys.* **31**: 147–156.

33. W. K. CHANDLER and A. L. HODGKIN, 1965. The effect of internal sodium on the action potential in the presence of different anions. *J. Physiol.* **181**: 594–611.

34. W. K. CHANDLER, A. L. HODGKIN, and H. MEVES, 1965. The effect of changing the internal solution on sodium inactivation and related phenomena in giant axons. *J. Physiol.* **180**: 821–836.

35. H. T. CHANG, 1952. Cortical and spinal neurons: cortical neurones with particular reference to the apical dendrites. *Cold Spring Harbor Symp. Quant. Biol.* **17**: 189–202.

36. C. CHARGAS, 1959. "Studies on the mechanism of curare fixation by cells," pp. 327–345, in: D. Bovet, F. Bovet-Nitti, and G. B. Marini-Bettolo (eds.), *Curare and Curare-like Agents.* Amsterdam: Elsevier Pub. Co.

37. L. W. CHU, 1954. A cytological study of anterior horn cells isolated from human spinal cord. *J. Comp. Neurol.* **100**: 381–413.

38. L. B. COHEN, R. D. KEYNES, and B. HILLE, 1968. Light scattering and birefringence changes during nerve activity. *Nature* **218**: 438–441.

39. K. S. COLE, 1968. *Membranes, Ions, and Impulses; A Chapter of Classical Biophysics.* Berkeley: University of California Press, 569 pp.

40. K. S. COLE and H. J. CURTIS, 1939. Electric impedence of the squid giant axon during activity. *J. Gen. Physiol.* **22**: 649–670.

41. K. S. COLE and A. L. HODGKIN, 1939. Membrane and protoplasm resistance in the squid giant axon. *J. Gen. Physiol.* **22**: 671–687.

42. C. M. CONNELLY, 1959. Recovery processes and metabolism of nerve. *Rev. Mod. Phys.* **31**: 475–484.

43. E. J. CONWAY, 1957. Nature and significance of concentration relation of potassium and sodium ions in skeletal muscle. *Physiol. Rev.* **37**: 84–132.

44. I. M. COOKE, J. M. DIAMOND, A. D. GRINNELL, S. HAGIWARA, and H. SAKATA, 1968. Supression of the action potential in nerve by nitrobenzene derivatives. *Proc. Natl. Acad. Sci. U. S.* **60**: 470–477.

45. J. S. COOMBS, D. R. CURTIS, and J. C. ECCLES, 1957. The interpretation of spike potentials of motoneurones. *J. Physiol.* **139**: 198–231.

46. J. S. COOMBS, D. R. CURTIS, and J. C. ECCLES, 1957. The generation of impulses in motoneurones. *J. Physiol.* **139**: 232–249.

47. J. S. COOMBS, J. C. ECCLES, and P. FATT, 1955. The electrical properties of the motoneurone membrane. *J. Physiol.* **130**: 291–325.

48. J. S. COOMBS, J. C. ECCLES, and P. FATT, 1955. The specific ionic conductances and the ionic movements across the motoneuronal membrane that produce the inhibitory postsynaptic potential. *J. Physiol.* **130**: 326–373.

49. J. S. COMBS, J. C. ECCLES, and P. FATT, 1955. Excitatory synaptic action in motoneurones. *J. Physiol.* **130**: 374–395.

50. H. J. CURTIS and K. S. COLE, 1942. Membrane resting and action potentials from the squid giant axon. *J. Cell. Comp. Physiol.* **19**: 135–144.

51. D. R. CURTIS and J. C. ECCLES, 1959. The time course of excitatory and inhibitory synaptic actions. *J. Physiol.* **145**: 529–546.

52. R. B. DEAN, 1941. Theories of electrolyte equilibrium in muscle. *Biol. Symp.* **3**: 331–348.

53. J. DEL CASTILLO and B. KATZ, 1954. Quantal components of the end-plate potential. *J. Physiol.* **124**: 560–573.

54. J. M. R. DELGADO, 1964. "Electrodes for extracellular recording and stimulation," pp. 88–143, in: W. L. Nastuk (ed.), *Physical Techiques in Biological Research, Vol. V, Electrophysiological Methods, Part A*, New York: Academic Press.

55. E. D. P. DE ROBERTIS, 1959. Submicroscopic morphology of the synapse. *Internat. Rev. Cytol.* **8**: 61–96.

56. W. D. DETTBARN, 1960. Effect of curare on conduction in myelinated, isolated nerve fibres of the frog. *Nature* **186**: 891–892.

57. W. D. DETTBARN, 1960. New evidence for the role of acetylcholine in conduction. *Biochim. Biophys. Acta* **41**: 377–386.

58. M. M. DEWEY and L. BARR, 1962. Intercellular connection between smooth muscle cells: the nexus. *Science* **137**: 670–672.

59. J. DIAMOND, J. A. B. GRAY, and D. R. INMAN, 1958. The depression of the receptor potential in Pacinian corpuscles. *J. Physiol.* **141**: 117–130.

60. J. DIAMOND, J. A. B. GRAY, and D. R. INMAN, 1958. The relation between receptor potentials and the concentration of sodium ions. *J. Physiol.* **142**: 382–394.

61. J. DIAMOND, J. A. B. GRAY, and M. SATO, 1956. The site of initiation of impulses in Pacinian corpuscles. *J. Physiol.* **133**: 54–67.

62. C. J. DUNCAN, 1965. Cation-permeability control and depolarization in excitable cells. *J. Theoret. Biol.* **8**: 403–418.

63. C. J. DUNCAN, 1967. *The Molecular Properties and Evolution of Excitable Cells.* Oxford: Pergamon Press, 253 pp.

64. J. C. ECCLES, 1946. An electrical hypothesis of synaptic and neuromuscular transmission. *Ann. N.Y. Acad. Sci.* **47**: 429–455.

65. J. C. ECCLES, 1953. *The Neurophysiological Basis of Mind: the Principles of Neurophysiology.* Oxford: Clarendon Press, 314 pp.

66. J. C. ECCLES, 1957. *The Physiology of Nerve Cells.* Baltimore: The Johns Hopkins Press, 270 pp.

67. J. C. ECCLES, 1964. *The Physiology of Synapses.* Berlin: Springer-Verlag, 316 pp.

68. J. C. ECCLES, R. M. ECCLES, and M. ITO, 1964. Effects of intracellular potassium and sodium injections on the inhibitory postsynaptic potential. *Proc. Roy. Soc. London B* **160**: 181–196.

69. J. C. ECCLES, R. M. ECCLES, and M. ITO, 1964. Effects produced on inhibitory postsynaptic potentials produced by the coupled injections of cations and anions into motoneurones. *Proc. Roy. Soc. London B* **160**: 197–210.

70. I. S. EDELMAN, 1961. Transport through biological membranes. *Ann. Rev. Physiol.* **23**: 37–70.

71. C. EDWARDS and S. HAGIWARA, 1959. Potassium ions and the inhibitory process in the crayfish stretch receptor. *J. Gen. Physiol.* **43**: 315–321.

72. C. EDWARDS and S. W. KUFFLER, 1959. The blocking effect of γ-aminobutyric acid (GABA) and the action of related compounds on single nerve cells. *J. Neurochem.* **4**: 19–30.

73. C. EDWARDS and D. OTTOSON, 1958. The site of impulse initiation in a nerve cell of a crustacean stretch receptor. *J. Physiol.* **143**: 138–148.

74. S. EHRENPREIS, 1960. Isolation and identification of the acetylcholine receptor protein of electric tissue. *Biochim. Biophys. Acta* **44**: 561–577.

75. S. EHRENPREIS and M. G. KELLOCK, 1960. Acetylcholine receptor protein and nerve action. I. Specific reaction of local anesthetics with the protein. *Biochem. Biophys. Research Commun.* **2**: 311–315.

76. M. Eigen and L. de Maeyer, 1958. Self-dissociation and protonic charge transport in water and ice. *Proc. Roy. Soc. London A* **247**: 505–533.

77. G. Falk and R. W. Gerard, 1954. Effect of microinjected salts and ATP on the membrane potential and mechanical response of muscle. *J. Cell. Comp. Physiol.* **43**: 393–403.

78. P. Fatt, 1961. "The change in membrane permeability during the inhibitory process," pp. 87–91, in: E. Florey (ed.), *Nervous Inhibition.* Oxford: Pergamon Press.

79. P. Fatt and B. L. Ginsborg, 1958. The ionic requirements for the production of action potentials in crustacean muscle fibres. *J. Physiol.* **142**: 516–543.

80. P. Fatt and B. Katz, 1951. An analysis of the end-plate potential recorded with an intra-cellular electrode. *J. Physiol.* **115**: 320–369.

81. W. Feder (ed.), 1968. Bioelectrodes. *Ann. N.Y. Acad. Sci.* **148**: 1–287.

82. A. Fessard, 1952. Diversity of transmission processes as exemplified by specific synapses in electric organs. *Proc. Roy. Soc. London B* **140**: 186–189.

83. A. Fessard and J. Posternak, 1950. Les mécanismes élémentaires de la transmission synaptique. *J. Physiol. Path. Gén.* **42**: 319–445.

84. E. Florey, 1961. Comparative physiology: transmitter substances. *Ann. Rev. Physiol.* **23**: 501–528.

85. K. Frank and M. C. Becker, 1964. "Microelectrodes for recording and stimulation," pp. 22–87, in: W. L. Nastuk (ed.), *Physical Techniques in Biological Research, Vol. V, Electrophysiological Methods, Part A.* New York: Academic Press.

86. R. Frater, S. E. Simon, and F. H. Shaw, 1959. Muscle: a three phase system. The partition of divalent ions across the membrane. *J. Gen. Physiol.* **43**: 81–96.

87. A. S. Frumento, 1965. The electrical effects of an ionic pump. *J. Theoret. Biol.* **9**: 253–262.

88. A. S. Frumento, 1965. Sodium pump: its electrical effects in skeletal muscle. *Science* **147**: 1442–1443.

89. F. A. Fuhrman, 1959. Transport through biological membranes. *Ann. Rev. Physiol.* **21**: 19–48.

90. M. G. F. Fuortes, K. Frank, and M. C. Becker, 1957. Steps in the production of motoneuron spikes. *J. Gen. Physiol.* **40**: 735–752.

91. E. J. Furshpan and T. Furukawa, 1962. Intracellular and extracellular responses of several regions of the Mauthner cell of the goldfish. *J. Neurophysiol.* **25**: 732–771.

92. E. J. Furshpan and D. D. Potter, 1959. Transmission at the giant motor synapses of the crayfish. *J. Physiol.* **145**: 289–325.

93. E. J. Furshpan and D. D. Potter, 1959. Slow post-synaptic potentials recorded from the giant motor fibre of the crayfish. *J. Physiol.* **145**: 326–335.

94. T. Furukawa and E. J. Furshpan, 1963. Two inhibitory mechanisms in the Mauthner neurons of goldfish. *J. Neurophysiol.* **26**: 140–176.

95. L. Girardier, J. P. Reuben, P. W. Brandt, and H. Grundfest, 1963. Evidence for anion permselective membrane in crayfish muscle fibers and its possible role in excitation-contraction coupling. *J. Gen. Physiol.* **47**: 189–214.

96. D. E. Goldman, 1943. Potential, impedance, and rectification in membranes. *J. Gen. Physiol.* **27**: 37–60.

97. J. A. B. Gray and M. Sato, 1953. Properties of the receptor potential in Pacinian corpuscles. *J. Physiol.* **122**: 610–636.

98. H. Grundfest, 1952. "Mechanisms and properties of bioelectric potentials," pp. 193–228, in: E. S. G. Barron (ed.), *Modern Trends in Physiology and Biochemistry.* New York: Academic Press.

99. H. GRUNDFEST, 1955. "The nature of the electrochemical potentials of bioelectric tissues," pp. 141–166, in: T. Shedlovsky (ed.), *Electrochemistry in Biology and Medicine.* New York: John Wiley & Sons, Inc.

100. H. GRUNDFEST, 1957. The mechanism of discharge of the electric organs in relation to general and comparative electrophysiology. *Prog. Biophys.* **7**: 1–85.

101. H. GRUNDFEST, 1957. Electrical inexcitability of synapses and some of its consequences in the central nervous system. *Physiol. Rev.* **37**: 337–361.

102. H. GRUNDFEST, 1963. "Impulse conducting properties of cells," pp. 227–322, in: D. Mazia and A. Tyler (eds.), *The General Physiology of Cell Specialization.* New York: McGraw Hill Book Co.

103. H. GRUNDFEST, 1966. Heterogeneity of excitable membrane: electrophysiological and pharmacological evidence and some consequences. *Ann. N.Y. Acad. Sci.* **137**: 901–949.

104. H. GRUNDFEST, 1966. Comparative electrobiology of excitable membranes. *Adv. Comp. Physiol. Biochem.* **2**: 1–116.

105. H. GRUNDFEST and M. V. L. BENNETT, 1961. "Studies on morphology and electrophysiology of electric organs. I. Electrophysiology of marine electric fishes," pp. 57–101, in: C. Chagas and A. Paes de Carvalho (eds.), *Bioelectrogenesis.* Amsterdam: Elsevier Pub. Co.

106. H. GRUNDFEST, C. Y. KAO, and M. ALTAMIRANO, 1954. Bioelectric effects of ions microinjected into the giant axon of *Loligo. J. Gen. Physiol.* **38**: 245–282.

107. S. HAGIWARA, S. CHICHIBU, and K.-I. NAKA, 1964. The effects of various ions on resting and spike potentials of barnacle muscle fibers. *J. Gen. Physiol.* **48**: 163–179.

108. S. HAGIWARA, K. KUSANO, and S. SAITO, 1960. Membrane changes in crayfish stretch receptor neuron during synaptic inhibition and under action of gamma-aminobutyric acid. *J. Neurophysiol.* **23**: 505–515.

109. S. HAGIWARA and H. MORITA, 1962. Electrotonic transmission between two nerve cells in leech ganglion. *J. Neurophysiol.* **25**: 721–731.

110. S. HAGIWARA and N. SAITO, 1959. Voltage-current relations in nerve cell membrane of *Onchidium verruculatum. J. Physiol.* **148**: 161–179.

111. S. HAGIWARA and I. TASAKI, 1958. A study of the mechaism of impulse transmission across the giant synapse of the squid. *J. Physiol.* **143**: 114–137.

112. S. HAGIWARA, A. WATANABE, and N. SAITO, 1959. Potential changes in syncytial neurons of lobster cardiac ganglion. *J. Neurophysiol.* **22**: 554–572.

113. E. J. HARRIS, 1957. Permeation and diffusion of K ions in frog muscle. *J. Gen. Physiol.* **41**: 169–195.

114. L. A. HEPPEL, 1940. The diffusion of radioactive sodium into the muscles of potassium-deprived rats. *Am. J. Physiol.* **128**: 449–454.

115. S. HESTRIN, 1950. Acylation reactions mediated by purified acetylcholine esterase II. *Biochim. Biophys. Acta* **4**: 310–321.

116. H. HIGMAN, T. R. PODLESKI, and E. BARTELS, 1963. Apparent dissociation constants between carbamylcholine, *d*-tubocurarine and the receptor. *Biochim. Biophys. Acta* **75**: 187–193.

117. A. V. HILL, 1933. The three phases of nerve heat production. *Proc. Roy. Soc. London B* **113**: 345–356.

118. D. K. HILL, 1950. Volume changes resulting from stimulation of a giant nerve fibre. *J. Physiol.* **111**: 304–327.

119. B. HILLE, 1967. The selective inhibition of delayed potassium currents in nerve by tetraethylammonium ion. *J. Gen. Physiol.* **50**: 1287–1302.

120. L. P. HINTERBUCHNER and D. NACHMANSOHN, 1960. Electrical activity evolked by a specific chemical reaction. *Biochim. Biophys. Acta* **44**: 554–560.

121. A. L. HODGKIN, 1951. The ionic basis of electrical activity in nerve and muscle. *Biol. Rev.* **26**: 339–409.

122. A. L. HODGKIN, 1958. Ionic movements and electrical activity in giant nerve fibres. *Proc. Roy. Soc. London B* **148**: 1–37.

123. A. L. HODGKIN, 1964. The ionic basis of nervous conduction. *Science* **145**: 1148–1154.

124. A. L. HODGKIN and P. HOROWICZ, 1960. The effect of sudden changes in ionic concentrations on the membrane potential of single muscle fibres. *J. Physiol.* **153**: 370–385.

125. A. L. HODGKIN and A. F. HUXLEY, 1952. Currents carried by sodium and potassium ions through the membrane of the giant axon of *Loligo*. *J. Physiol.* **116**: 449–472.

126. A. L. HODGKIN and A. F. HUXLEY, 1952. The components of membrane conductance in the giant axon of *Loligo*. *J. Physiol.* **116**: 473–496.

127. A. L. HODGKIN and A. F. HUXLEY, 1952. The dual effect of membrane potential on sodium conductance in the giant axon of *Loligo*. *J. Physiol.* **116**: 497–506.

128. A. L. HODGKIN and A. F. HUXLEY, 1952. A quantitative description of membrane current and its application to conduction and excitation in nerve. *J. Physiol.* **117**: 500–544.

129. A. L. HODGKIN, A. F. HUXLEY, and B. KATZ, 1952. Measurement of current-voltage relations in the membrane of the giant axon of *Loligo*. *J. Physiol.* **116**: 424–448.

130. A. L. HODGKIN and B. KATZ, 1949. The effect of sodium ions on the electrical activity of the giant axon of the squid. *J. Physiol.* **108**: 37–77.

131. A. L. HODGKIN and R. D. KEYNES, 1953. The mobility and diffusion coefficient of potassium in giant axons from *Sepia*. *J. Physiol.* **119**: 513–528.

132. A. L. HODGKIN and R. D. KEYNES, 1955. Active transport of cations in giant axons from *Sepia* and *Loligo*. *J. Physiol.* **128**: 28–60.

133. A. L. HODGKIN and R. D. KEYNES, 1956. Experiments on the injection of substances into squid giant axons dy means of a micro-syringe. *J. Physiol.* **131**: 592–616.

134. A. L. HODGKIN and R. D. KEYNES, 1958. The potassium permeability of a giant nerve fibre. *J. Physiol.* **128**: 61–88.

135. M. E. HOLMAN, 1958. Membrane potentials recorded with high-resistance microelectrodes; and the effects of changes in ionic environment on the electrical and mechanical activity of the smooth muscle of the *taenia coli* of the guinea pig. *J. Physiol.* **141**: 464–488.

136. G. HOYLE, 1962. Neuromuscular physiology. *Adv. Comp. Physiol. Biochem.* **1**: 177–216.

137. F. HUNEEUS-COX, 1964. Electrophoretic and immunological studies of squid axoplasm proteins. *Science* **143**: 1036–1037.

138. F. C. HUNEEUS-COX and H. L. FERNANDEZ, 1967. Effect of specific antibodies on the excitability of internally perfused squid axons. *J. Gen. Physiol.* **50**: 2407–2420.

139. F. HUNEEUS-COX, H. L. FERNANDEZ, and B. H. SMITH, 1966. Effects of redox and sulfhydral reagents on the bioelectric properties of the giant axon of the squid. *Biophys. J.* **6**: 675–689.

140. C. C. HUNT and A. TAKEUCHI, 1962. Responses of the nerve terminal of the Pacinian corpuscle. *J. Physiol.* **160**: 1–21.

141. W. P. HURLBUT, 1961. "Sodium and potassium distribution of nerve fibers," pp. 97–118, in: A. M. Shanes (ed.), *Biophysics of Physiological and Pharmacological Actions*. Washington D.C.: AAAS.

142. A. F. HUXLEY, 1959. Ionic movements during nerve activity. *Ann. N.Y. Acad. Sci.* **81**: 221–246.

143. A. F. Huxley, 1964. Excitation and conduction in nerve; quantitative analysis. *Science* **145**: 1154–1159.

144. A. F. Huxley and R. Stämpfli, 1951. Effect of potassium and sodium on resting and action potentials of single myelinated nerve fibres. *J. Physiol.* **112**: 496–508.

145 N. Ishiko and W. R. Loewenstein, 1961. Effects of temperature on the generator and action potentials of a sense organ. *J. Gen. Physiol.* **45**: 105–124.

146. M. Ito, P. G. Kostyuk, and T. Oshima, 1962. Further study on anion permeability in cat spinal motoneurones. *J. Physiol.* **164**: 150–156.

147. L. L. Iversen, 1965. The uptake of catecholamines at high perfusion concentrations in the rat isolated heart: a novel catecholamine uptake process. *Brit. J. Pharmacol.* **25**: 18–33.

148. T. L. Jahn, 1962. A theory of electronic conduction through membranes and of active transport of ions based on redox transmembrane potentials. *J. Theoret. Biol.* **2**: 129–138.

149. C. Y. Kao and H. Grundfest, 1957. Postsynaptic electrogenesis in septate giant axons. I. Earthworm median giant axon. *J. Neurophysiol.* **20**: 554–573.

150. A. Karlin and M. Winnik, 1968. Reduction and specific alkylation of the receptors for acetylcholine. *Proc. Natl. Acad. Sci. U. S.* **60**: 666–674.

151. B. Katz, 1950. Action potentials from a sensory nerve ending. *J. Physiol.* **111**: 248–260.

152. B. Katz, 1950. Depolarization of sensory terminals and the initiation of impulses in the muscle spindle. *J. Physiol.* **111**: 261–282.

153. B. Katz, 1962. The transmission of impulses from nerve to muscle, and the subcellular unit of synaptic action. *Proc. Roy. Soc. London B* **155**: 455–479.

154. B. Katz, 1966. *Nerve, Muscle, and Synapse.* New York: McGraw-Hill Book Co., 193 pp.

155. B. Katz and R. Miledi, 1965. "The quantal release of transmitter substance," pp. 118–125, in: D. R. Curtis and A. K. McIntyre (eds.), *Studies in Physiology.* New York: Springer Pub. Co.

156. G. A. Kerkut and R. C. Thomas, 1963. Acetylcholine and the spontaneous inhibitory postsynaptic potentials in the snail neurone. *Comp. Biochem. Physiol.* **8**: 39–45.

157. G. Kewitz, 1957. A specific antidote against lethal alkyl phosphate intoxication. III. Repair of chemical lesion. *Arch. Biochem. Biophys.* **66**: 263–270.

158. H. Kewitz and I. B. Wilson, 1956. A specific antidote against lethal alkyl phosphate intoxication. *Arch. Biochem. Biophys.* **60**: 261–263.

159. H. Kewitz, I. B. Wilson, and D. Nachmansohn, 1956. A specific antidote against lethal alkyl phosphate intoxication. II. Antidotal properties. *Arch. Biochem. Biophys.* **64**: 456–465.

160. R. D. Keynes, 1963. Chloride in the squid giant axon. *J. Physiol.* **169**: 690–705.

161. R. D. Keynes, 1963. Book review: G. N. Ling, A Physical Theory of the Living State. *Comp. Biochem. Physiol.* **9**: 261–262.

162. R. D. Keynes and G. W. Maisel, 1954. The energy requirement for sodium extrusion from a frog muscle. *Proc. Roy. Soc. London B* **142**: 383–392.

163. R. D. Keynes and H. Martins-Ferreira, 1953. Membrane potentials in the electroplates of the electric eel. *J. Physiol.* **119**: 315–351.

164. K. Koketsu, J. A. Cerf, and S. Nishi, 1959. Effect of quaternary ammonium ions on electrical activity of spinal ganglion cells in frog. *J. Neurophysiol.* **22**: 177–194.

165. K. Koketsu and Y. Kimura, 1960. The resting potential and intracellular potassium of skeletal muscle in frogs. *J. Cell. Comp. Physiol.* **55**: 239–244.

166. I. J. KOPIN, 1967. "The adrenergic synapse," pp. 427–432, in: G. C. Quarton, T. Melnechuk, and F. O. Schmitt (eds.), *The Neurosciences—A Study Program.* New York: Rockefeller University Press.

167. S. W. KUFFLER and C. EDWARDS, 1958. Mechanisms of gamma-aminobutyric acid (GABA) action and its relation to synaptic inhibition. *J. Neurophysiol.* **21**: 589–610.

168. H. C. LAWLER, 1961. Turnover time of acetylcholinesterase. *J. Biol. Chem.* **236**: 2296–2301.

169. K. LINDERSTRÖM-LANG, 1923. On the salting-out effect. *C. R. Trav. Lab. Carlsberg* (No. 4) **15**: 1–65.

170. K. LINDERSTRÖM-LANG, 1924. On the ionization of proteins. *C. R. Trav. Lab. Carlsberg* (No. 7) **15**: 1–30.

171. G. N. LING, 1960. The interpretation of selective ionic permeability and cellular potentials in terms of the fixed charge-induction hypothesis. *J. Gen. Physiol. Suppl.* **43**: 149–174.

172. G. N. LING, 1962. *A Physical Theory of the Living State: the Association-Induction Hypothesis.* New York: Blaisdell Pub. Co., 680 pp.

173. G. LING and R. W. GERARD, 1949. The normal membrane potential of frog sartorius fibers. *J. Cell. Comp. Physiol.* **34**: 383–396.

174. G. N. LING and R. W. GERARD, 1949. The membrane potential and metabolism of muscle fibers. *J. Cell. Comp. Physiol.* **34**: 413–438.

175. G. N. LING and R. W. GERARD, 1950. External potassium and the membrane potential of single muscle fibres. *Nature* **165**: 113–114.

176. G. N. LING and J. W. WOODBURY, 1949. Effect of temperature on the membrane potential of frog muscle fibers. *J. Cell. Comp. Physiol.* **34**: 407–412.

177. W. R. LOEWENSTEIN, 1959. The generation of electrical activity in a nerve ending. *Ann. N.Y. Acad. Sci.* **81**: 367–387.

178. W. R. LOEWENSTEIN, 1961. Excitation and inactivation in a receptor membrane. *Ann. N.Y. Acad. Sci.* **94**: 510–534.

179. W. R. LOEWENSTEIN and S. COHEN, 1959. I. After-effects of repetitive activity in a nerve ending. *J. Gen. Physiol.* **43**: 335–345.

180. W. R. LOEWENSTEIN and S. COHEN, 1959. II. Post-tetanic potentiation and depression of generator potential in a single non-myelinated nerve ending. *J. Gen. Physiol.* **43**: 347–376.

181. W. R. LOEWENSTEIN and N. ISHIKO, 1959. Properties of a receptor membrane. Spatial summation of electrical activity in a nerve ending. *Nature* **183**: 1724–1726.

182. W. R. LOEWENSTEIN and R. RATHKAMP, 1958. The sites for mechano-electric conversion in a Pacinian corpuscle. *J. Gen. Physiol.* **41**: 1245–1265.

183. W. R. LOEWENSTEIN, C. A. TERZUOLO, and Y. WASHIZU, 1963. Separation of transducer and impulse generating processes in sensory receptors. *Science* **142**: 1180–1181.

184. R. LORENTE DE NÓ, 1947. A Study of Nerve Physiology. *Studies from the Rockefeller Institute for Medical Research* **131**: 1–496; **132**: 1–548.

185. E. J. LUND, 1928. Relation between continuous bio-electric currents and cell respiration. *J. Exptl. Zool.* **51**: 265–307.

186. T. MAÉNO, 1959. Electrical characteristics and activation potential of *Bufo* eggs. *J. Gen. Physiol.* **43**: 139–157.

187. G. MARSH, 1935. Redox theory of membrane potentials. *Plant Physiol.* **10**: 681–697.

188. H. McLENNAN, 1970. *Synaptic Transmission.* Philadelphia: W. B. Saunders Co., 178 pp.

189. L. M. MICHAELIS and W. A. PERLZWEIG, 1927. Studies on permeability of membranes. I. Introduction and the diffusion of ions across the dried collodion membrane. *J. Gen. Physiol.* **10**: 575–598.

190. R. MILEDI, 1962. "Induction of receptors," pp. 220–235, in: J. L. Mongar and A. V. S. DeReuck (eds.), *Enzymes and Drug Action*. Boston: Little, Brown and Co.

191. R. MILEDI and C. R. SLATER, 1968. Electrophysiology and electron microscopy of rat neuromuscular junctions after nerve degeneration. *Proc. Roy. Soc. London B* **169**: 289–306.

192. J. W. MOORE, M. P. BLAUSTEIN, N. C. ANDERSON, and T. NARAHASHI, 1967. Basis of tetrodotoxin's selectivity in blockage of squid axons. *J. Gen. Physiol.* **50**: 1401–1411.

193. J. W. MOORE and K. S. COLE, 1963. "Voltage clamp techniques," pp. 262–321, in: W. L. Nastuk (ed.), *Physical Techniques in Biological Research, Vol. VI, Electrophysiological Methods, Part B*. New York: Academic Press.

194. J. W. MOORE, T. NARAHASHI, and T. I. SHAW, 1967. An upper limit to the number of sodium channels in nerve membrane? *J. Physiol.* **188**: 99–105.

195. L. J. MULLINS, 1959. The penetration of some cations into muscle. *J. Gen. Physiol.* **42**: 817–829.

196. D. NACHMANSOHN, 1959. *Chemcial and Molecular Basis of Nerve Activity*. New York: Academic Press, 235 pp.

197. D. NACHMANSOHN, 1961. Chemical factors controlling nerve activity. *Science* **134**: 1962–1968.

198. D. NACHMANSOHN, 1966. Chemical control of the permeability cycle in excitable membranes during electrical activity. *Ann. N.Y. Acad. Sci.* **137**: 877–900.

199. D. NACHMANSOHN, 1970. Proteins in excitable membranes. *Science* **160**: 1059–1066.

200. D. NACHMANSOHN and I. B. WILSON, 1951. The enzymatic hydrolysis and synthesis of acetylcholine. *Adv. Enzymol.* **12**: 259–339.

201. S. NAKAJIMA, 1966. Analysis of K inactivation and TEA action in the supramedullary cells of puffer. *J. Gen. Physiol.* **49**: 629–640.

202. S. NAKAJIMA and K. KUSANO, 1966. Behavior of delayed current under voltage clamp in the supramedullary neurons of puffer. *J. Gen. Physiol.* **49**: 613–628.

203. Y. NAKAMURA, S. NAKAJIMA, and H. GRUNDFEST, 1965. The action of tetrodotoxin on electrogenic components of squid giant axons. *J. Gen. Physiol.* **48**: 985–996.

204. Y. NAKAMURA, S. NAKAJIMA, and H. GRUNDFEST, 1965. Analysis of spike electrogenesis and depolarizing K inactivation in electroplaques of *Electrophorus electricus* L. *J. Gen. Physiol.* **49**: 321–349.

205. T. NAMBA and K. HIRAKI, 1958. PAM (pyridine-2-aldoxime methiodide) therapy for alkyl phosphate poisoning. *J. Am. Med. Assoc.* **166**: 1834–1839.

206. T. NARAHASHI, 1963. Dependence of resting and action potentials on internal potassium in perfused squid axons. *J. Physiol.* **169**: 91–115.

207. T. NARAHASHI, T. DEGUCHI, N. URAKAWA, and Y. OHKUBO, 1960. Stabilization and rectification of muscle fiber membrane by tetrodotoxin. *Am. J. Physiol.* **198**: 934–938.

208. T. NARAHASHI, J. W. MOORE, and W. R. SCOTT, 1964. Tetrodotoxin blockage of sodium conductance increase in lobster giant axons. *J. Gen. Physiol.* **47**: 965–974.

209. D. NOBEL, 1966. Applications of Hodgkin-Huxley equations to excitable tissue. *Physiol. Rev.* **46**: 1–50.

210. W. J. V. OSTERHOUT, 1933. The electrical behavior of large plant cells. *Cold Spring Harbor Symp. Quant. Biol.* **1**: 125–130.

211. M. OZEKI, A. R. FREEMAN, and H. GRUNDFEST, 1966. The membrane components

of crustacean neuromuscular systems. I. Immunity of different electrogenic components to tetrodotoxin and saxitoxin. *J. Gen. Physiol.* **49**: 1319–1334.

212. M. OZEKI, A. R. FREEMAN, and H. GRUNDFEST, 1966. The membrane components of crustacean neuromuscular systems. II. Analysis of interaction among the electrogenic components. *J. Gen. Physiol.* **49**: 1335–1349.

213. M. OZEKI and M. SATO, 1965. Changes in the membrane potential and membrane conductance associated with a sustained compression of the non-myelinated nerve terminal in Pacinian corpuscles. *J. Physiol.* **180**: 186–208.

214. W. B. PATTERSON and D. STETTEN, 1949. A study of gastric HCl formation. *Science* **109**: 256–258.

215. D. C. PEASE and T. A. QUILLIAM, 1957. Electron microscopy of the Pacinian corpuscle. *J. Biophys. Biochem. Cytol.* **3**: 331–342.

216. J. W. S. PRINGLE, 1962. Prologue: the input element. *Symp. Exptl. Biol.* **16**: 1–11.

217. T. A. QUILLIAM and M. SATO, 1955. The distribution of myelin on nerve fibres from Pacinian corpuscles. *J. Physiol.* **129**: 167–176.

218. W. S. REHM and W. H. DENNIS, 1957. "A discussion of theories of hydrochloric acid formation in the light of electrophysiological findings," pp. 303–330, in: Q. R. Murphy (ed.), *Metabolic Aspects of Transport across Cell Membranes*. Madison: University of Wisconsin Press.

219. P. ROSENBERG and S. EHRENPREIS, 1961. Reversible block of axonal conduction by curare after treatment with cobra venom. *Biochem. Pharmacol.* **8**: 192–206.

220. P. ROSENBERG and H. HIGMAN, 1960. An improved isolated single electroplax preparation. II. Compounds acting on the conducting membrane. *Biochim. Biophys. Acta* **45**: 348–354.

221. M. A. ROTHENBERG, D. B. SPRINSON, and D. NACHMANSOHN, 1948. Site of action of acetylcholine. *J. Neurophysiol.* **11**: 111–116.

222. M. SATO, 1961. Response of the Pacinian corpuscle to sinusoidal vibration. *J. Physiol.* **159**: 391–409.

223. M. SATO and M. OZEKI, 1963. Response of the non-myelinated nerve terminal in Pacinian corpuscle to mechanical and antidromic stimulation and the effect of procaine, choline, and cooling. *Jap. J. Physiol.* **13**: 564–582.

224. F. O. SCHMITT, 1966. Molecular and ultrastructural correlates of function in neurons, neuronal nets, and the brain. *Naturwissenschaften* **53**: 71–79.

225. F. O. SCHMITT and P. F. DAVISON, 1965. Role of protein in neural function, an essay. *Neurosci. Res. Prog. Bull.* **3**: 55–76.

226. E. SCHOFFENIELS and D. NACHMANSOHN, 1959. Ion movements studied with single isolated electroplax. *Ann. N.Y. Acad. Sci.* **81**: 285–306.

227. A. M. SHANES, 1958. Electrochemical aspects of physiological and pharmacological action in excitable cells. Part I. The resting cell and its alteration by extrinsic factors. *Pharmacol. Rev.* **10**: 1–164.

228. A. M. SHANES, 1958. Electrochemical aspects of physiological and pharmacological action in excitable cells. Part II. The action potential and excitation. *Pharmacol. Rev.* **10**: 165–273.

229. A. M. SHANES, W. H. FREYGANG, H. GRUNDFEST, and E. AMATNIEK, 1959. Anesthetic and calcium action in the voltage clamped squid giant axon. *J. Gen. Physiol.* **42**: 793–802.

230. F. H. SHAW, S. E. SIMON, and B. M. JOHNSTONE, 1956. The non-correlation of bio-electric potentials with ionic gradients. *J. Gen. Physiol.* **40**: 1–17.

231. T. Shedlovsky, 1952. Electromotive force from proton transfer reactions: a model for bioelectric phenomena. *Cold Spring Harbor Symp. Quant. Biol.* **17**: 97–102.

232. C. S. Sherrington, 1906. *The Integrative Action of the Nervous System.* New Haven: Yale University Press.

233. S. E. Simon, B. M. Johnstone, K. H. Sharkley, and F. H. Shaw, 1959. Muscle: a three phase system. The partition of monovalent ions across the cell membrane. *J. Gen Physiol.* **43**: 55–79.

234. K. Sollner, I. Abrams, and C. W. Carr, 1941. The structure of the collodion membrane and its electrical behavior. I. The behavior and properties of commercial collodion. *J. Gen. Physiol.* **24**: 467–482.

235. K. Sollner, I. Abrams, and C. W. Carr, 1941. The structure of the collodion membrane and its electrical behavior. II. The activated collodion membrane. *J. Gen. Physiol.* **25**: 7–27.

236. R. Stämpfli, 1954. A new method for measuring membrane potentials with external electrodes. *Experientia* **10**: 508–509.

237. M. Takata, J. W. Moore, C. Y. Koa, and F. A. Fuhrman, 1966. Blockage of sodium conductance increase in lobster giant axon by tarichatoxin (tetrodotoxin). *J. Gen. Physiol.* **49**: 977–988.

238. A. Takeuchi and N. Takeuchi, 1960. On the permeability of the end-plate membrane during the action of transmitter. *J. Physiol.* **154**: 52–67.

239. A. Takeuchi and N. Takeuchi, 1960. Further analysis of relationship between end-plate potential and end-plate current. *J. Neurophysiol.* **23**: 397–402.

240. I. Tasaki, 1968. *Nerve Excitation. A Macromolecular Approach.* Springfield, Ill.: Charles C Thomas, Pub., 201 pp.

241. I. Tasaki and I. Singer, 1966. Membrane macromolecules and nerve excitability: a physico-chemical interpretation of excitation in squid giant axon. *Ann. N.Y. Acad. Sci.* **137**: 792–806.

242. I. Tasaki and I. Singer, 1968. Some problems involved in electric measurements of biological systems. *Ann. N.Y. Acad. Sci.* **148**: 36–53.

243. I. Tasaki, I. Singer, and T. Takenaka, 1965. Effects of internal and external ionic environment on excitability of squid giant axon. A macromolecular approach. *J. Gen. Physiol.* **48**: 1095–1123.

244. I. Tasaki, T. Teorell, and C. S. Spyropoulos, 1961. Movement of radioactive tracers across squid axon membrane. *Am. J. Physiol.* **200**: 11–22.

245. L. Tauc, 1958. Processus post-synaptique d'excitation et d'inhibition dans le soma neuronique de l'Aplysie et de l'Escargot. *Arch. Ital. Biol.* **96**: 78–110.

246. L. Tauc and G. M. Hughes, 1963. Modes of initiation and propagation of spikes in the branching axons of molluscan central neurons. *J. Gen. Physiol.* **46**: 533–549.

247. R. E. Taylor, 1959. Effect of procaine on electrical properties of squid axon membrane. *Am. J. Physiol.* **196**: 1071–1078.

248. R. E. Taylor, 1963. "Cable theory," pp. 219–262, in: W. L. Nastuk (ed.), *Physical Techniques in Biological Research, Vol. VI, Electrophysiological Methods, Part B.* New York: Academic Press.

249. T. Teorell, 1953. Transport processes and electrical phenomena in ionic membrane. *Prog. Biophys.* **3**: 305–369.

250. T. Teorell, 1956. Transport phenomena in membrane. *Disc. Faraday Soc.* **21**: 9–26.

251. W. TRAUTWEIN and J. DUDEL, 1958. Zum Mechanisms der Membranwirkung des Acetylcholin an der Herzmuskelfaser. *Pflüger's Arch. Ges. Physiol.* **266**: 324–334.

252. W. TRAUTWEIN, S. W. KUFFLER, and C. EDWARDS, 1956. Changes in membrane characteristics of heart muscle during inhibition. *J. Gen. Physiol.* **40**: 135–145.

253. R. R. WALSH and S. E. DEAL, 1959. Reversible conduction block produced by lipid-insoluble quaternary ammonium ions in cetyltrimethylammonium bromide-treated nerves. *Am. J. Physiol.* **197**: 547–550.

254. A. WATANABE and H. GRUNDFEST, 1961. Impulse propagation at the septal and commissural junctions of crayfish lateral giant axons. *J. Gen. Physiol.* **45**: 267–308.

255. R. WERMAN, 1966. Criteria for identification of a central nervous system transmitter. *Comp. Biochem. Physiol.* **18**: 745–766.

256. R. WERMAN and H. GRUNDFEST, 1961. Graded and all-or-none electrogenesis in arthropod muscle. II. The effect of alkali-earth and onium ions on lobster muscle fibers. *J. Gen. Physiol.* **44**: 997–1027.

257. R. WERMAN, F. V. MCCANN, and H. GRUNDFEST, 1961. Graded and all-or-none electrogenesis in arthropod muscle. I. The effects of alkali-earth cations on the neuromuscular system of *Romalea microptera. J. Gen. Physiol.* **44**: 979–995.

258. V. P. WHITTAKER and E. G. GRAY, 1962. The synapse: biology and morphology. *Brit. Med. Bull.* **18**: 223–228.

259. I. B. WILSON, 1952. Acetylcholinesterase. XII. Further studies of binding forces. *J. Biol. Chem.* **197**: 215–225.

260. I. B. WILSON, 1952. Acetylcholinesterase. XIII. Reactivation of alkyl phosphate-inhibited enzyme. *J. Biol. Chem.* **199**: 113–120.

261. I. B. WILSON, F. BERGMANN, and D. NACHMANSOHN, 1950. Acetylcholinesterase. X. Mechanism of the catalysis of acylation reactions. *J. Biol. Chem.* **186**: 781–790.

262. I. B. WILSON and M. COHEN, 1953. The essentiality of acetylcholinesterase in conduction. *Biochim. Biophys. Acta* **11**: 147–156.

263. I. B. WILSON and S. GINSBURG, 1955. A powerful reactivator of alkyl phosphate-inhibited acetylcholinesterase. *Biochim. Biophys. Acta* **18**: 168–170.

264. I. B. WILSON, S. GINSBURG, and C. QUAN, 1958. Molecular complementariness as basis for reactivation of alkyl phosphate-inhibited enzyme. *Arch. Biochem. Biophys.* **77**: 286–296.

265. R. W. G. WYCKOFF and J. Z. YOUNG, 1956. The motoneurone surface. *Proc. Roy. Soc. London B* **144**: 440–450.

ARTIFICIAL
MEMBRANE SYSTEMS

chapter five

To simplify the study of any complex system, a useful approach is to control vigorously many of the parameters or to use only a small portion of the events. Another approach that oscillates between being helpful and being misleading is to build a model system which possesses many of the features but which is easier to manipulate, control, measure, and—most important—understand what is happening. Model systems for studies on the properties of biological membranes have yielded intriguing results.

One of the early models for biological membranes was proposed by Lillie [45] who showed that an iron wire can be made to exhibit some of the properties of a propagating impulse and, when covered with regular segments of glass tubing, can even show iterative behavior similar to myelinated neurons. Teorell [83] and Franck [26] obtained slow oscillatory electrical signals from glass membranes. Other model systems were developed to promote understanding of the mechanisms of membrane transport and permeation using guaiacol membranes, collodion membranes, and various fixed-charge or ion-exchange membranes.

In more recent years, partly as a result of the lipid bilayer postulate in the Davson-Danielli concept of membrane structure investigators have attended to lipid membrane systems of less that 100 Å thickness [68, 69]. Mueller and Rudin achieved electrical excitability in their bilipid layers but only when certain additives were present in the system, such as alamethicin and other cyclic antibiotics or the polymeric excitability inducing materials (EIM) extracted from living membranes. Many of the physical properties of this class of artificial membranes and their permeabilities have been measured in a number of different laboratories.

Protein is also beginning to receive proper attention. For example, Shashoua [76, 77] obtained spikes from lipid-free membranes made from a variety of proteins, polynucleic acids, and synthetic polymers. In addition to the experiments of Mueller and Rudin mentioned above, other investigators have found that lipid bilayers, when treated with bacterial protein, protamine, and the like, display electrical phenomena that, when the proper additives are selected and adjusted, mimic the behavior of naturally excitable membranes: action potentials, resting potentials, thresholds, delayed rectification, spontaneous and rhythmic firing, and reversible block by anesthetics [42]. A protein such as albumen can increase the conductance in artificial lipid membranes but yields no selectivity between potassium and sodium [98] unlike the selectivity conferred by the polypeptide antibiotics alamethicin and valinomycin [54].

Still other investigators have presented theoretical models for the operation of biological membranes [9, 13, 47, 59, 100–103]. The number of theoretical papers published conveys both the intensity of serious interest and the need for far more experimental data.

5.1 The Lipid Bilayer

The involved electrochemical functions of living plasma membrane obviously result from their physicochemical microstructure. The accumulating evidence favors the existence of many kinds of biological membranes; the different molecular structures perform the different functions. Some of this evidence is contrary to the Davson-Danielli concept and contrary to the uniformity implied in Robertson's early descriptions of his unit-membrane hypothesis. Complex membranes composed of materials of different properties can be assembled into either lamellar or mosaic structures. The present view is that the plasma membrane contains mosaic structures, although the extent and variation in the patchiness is still argued. Consequently, the lipid bilayer, so important in the early descriptions of membrane structure, is no longer expected to be ubiquitous and to possess uniform properties. The phospholipid bilayer is just one component—but still an important one—among the many in different types of biological membranes. Many of the properties of living membranes can be explained by a phospholipid bilayer so that, in spite of recent difficulties, it probably represents a major portion of most membranes. The other physiological properties may belong to other components, e.g., proteins in the mosaic. The approach to the study of membrane construction recommended by many investigators is to use a phospholipid bilayer as a matrix for an artificial membrane system and then to add materials (proteins, etc.) in order to add specific properties that parallel the biological ones (e.g., facilitated diffusion, glucose transport, action potentials).

5.2 Artificial Lipid
Membranes

Most of the artificial lipid membranes have been formed from solutions containing different phospholipids and an excess of aliphatic hydrocarbon or DL-α-tocopherol dissolved in chloroform-methanol mixtures [37, 57] or the hydrocarbon may act as the solvent [30]. The phospholipid used in these model systems have included component mixtures from brain white matter [57], red cell membranes [1], or purified preparations of lecithin [30, 37], sphingomyelin [52], phosphatidyl inositol [44], and synthetic dioleolecithin [42]. The films can be made more stable by adding cholesterol to the membrane-forming solutions, and amphipathic molcules (glycerol distearate, oxidized cholesterol, etc.) can be substituted for phospholipid [89].

A number of investigators have simply painted the membrane-forming solution across an opening (between 0.5 and 5 mm^2) in a plastic (polyethylene or teflon) partition separating identical salt solutions [57, 88]. The formed film thins to a lipid bilayer—the "black" membrane so named because of its optical properties—in a process analogous to the formation and thinning of soap films. These studies have varied many pertinent conditions: the concentration in the bathing solutions of NaCl and KCl ($0.001-1M$), other ions (Mg, Ca), the pH ($3.0-9.5$), and the temperature ($20-45°C$). The membranes formed remain fairly stable for hours or days. The exclusion of oxygen improves the life of lecithin-containing films, which implies that oxidation of lipid leads to film rupture. It has not been possible, however, to prepare phospholipid membranes free from traces of heavy metals (e.g., Ca, Mg, Fe) and these may, in fact, be important as counterions to hold the phospholipid molecules together [85].

It is generally assumed that these black membranes are bilayer structures of the Davson-Danielli type; the aliphatic hydrocarbon chains lie between and parallel to the phospholipid chains. Capacitance [30], optical properties [35, 87], and electron micrographs [34, 58] all measure a film thickness near 70 Å, which is consistent with the space occupied by two extended phospholipid molecules. For example, phosphatidylcholine molecules with 18-carbon chains are 34 Å long [42]. Phospholipids can form colloidal structures in water, however. According to X-ray data [46], these structures are liquid crystals and may be mosaic structures in the form of hexagonal arrays of cylinders (either hydrocarbon chains fill the interior of the cylinders and water fills the outside or water fills the cylinders and hydrocarbon fills the gaps) or they may be lamellar structures, normally bilayers, of an alternating sequence of planar lipids and water. Transitions occur that allow lipid structures to open and close randomly to form pores, and in one of the possible forms, narrow water channels lined with polar groups exist. According to Overbeek [62], however, a priori energy considerations predict that hydrocarbon is squeezed out of the thick films until the limiting bilayer is reached.

Other preparations have been developed to obtain the larger areas of membrane that permit more precise permeability and flux measurements. Van den Berg [104] built larger membranes but they contained big islands of thick lipid, were fragile, and required careful shock mounting. Thompson [85] and his colleagues developed a technique to form large spherical vesicles that were entirely black membrane except for the cap at the top, which contained the excess lipid. Mueller and Rudin [54] made spherules, bounded by a 60–100 Å thick membrane, in large numbers by shaking a 0.1–1 % solution of total brain or beef heart lipids dissolved in hexane or octane with a small volume of protein. Bangham, Standish, and Watkins [5] formed myelinic structures by allowing dry phospholipids to swell in aqueous salt solutions with gentle agitation. The heterogeneous structures formed consisted of concentric bimolecular layers of lipid, each separated by an aqueous compartment. Papahadjopoulos and Miller [64] ultrasonicated such suspensions to produce a rather uniform population of sac-like vesicles, approximately 500 Å in diameter, each surrounded by one or more bilayers.

Ambiguities occur in the interpretations of data obtained from all of these systems. It is difficult to ascribe a specific property to the thin (black) region of a model film since areas of thick residual lipid also are present, especially at the edges. The most convincing evidence that a given phenomenon occurs in black membrane is that it is not obtainable in films prior to thinning or that, in permeability measurements, a direct proportionality with the area exists. Moreover, it is not easy to identify exact ratios of components in black membrane systems. The ratios can be quite different from those in the membrane-forming solutions; hence, the precise composition of any of the black membranes is just not known. In contrast, the membrane composition is known in lipid dispersion systems. The dispersion systems possess a multi-compartment complexity, however, that makes interpretation of solute-flux data difficult. Although considerable evidence has been presented to indicate that the fluxes occur across lamellae [65], the possibility exists for fluxes of materials along the aqueous channels between lamellae. Moreover, estimates of surface areas for disperion systems are difficult [6], which complicates the determination of the actual permeability values. As mentioned later, the two model systems—films and dispersions—also show differences in permeability properties although, according to resistance measurements, both are essentially ion-impermeable systems. The films have aliphatic hydrocarbons or DL-α-tocopherol in addition to phospholipid; so membranes formed from phospholipids alone may contain channels that can be occluded by neutral lipid or hydrocarbon, and dispersions formed with equal molar ratios of phosphatidylcholine and cholesterol have been found to be less permeable to chloride than dispersions with phosphatidylcholine alone [65].

5.3 Properties
of Lipid Films

Black membranes formed from phospholipids possess a surface tension near 1 dyne/cm [84, 85, 89, 90], which is equal to that of living membranes [32]. Originally, the low surface tension of plasma membranes was attributed to the protein adsorbed on the surface but, in reality, a thin phospholipid film alone possesses this characteristic. The electrical properties of black films are equivalent to a parallel circuit of resistance ($R = 10^6 - 10^9$ ohm/cm^2) [30, 37, 58] and capacitance ($C = 0.3 - 1.3\ \mu F/cm^2$) [31, 58]. During the application of an electric field across the film, the resistance remains linear up to a potential difference of 150 mV and above 200 mV the films generally break [37, 58]. These properties are consistent with a thin (about 50 Å) hydrocarbon film with a dielectric constant of approximately 2; rupture at 200 mV is attributed to dielectric breakdown in an intense electric field (about 4×10^5 V/cm). The large resistance results from a virtual impermeability to ions.

These films generally show a significant permeability to water. A water permeability coefficient of 0.5×10^{-3} to 1×10^{-2} cm/sec has been measured by both isotopic water flux and by osmotic flow experiments [29, 36, 86]. Early reports of differences between these two types of measurements came from underestimates in the water tracer experiments due to unstirred layers [12]. Water appears to traverse the films simply by dissoving into and diffusing through the hydrocarbon phase [28]. No aqueous pores need to be postulated. An increase in the cholesterol concentration in the membrane-forming solution reduces the water permeability probably by increasing the viscosity of the hydrocarbon phase and, therefore, lowering the diffusion coefficient of water within it [24].

The water permeability of the films is about the same as in biological membranes but the electrical resistance of the films is too big. The implication of this observation is that the magnitude of water permeability of biological membranes is simply a consequence of the presence of a thin lipid film similar to that formed in the model systems—the models provide no evidence for pores—whereas permeability to ions and solutes which are not soluble in lipids (e.g., sugar) and which do not penetrate the lipid films well [12] is a consequence of modifications in this structure by other components. The importance of nonlipid components is also suggested by the experiments that show that protein, adsorbed onto phospholipid membranes, lowers the electrical resistance by a factor of 1000 or so but has no effect on water permeability [12, 25].

The black phospholipid films generally show a greater permeability to cations than to anions [1] but membranes formed in dispersion systems are almost thirty times more permeable to chloride ions than to sodium or potassium [5, 63]. In the latter systems, acidic lipids, phosphatidylinositol, and phosphatidic acid are slightly more permeable to cations than to anions [65]. Cation permeability can be reduced by including in the dispersions some long-chain cations (e.g., steryl amine) and increased by long-chain anions (e.g., dicetylphosphoric acid) [5]. The lipid dispersions are highly permeable to erythritol, urea, and glycerol [6, 67] but the thin lipid films are virtually impermeable to them [68]. The addition of calcium causes phosphatidyl-serine crystals to increase their ability to discriminate between potassium and sodium [63]. Calcium does not change the amount of water inside the particles but does decrease the amount of water between particles presumably by altering the surface charge. The lipid particles behave as semipermeable membranes toward water; they swell and shrink in various concentrations of sucrose [67].

The behavior of phospholipid films is governed by London-van der Waal's interactions, configurational and entropy effects, and alterations in the structure of water adjacent to the lipid layers. These factors depend on chain length and degree of unsaturation of the lipid molecules [20, 75]. Cholesterol [3], calcium [65], local anesthetics [4], steroids [7], and polyene antibiotics [74, 107] all produce profound effects on the ion permeabilities of these systems. The macrocyclic antibiotics, such as valinomycin, enniatins, nactins, and their analogs, form a family of compounds that make membranes, biological and artificial, more permeable to alkali metal ions, especially to potassium. These compounds, which are all macrocyclic structures of peptides, depsides, or depsipeptides, complex with the cation by ion dipole interaction with the ordered system of polar groups within the central cavity of the compound [79]. The cation is thought to be sequestered within the molecular cage of the antibiotic so that the polar ion is covered by a nonpolar coat. This complex can dissolve in the hydrophobic regions of membranes. The chemistry of these antibiotics, then, explains their ability to induce selective cation permeability changes in biological and model membrane systems. The polyene antibiotics (e.g., filipin, nistatin, amphotericins) augment membrane permeability in a nonspecific fashion; these compounds do not interact with the permeating ions or molecules [79]. These antibiotics appear to produce injury to the membrane so that the type of molecules capable of passing through the antibiotic-treated membrane is determined solely by the magnitude of the holes or pores formed in the membrane, which is, then, functioning as a molecular sieve.

5.4 Ion-Exchange Membranes

A general impression of a biological membrane is that it is an ultrathin structure composed of a lipid bilayer with polar head groups oriented toward the aqueous phases and covered by adherent layers of protein or polysaccharide or both. In view of the unsaturation of a number of the fatty acids thought to be present, the hydrocarbon region of this structure is likely to be more [80] or less [66] liquid. The physical mechanisms by which ions cross these membranes are yet to be satisfactorily described [14] but the postulated mechanisms include transmembrane pores lined with fixed charges [40, 82], lipid-soluble carrier molecules [108], diffusion through a homogeneous dielectric substance but with rate limiting gating mechanisms at the surfaces [27], and translocation of vesicles [8]. In general, the possibilities for ion migration across a barrier can be classified as a collision mechanism, a jump mechanism from site to site, a carrier mechanism, and a solvent-drag mechanism. At the boundaries, ions may be distributed according to Gibbs-Donnan equilibria, specific ion-exchange equilibria, and partition coefficients and they may be affected by unstirred layers, high-energy barriers, and particular macromolecular structures. According to Eisenman, Sandblom, and Walker [22], a homogeneous membrane may be classified according to whether it contains ion-exchange sites or it is site-free. If a membrane is an ion exchanger, it may be further categorized according to whether the sites are fixed within the membrane or mobile and free to migrate, and according to whether the counterions are associated with or dissociated from the ion-exchange sites.

It is not at all unusual for workers to consider biological membranes as ion-exchange membranes as evidenced by the vigor of Tasaki's [81] arguments. Cole [15] notes that biological membranes have more complex behavior than can be accounted for by simple electrodiffusion processes; the Nernst-Planck-Einstein electrodiffusion equations are not sufficient to explain the electrical characteristics of the squid giant axon membrane. At the same time, Cole is not willing to abandon this approach until a clearly better substitute is presented. Ion-exchange membranes have yet to be so carefully examined that their suitability can be assessed as that "clearly better substitute" for the electrodiffusion models described by Cole above. In their extensive review, Eisenman, Sandblom, and Walker [22] relate their studies on the structure and function of ion-exchange membranes to analyses of biological mechanisms for ion permeation. If biological membranes are continuous lamellar structures, as their cross sections seen in electron micrographs would have you believe, ion permeation is likely to occur either by diffusion through a site-free interior or by a mechanism involving mobile sites for ion exchange and associated counterions. Lipid-soluble polar molecules serve as mobile carriers. A system with mobile sites but dissociated counterions appears unlikely because of the low dielectric constant of the lipid interior. If biological membranes have a micellar substructure with transverse pores lined with polar head groups, however, the permeation mechanism may be a fixed-site ion-exchange system with either associated or dissociated counterions. Solid ion exchangers are models for pores lined by fixed charges and liquid ion exchangers are models for lipid-soluble carriers.

Functionally, the sequence of cation selectivity observed in biological systems, as well as their magnitudes, are consistent with those known to exist in narrow-pore, fixed-site systems [22] but they may also be consistent with the selectivity properties of mobile-site systems, which are yet to be sufficiently well characterized for proper comparison. The large values for the temperature coefficients of ionic mobilities and ionic concentrations in biological membranes and the low mobilities (10^{-8} and 10^{-9} cm^2/sec V for Na and K, respectively [14]) are not consistent with a fixed-site system with dissociated counterions but they are consistent with the possibilities of systems with counterions associated with either fixed or mobile ion-exchange sites and with counterions dissociated from mobile sites.

The electrical properties of model ion-exchange membranes show many interesting correlations with the behavior of living membranes. Metastability of electrical characteristics has been observed in ion-exchange membranes with fixed sites and associated counterions [22]. This implies that metastable characteristics may be general features of simple ion-exchange membranes so that highly specialized structures may not be required for their existence. The steady state current-voltage characteristics of fixed-site and mobile-site membranes are significantly different: In high electrical fields, fixed-site systems approach a finite limiting conductance, whereas mobile-site systems approach a finite limiting current [16–18, 70, 71, 105, 106]. Ion-exchange membranes with charged pores yield kinetics for ion movements that resemble the carrier system proposed for biological membranes including saturation phenomena, competition, specificity, and poisoning [78], but a number of discrepancies have been observed between the electrical properties and the fluxes of biological membranes. For example, exchange diffusion, in which measured ion fluxes are larger than those expected on the basis of electrical conductance [99, 109], can be explained in a mobile-site system with associated counterions but not in either a mobile-site system with dissociated counterions or a simple fixed-site membrane. In contrast, the electrical conductances of some plant cells have been found to be some ten times larger than that predicted by flux measurements [19], an effect that can be explained in a fixed-site ion-exchange system with associated counterions. Increased association of the counterions to the ion-exchange sites makes a mobile-site membrane progressively more difficult to distinguish from a fixed-site membrane on the basis of its purely electrical properties but progressively easier to distinguish on the basis of a comparison of electrical and flux properties [22].

Membranes with fixed or mobile ion-exchange sites and associated counterions can have more complex properties than the idealized models have previously suggested. This means that considerably more must be learned about the details of ion-permeation mechanisms in many ion-exchange membranes before all the properties of these membranes can be defined. In any case, the properties of such ion-exchange membranes are now known to be sufficiently complex to offer models for biological phenomena such as excitability and diverse types of permeation.

5.5 A Hypothesis for Excitation—the Work of Tobias

Prior to his untimely death, Julian Tobias and his co-workers studied molecular events in the excitable systems of cellular surfaces and especially the role of calcium in this activity. Their initial approaches with chemically modified axons led to their later approaches with artificial lipid membranes.

Work during the 1950's on various optical and mechanical correlates of nerve activity [91–97] led Tobias to look for the accompanying alterations in molecular structure or changes in state. The experiments were only able to suggest reversible changes in hydration and spatial dispersion of unknown surface molecules, however. A second approach tried by Tobias [92, 93] was enzymatic digestion. The integrity of the lipid components but not of the protein components is required to retain axon excitability as well as other parameters (e.g., membrane resistance and capacitance, resting and action potentials, threshold, and conduction velocity). Axons function normally despite digestion of membrane proteins but axons loose their excitability when treated with phospholipases or digitonin, which disrupt phospholipids and cholesterol, respectively.

In order to examine pertinent properties of the lipid components of biological membranes more directly, Tobias [95] developed a phospholipid-cholesterol membrane by impregnating Millipore filters with animal cephalin (primarily phosphatidylserine) and cholesterol. The electrical conductance, water content, and water permeability of these membranes can be lowered by calcium or raised by potassium [43, 95], which mimics behavior often attributed to living membranes. There exists ample evidence to indicate that this model membrane contains negatively charged fixed sites and that the predominant cation within the membrane determines its electrical resistance [43, 60, 95]. This membrane model, however, is not able to distinguish between K and Na. The conclusion is that the model membrane is a cation exchanger [49] since (1) the ionic distribution between the membrane and its bathing solutions can be described in terms of either a Gibbs-Donnan equilibrium, in which the membrane salts are completely dissociated, or the formation of a weak salt of fixed anion and counter cation; (2) the resistance, which can be modified by externally applied salt, is actually dependent on the amount of Ca and K within the membrane; and (3) the permeability to NaCl and KCl is decreased by Ca. Fragments of red blood cell ghosts exhibit comparable ion uptake and Ca-dependent water uptake suggesting a similar ion exchanger nature [11, 49, 72, 73].

These data caused Tobias [92, 94, 96] to speculate upon the implications. The resistance of these membranes increases as calcium displaces potassium on lipid acidic groups, an association dependent on pH. Since calcium dehydrates but potassium rehydrates the membrane barrier, the resultant hydration changes may similarly occur in plasma membranes to explain the light scattering, dimensional, rigidity, and surface contour changes seen in nerve fibers during activity. The excitability system in the cell surface contains calcium associated with phospholipid acid groups, for example, the phosphate oxygen in cephalin. According to Tobias, the fundamental event in excitation, then, is the displacement of calcium from negative binding sites by potassium; the exchange of K for Ca may be forced by the exciting electrical potential. The resulting hydration of lipid permits a tangential spreading of its protein companion to produce the changes in the transmembrane ion fluxes that are characteristic of nerve activity.

5.6 Electrical Excitability of
Artificial Lipid Membranes

Mueller and Rudin [56, 57] built stable bimolecular lipid membranes from solubilized lipids extracted from brain white matter. In aqueous solutions, these experimental membranes possess many of the mechanical and electrical properties of living membranes. Mueller [55] actually suggested that they were reconstituted cellular membranes because they were constructed from extracted materials and because their theoretical, mechanical, electron microscopic, and electrical properties were like those attributed to biological membranes.

What makes these preparations different from the artificial lipid systems described earlier is that Mueller and Rudin [56] permitted them to adsorb appropriate, although unidentified, macromolecules obtained from biological

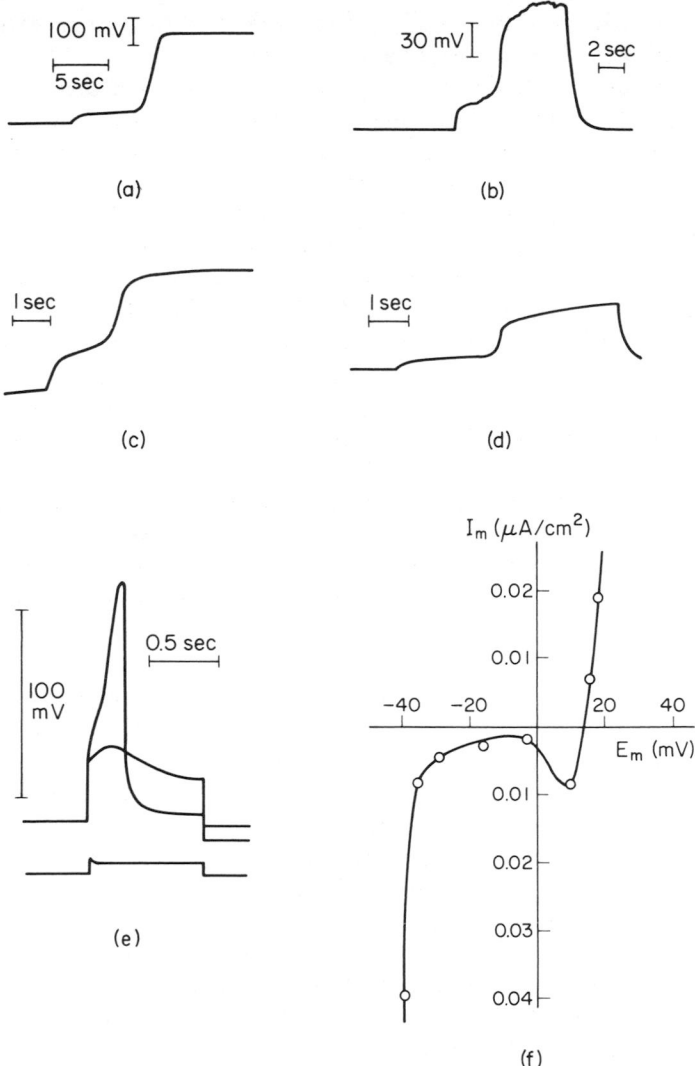

Figure 59. *A sample of responses of Mueller-Rudin membranes:* (a)-(e) are examples of potential changes across membranes in response to applied rectangular currents. Compare the response of a reconstituted biological membrane (b) to that of: a tunnel diode (a), frog nerve in isomolar KCl (c), and a *Valonia* algal cell (d). (e) and (f) illustrate the behavior of bimolecular lipid membranes in the presence of 10^{-7} g/cm^3 alamethicin. (e) shows a subthreshold response and a monostable action potential (upper traces) during an applied depolarizing pulse (lower trace). (f) is a current-voltage plot of responses of this membrane showing both asymmetric rectification and negative resistance characteristics.

sources (e.g., proteinaceous extracts from retina, white matter, fermented egg white) and they became electrically excitable (Fig. 59).

One particularly effective additive is a proteinaceous material obtained from the culture medium of *Enterobacter cloacae*. This excitability-inducing material (EIM) has a molecular weight less than 70,000 and can lower the electrical resistance of a lipid membrane from a value of 10^8 ohm cm^2, common for pure lipid films, to values between 10^3 and 10^5 ohm cm^2, values comparable to those measured across plasma membranes. This effect is concentration dependent. An EIM-treated film in a steady applied electric field behaves as a passive resistance at lower voltages but at voltages higher than some threshold level the membrane shows a reversible and regenerative rise in conductance; at suprathreshold voltages the membrane resistance changes between two definite values to appear as half of an action potential (Fig. 59). The detailed kinetics of these resistance changes are similar to the action potentials of frog nerve in 0.1 M KCl [50]; frog skin [23]; and *Valonia*, a marine alga [10]. The addition of EIM to a lipid membrane also confers a cation selectivity but no discrimination between K and Na. There is no refractoriness. Mueller and Rudin [51] attribute the lowering of the membrane resistance by EIM to a perturbation of the regularity of the lipid bilayer to form a localized pathway for the passage of electrolytes.

Mueller and Rudin [51–54] were also able to mimic inactivation by adding to their membranes a second molecular species (protamine, polyserine, or spermine) that possesses current-voltage characteristics the inverse of those due to EIM. Therefore, in a black membrane that contains sphingomyelin, EIM, and protamine, a complete action potential can be produced that possesses the same time course and magnitude as that recorded from biological systems (Fig. 59). With careful adjustment of the salt gradient and the concentration of EIM and protamine, the artificial membranes show repetitive and rhythmic electrical firing. These action potentials can be reversibly blocked by local anesthetics at physiological concentrations.

Alamethicin, an antibiotic obtained from *Trichoderma viride*, is a cyclopeptide of known amino acid composition [48]. Alamethicin can replace EIM in a variety of lipid membrane preparations in its ability to evoke membrane characteristics of negative slope resistance, delayed rectification, bistable changes of EMF, and simple or rhythmic spikes [53].

Other antibiotics also bind to these lipid films to produce interesting results. The addition of the cyclic peptide valinomycin confers ionic selectivity to these films in the order of Rb > K > Cs > Na = Li; the films are made some 300 times more permeable to K than to Na, like some biological membranes [7]. Nonactin is similar in action but gramicidin is not able to discriminate among the cations [52]. The polyene antibiotics amphotericin A and B and nystatin produce anion specificity [25]; their ability to interact with and to break lipid films is dependent on the cholesterol content of the films and these antibiotics are known to attack only sterol-containing membranes [41]. All these compounds are ring structures except gramicidin, which can form a ring. Altering the structure of valinomycin—increasing the ring structure by three of four units or decreasing it by two units—depresses the electrogenic action [52]. Increasing the lipophilic balance by substituting leucine for alanine increases the effectiveness and, conversely, cyclic compounds with polar groups on the outer part of the ring (e.g., filipin, colimycin, amphotericin) are without effect [52]. Hence, speculation centers on the ability of ions, with or without their hydration shells, to fit into the rings.

To explain the origin of the ion selectivity conferred upon the experimental membranes by these cyclic compounds, Mueller and Rudin [54] proposed that the carbonyl oxygen atoms, which are proton acceptors, project toward the center of the ring to complex with either the naked or the hydrated cation. This orientation, with the cation hidden within the ring and the hydrocarbon side chains of the ring exposed to the lipid outside, allows the entire charge complex to be soluble in the lipid membrane. The poor intracationic selectivity displayed by alamethicin results from the larger opening within the ring structure (13 Å internal diameter) as compared to that of valinomycin (7 Å), which shows good selectivity. The more basic protamine inverts the sign of this conductance by increasing the ratio of proton donors (and positive charges) to proton acceptors (and negative charges). Mueller and Rudin interpret the action potentials as originating from two gated conducting systems (e.g., alamethicin and protamine) which have appropriate time constants and different ionic selectivities and which act as parallel EMF's and variable resistive loads on each other.

The nature of the underlying gating mechanisms have not been resolved. A simple monovalent carrier mechanism is excluded as inconsistent with the highly nonlinear cooperative phenomena shown by EIM and alamethicin. The electrokinetic and chemical data suggest that six or more alamethicin molecules form either a carrier molecule or a channel for ion permeation but such a complex would have a molecular weight in excess of 11,000 and would be bulky. If aggregates form with protamine or polyserine, molecular weights up to 50,000 are possible and it is doubtful that such a molecule could retain the essential mobility of a carrier. Alternatively, Mueller and Rudin [54] suggest that alamethicin may cause a reversible micellarization of the lipid bilayer structure to provide a series of sites for ions to bridge the membrane. Indeed, further speculation leads Mueller and Rudin to suggest that excitability phenomena, active transport processes, and key steps in energy utilization all possess a common mechanism located within cellular membranes.

5.7 Electrically Active
Polyelectrolyte Membranes

To more closely mimic the behavior of biological membranes, and especially electrogenic ones, it is necessary to modify the insulator properties of the artificial lipid membrane with protein or peptide additives, such as EIM and alamethicin in the Mueller and Rudin preparations. The important factor for electrogenesis, then, is the protein alone or the protein-lipid complex. To more carefully examine the contributions of the nonlipid components to the behavior of artificial membranes, Shashoua [76, 77] prepared lipid-free membranes made with proteins, nucleic acids, and synthetic polymers. When these polyelectrolyte membranes are placed between two equal NaCl solutions and in a steady electric field, they spontaneously generate electrical transients with time constants and amplitudes analogous to the spike potentials of neuronal membranes (Fig. 60).

Shashoua prepared his membranes by permitting a polyacid to react with a polybase at an interface. Some of his preparations were mechanically stable, while others required the support of a mesh matrix formed by inert polymers. These membranes are thin and their manner of preparation makes it possible to assume that the surface facing the polyacid solution is negatively charged, while the polybase side is positively charged (Fig. 60). The polyelectrolytes used in these membrane systems included polyacrylamide, polyglutamate, dextran sulfate, methylacrylate, yeast RNA, cytochrome c, pancreatic RNAse, and poly-L-lysine.

The conditions necessary for exposing the electrogenic properties of these membranes include (1) an anisotropy of chemical composition, a sandwich-like structure in which a cation exclusion barrier is separated from an anion exclusion barrier by a neutral polyampholyte zone; (2) the presence of electrolytes on both sides of the membrane to act as the current carrying species; and (3) a polarization across the membrane such that the current drives cations into the polyanion phase and anions into the polycation phase. When these conditions are met, current flowing through the membrane results in a sequence of mechanical and electrical events as a function of the applied voltage. At a critical voltage, the recording electrodes become sensitive to mechanical vibrations. A further increase in the applied voltage results in a loss of the mechanical instability, the generation of spike-like activity, and a region of negative resistance in the current-voltage curves. Higher voltages cause a decrease and, finally, a complete loss of these properties.

Katchalsky [38, 39] proposed a molecular mechanism to explain these properties based upon a phase transition in the polyelectrolytes at certain salt concentrations. Current flow through the experimental membrane causes an accumulation of salt in the swollen polyelectrolyte at the interface region;

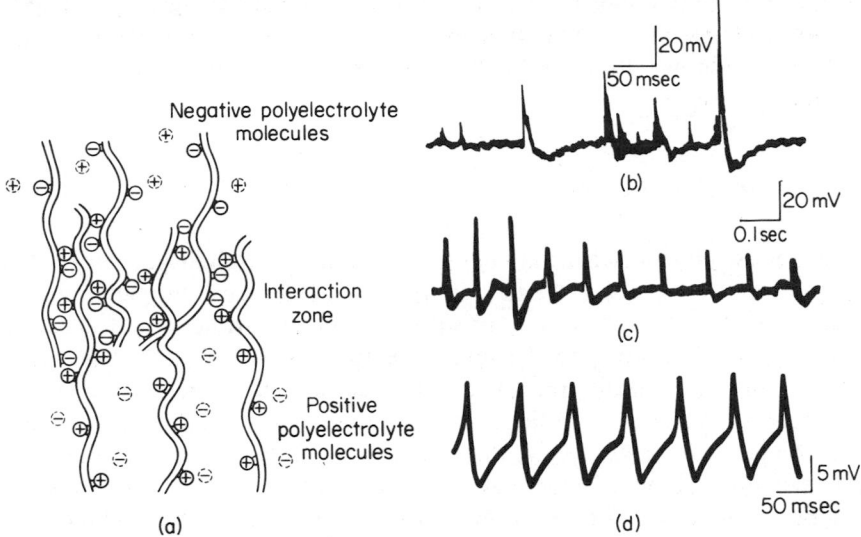

Figure 60. *Electrogenic responses of Shashoua's artificial membranes:* the membranes formed by the interaction at the interface of two polyelectrolytes are schematically represented by the double layer composed of two polyelectrolyte monolayers (a). The closed circles represent charges fixed on the macromolecules and the dashed circles represent free mobile counterions. Oscilloscope traces of spikes generated by such polyelectrolyte membranes in 0.15N NaCl during an applied steady voltage are given in (b)-(d). The membranes are constructed by the interaction of: (b) polyacrylic/acrylamide with dimethylaminoethyl acrylate, (c) RNA with Ca^{2+}, and (d) methylacrylate/acrylic acid with Ca^{2+}.

cations and anions arrive there from opposite sides. The local increase in salt concentration sharply raises the conductance and, hence, the amount of current passing through the membrane, The accumulation of salt also has an antagonistic effect. It is known that polyelectrolyte molecules are stretched at low ionic strengths but contract or collapse at high ionic strengths. Polyelectrolyte membranes become more permeable to salt with increasing current. Thus, when the peak of the spike is reached, the permeability breaks down and water flushes away the accumulated intramembrane salt. The membrane then returns to its initial state.

Katchalsky considers it reasonable to attribute stimulation phenomena to changes in intramembrane ion concentrations. The energy input for each charged group produced by increasing the internal potential of a cell by some 20 mV—sufficient to fire most neurons—is 2×10^{-2} electron volts, which is an amount lower than that required to change any chemical structure or to open even weak bonds. The conclusion is that the stimulating potential is only a trigger mechanism that releases other forms of energy, for example, an accumulation of ions that, upon interacting with the macromolecular components of the membrane, leads to phase transitions and the resulting permeability changes.

5.8 Conclusions

The studies on artificial membrane systems have uncovered properties that are similar to the characterstics of natural membranes; but attempts to identify molecular mechanisms of operation in the artificial systems have been equally frustrating to the same attempts on biological systems. It is always difficult to identify the relations between a model system and its living counterpart. While many of the model membranes have behaved in a manner too complex to analyze easily, the situation is comforting in that the complicated functioning of membranes, dead or alive, may not result from a unique and special structural organization but may result spontaneously as the combined properties of the classes of molecules found in them are expressed. There may be many ways to build membranes that function alike. At the same time, studies on model systems show how small alterations in the balance between many of the factors can alter behavior greatly; hence, the basic plan for membranes may not be so different among the diverse types available in nature. Curiously, the features (e.g., spike) once attributed only to nerve and muscle are now being found in such cells as different as those in liver and pancreas. Alterations in the proportions of components in the mosaic arrangements may be sufficient to convert pancreas membrane to nerve membrane.

The evidence now available suggests that lipid is important in membranes to provide the insulating barrier between cytoplasm and environment. Films of phospholipid are especially able to mimic so many of the normal properties of plasma membranes and especially the permeability properties. But lipid alone makes an insufficient model; other molecules are required to be present. Some proteins and peptides adsorbed to the lipid film, and sometimes alone, elicit the distinctive properties of excitation and electrogenesis, ionic discrimination and permeation, and so on. The conclusion is that the use of model membranes shows many promising approaches to the elucidation of mechanisms underlying membrane phenomena without the complicating presence of other operations found in multifunctioning living membranes. The literature on model membrane systems is characterized by much theory and speculation. There is a need for sound experimental data to seriously challenge the diverse suggestions presented to explain phenomena in order to accept, modify, or reject them. Recent progress is too young to clearly see the proper directions of future work. Although already fruitful, a still better understanding of the behavior of phospholipids, ion exchangers, and poly-electrolytes is necessary before the resulting information can be applied to plasma membranes with confidence.

There is much to do.

REFERENCES CITED

1. T. E. ANDREOLI, J. A. BANGHAM, and D. C. TOSTESON, 1967. The formation and properties of thin lipid membranes from HK and LK sheep red cell lipids. *J. Gen. Physiol.* **50**: 1729–1749.

2. T. E. ANDREOLI, M. TIFFENBERG, and D. C. TOSTESON, 1967. The effect of valinomycin on the ionic permeability of thin lipid membranes. *J. Gen. Physiol.* **50**: 2527–2547.

3. A. D. BANGHAM and D. PAPAHADJOPOULOS, 1966. Biophysical properties of phospholipids. I. Interaction of phosphatidylserine monolayers with metal ions. *Biochim. Biophys. Acta* **126**: 181–184.

4. A. D. BANGHAM, M. M. STANDISH, and N. MILLER, 1965. Cation permeability of phospholipid model membranes: effect of narcotics. *Nature* **208**: 1295–1297.

5. A. D. BANGHAM, M. M. STANDISH, and J. C. WATKINS, 1965. Diffusion of univalent ions across the lamellae of swollen phospholipids. *J. Mol. Biol.* **13**: 238–252.

6. A. D. BANGHAM, M. M. STANDISH, J. C. WATKINS, and G. WEISSMANN, 1967. The diffusion of ions from a phospholipid model membrane system. *Protoplasma* **63**: 183–187.

7. A. D. BANGHAM, M. M. STANDISH, and G. WEISSMANN, 1965. The action of steroids and streptolysin S on the permeability of phospholipid structures to cations. *J. Mol. Biol.* **13**: 252–259.

8. H. S. BENNETT, 1956. The concept of membrane flow and membrane vesiculation as mechanisms for active transport and ion pumping. *J. Biophys. Biochem. Cytol. Suppl.* **2**: 99–103.

9. M. BLANK, 1966. Physical models in research on biological membranes. *Ann. N.Y. Acad. Sci.* **137**: 755–758.

10. L. R. BLINKS, 1936. Effects of current flow on bioelectric potential. I. *Valonia. J. Gen. Physiol.* **19**: 633–672.

11. A. P. CARVALHO, H. SANUI, and N. PACE, 1963. Calcium and magnesium binding properties of cell membrane materials. *J. Cell. Comp. Physiol.* **62**: 311–317.

12. A. CASS and A. FINKELSTEIN, 1967. Water permeability of thin lipid membranes. *J. Gen. Physiol.* **50**: 1765–1784.

13. J.-P. CHANGEUX, J. THIERY, Y. Tung, and C. KITTEL, 1967. On the cooperativity of biological membranes. *Proc. Natl. Acad. Sci. U. S.* **57**: 335–341.

14. K. S. COLE, 1965. Electrodiffusion models for the membrane of squid giant axon. *Physicl. Rev.* **45**: 340–379.

15. K. S. COLE, 1968. *Membranes, Ions, and Impulses. A Chapter of Classical Biophysics.* Berkeley: University of California Press, 569 pp.

16. F. CONTI and G. EISENMAN, 1965. The non-steady state membrane potential of ion exchangers with fixed sites. *Biophys. J.* **5**: 247–256.

17. F. CONTI and G. EISENMAN, 1965. The steady-state properties of ion-exchange membranes with fixed sites. *Biophys. J.* **5**: 511–530.

18. F. Conti and G. Eisenman, 1966. The steady-state properties of an ion-exchange membrane with mobile sites. *Biophys. J.* **6**: 227–246.

19. J. Dainty, 1962. Ion transport and electrical potentials in plant cells. *Ann. Rev. Plant Physiol.* **13**: 379–402.

20. R. A. Demel, L. L. M. van Deenen, and B. A. Pethica, 1967. Monolayer interactions of phospholipids and cholesterol. *Biochim. Biophys. Acta* **135**: 11–19.

21. G. Eisenman and F. Conti, 1965. Some implications for biology of recent theoretical and experimental studies of ion permeation in model membranes. *J. Gen. Physiol.*, Part II, **48**: 65–73.

22. G. Eisenman, J. P. Sandblom, and J. L. Walker, Jr., 1967. Membrane structure and ion permeation. *Science* **155**: 965–974.

23. A. Finkelstein, 1961. Electrical excitability of isolated frog skin. *Nature* **190**: 1119–1120.

24. A. Finkelstein and A. Cass, 1967. Effect of cholesterol on water permeability of thin lipid membranes. *Nature* **216**: 717–718.

25. A. Finkelstein and A. Cass, 1968. Permeability and electrical properties of thin lipid membranes. *J. Gen. Physiol. Suppl.* **52**: 145–173.

26. U. F. Franck, 1956. Models for biological excitation processes. *Prog. Biophys.* **6**: 171–206.

27. D. E. Goldman, 1964. A molecular structural basis for the excitation properties of axons. *Biophys. J.* **4**: 167–188.

28. T. Hanai and D. A. Haydon, 1966. The permeability to water of bimolecular lipid membranes. *J. Theoret. Biol.* **11**: 370–382.

29. T. Hanai, D. A. Haydon, and W. R. Redwood, 1966. The water permeability of artificial bimolecular leaflets: a comparison of radio-tracer and osmotic methods. *Ann. N.Y. Acad. Sci.* **137**: 731–739.

30. T. Hanai, D. A. Haydon, and J. Taylor, 1964. An investigation by electrical methods of lecithin-in-hydrocarbon films in aqueous solutions. *Proc. Roy. Soc. London A* **281**: 377–391.

31. T. Hanai, D. A. Haydon, and J. Taylor, 1965. Some further experiments in bimolecular lipid membranes. *J. Gen. Physiol. Suppl.* **48**: 59–63.

32. E. N. Havey and J. F. Danielli, 1938. Properties of the cell surface. *Biol. Rev.* **13**: 319–341.

33. D. A. Haydon and J. Taylor, 1963. The stability and properties of bimolecular lipid leaflets in aqueous solutions. *J. Theoret. Biol.* **4**: 281–296.

34. F. A. Henn, G. L. Decker, J. W. Greenawalt, and T. E. Thompson, 1967. Properties of lipid bilayer membranes separating two aqueous phases: electron microscope studies. *J. Mol. Biol.* **24**: 51–58.

35. C. Huang and T. E. Thompson, 1965. Properties of lipid bilayer membranes separating two aqueous phases: determination of membrane thickness. *J. Mol. Biol.* **13**: 183–193.

36. C. Huang and T. E. Thompson, 1966. Properties of lipid bilayer membranes separating two aqueous phases: water permeability. *J. Mol. Biol.* **15**: 539–554.

37. C. Huang, L. Wheeldon, and T. E. Thompson, 1964. The properties of lipid bilayer membranes separating two aqueous phases: formation of a membrane of simple composition. *J. Mol. Biol.* **8**: 148–160.

38. A. Katchalsky, 1967. "Membrane thermodynamics," pp. 326–343, in: G. C. Quarton, T. Melnechuk, and F. O. Schmitt (eds.), *The Neurosciences, A Study Program.* New York: Rockefeller University Press.

39. A. KATCHALSKY and R. SPANGLER, 1968. Dynamics of membrane processes. *Quart. Rev. Biophys.* **1**: 127–175.

40. J. L. KAVANAU, 1965. *Structure and Function in Biological Membranes*, Vol. 1. San Francisco: Holden-Day, Inc., Pub., 322 pp.

41. J. O. LAMPEN, 1966. "Interference by polyenic antifungal antibiotics (especially nystatin and filipin) with specific membrane functions," pp. 111–130, in: B. A. Newton and P. E. Reynold (eds.), *Biochemical Studies of Antimicrobial Drugs*. 16th Symp. Soc. Gen. Microbiol. Cambridge: University Press.

42. P. LAÜGER, W. LESSLAUER, E. MARTI, and J. RICHTER, 1967. Electrical properties of bimolecular phospholipid membranes. *Biochim. Biophys. Acta* **135**: 20–32.

43. G. T. LEITCH and J. M. TOBIAS, 1964. Phospholipid-cholesterol membrane model: effect of calcium, potassium, and protamine on membrane hydration, water permeability, and electrical resistance. *J. Cell. Comp. Physiol.* **63**: 225–232.

44. W. LESSLAUER, J. RICHTER, and P. LAÜGER, 1967. Some electrical properties of bimolecular phosphatidyl inositol membranes. *Nature* **213**: 1224–1226.

45. R. S. LILLIE, 1925. Factors affecting transmission and recovery in the passive iron nerve model. *J. Gen. Physiol.* **7**: 473–507.

46. V. LUZZATTI, F. REISS-HUSSON, E. RIVAS, and T. GALIK-KRZYWICKI, 1966. Structure and polymorphism in lipid-water systems and their possible biological implications. *Ann. N.Y. Acad. Sci.* **137**: 409–413.

47. A. MAURO and A. FINKELSTEIN, 1958. Realistic model of a fixed-charge membrane according to the theory of Teorell, Meyer, and Sievers. *J. Gen. Physiol.* **42**: 385–391.

48. C. E. MEYER and F. REUSSER, 1967. A polypeptide antibacterial agent isolated from *Trichoderma veride*. *Experientia* **23**: 85–86.

49. D. C. MIKULECKY and J. M. TOBIAS, 1964. Phospholipid-cholesterol membrane model. I. Correlation of resistance with ion content. II. Cation exchange properties. III. Effect of Ca on salt permeability. IV. Ca-K uptake by sonically fragmented erythrocyte ghosts. *J. Cell. Comp. Physiol.* **64**: 151–164.

50. P. MUELLER, 1958. Prolonged action potentials from single nodes of Ranvier. *J. Gen. Physiol.* **42**: 137–162.

51. P. MUELLER and D. O. RUDIN, 1963. Induced excitability in reconstituted cell membrane structure. *J. Theoret. Biol.* **4**: 268–280.

52. P. MUELLER and D. O. RUDIN, 1967. Action potential phenomena in experimental bimolecular lipid membranes. *Nature* **213**: 603–604.

53. P. MUELLER and D. O. RUDIN, 1968. Action potentials induced in bimolecular lipid membranes. *Nature* **217**: 713–719.

54. P. MUELLER and D. O. RUDIN, 1968. Resting and action potentials in experimental bimolecular lipid membranes. *J. Theoret. Biol.* **18**: 222–258.

55. P. MUELLER, D. O. RUDIN, H. T. TIEN, and W. C. WESCOTT, 1962. Reconstitution of excitable cell membrane structure *in vitro*. *Circulation* **26**: 1167–1170.

56. P. MUELLER, D. O. RUDIN, H. T. TIEN, and W. C. WESCOTT, 1962. Reconstituion of cell membrane structure *in vitro* and its transformation into an excitable system. *Nature* **194**: 979–980.

57. P. MUELLER, D. O. RUDIN, H. T. TIEN, and W. C. WESCOTT, 1963. Methods for the formation of single bimolecular lipid membranes in aqueous solution. *J. Phys. Chem.* **67**: 534–535.

58. P. MUELLER, D. O. RUDIN, H. T. TIEN, and W. C. WESCOTT, 1964. "Formation and properties of bimolecular lipid membranes," pp. 379–393, in: J. F. Danielli, K. G. A.

Pankhurst, and A. C. Riddiford (eds.), *Recent Progress in Surface Science*, Vol. I. New York: Academic Press.

59. L. J. Mullins, 1961. The macromolecular properties of excitable membrane. *Ann. N.Y. Acad. Sci.* **94**: 390–404.

60. H. A. Nash and J. M. Tobias, 1964. Phospholipid membrane model: importance of phosphatidylserine and its cation exchange nature. *Proc. Natl. Acad. Sci. U. S.* **51**: 476–480.

61. W. J. V. Osterhout, 1940. Some models of protoplasmic surfaces. *Cold Spring Harbor Symp. Quant. Biol.* **8**: 51–62.

62. J. T. G. Overbeek, 1960. Black soap films. *J. Phys. Chem.* **64**: 1178–1183.

63. D. Papahadjopoulos and A. D. Bangham, 1966. Biophysical properties of phospholipids. II. Permeability of phosphatidylserine liquid crystals to univalent ions. *Biochim. Biophys. Acta* **126**: 185–188.

64. D. Papahadjopoulos and N. Miller, 1967. Phospholipid model membranes. I. Structural characteristics of hydrated liquid crystals. *Biochim. Biophys. Acta* **135**: 624–638.

65. D. Papahadjopoulos and J. C. Watkins, 1967. Phospholipid model membranes. II. Permeability properties of hydrated liquid crystals. *Biochim. Biophys. Acta* **135**: 639–652.

66. B. A. Pethica, 1967. Structure and physical chemistry of membranes. *Protoplasma* **63**: 147–156.

67. R. Rendi, 1967. Water extrusion in isolated subcellular fractions. VI. Osmotic properties of swollen phospholipid suspensions. *Biochim. Biophys. Acta* **135**: 333–346.

68. L. Rothfield and A. Finkelstein, 1968. Membrane biochemistry. *Ann. Rev. Biochem.* **37**: 463–496.

69. A. Rothstein, 1968. Membrane phenomena. *Ann. Rev. Physiol.* **30**: 15–72.

70. J. P. Sandblom, 1967. A method to relate steady-state ionic currents, conductances, and membrane potential in ion exchange membranes with unknown thermodynamic properties. *Biophys. J.* **7**: 243–265.

71. J. P. Sandblom and G. Eisenman, 1967. Membrane potentials at zero current: the significance of a constant ionic permeability ratio. *Biophys. J.* **7**: 217–242.

72. H. Sanui, A. P. Carvalho, and N. Pace, 1962. Relationship of hydrogen ion binding to sodium and potassium binding by rat liver cell microsomes and human erythrocyte ghosts. *J. Cell. Comp. Physiol.* **59**: 241–250.

73. H. Sanui and N. Pace, 1962. Sodium and potassium binding by erythrocyte ghosts. *J. Cell. Comp. Physiol.* **59**: 251–257.

74. G. Sessa and G. Weissmann, 1967. Effect of polyene antibiotics on phospholipid spherules containing varying amounts of charged components. *Biochim. Biophys. Acta* **135**: 416–426.

75. D. O. Shah and J. H. Schulman, 1967. The ionic structure of sphingomyelin monolayers. *Biochim. Biophys. Acta* **135**: 184–187.

76. V. E. Shashoua, 1967. Electrically active polyelectrolyte membranes. *Nature* **215**: 846–847.

77. V. E. Shashoua, 1969. "Electrically active protein and polynucleic acid membranes," pp. 147–159, in: D. C. Tosteson (ed.), *The Molecular Basis of Membrane Function*. Englewood Cliffs, N.J.: Prentice-Hall, Inc.

78. G. M. Shean and K. Sollner, 1966. Carrier mechanisms in the movement of ions across porous and liquid ion exchanger membranes. *Ann. N.Y. Acad. Sci.* **137**: 759–776.

79. M. M. Shemyakin, V. K. Antonov, L. D. Bergelson, V. T. Ivanov, G. G. Malenkov,

Y. A. OVCHINNIKOV, and A. M. SHKROB, 1969. "Chemistry of membrane-affecting peptides, depsipeptides, and depsides (structure-function relations)," pp. 173–210, in: D. C. Tosteson (ed.), *The Molecular Basis for Membrane Function*. Englewood Cliffs, N.J.: Prentice-Hall, Inc.

80. W. STOECKENIUS, 1962. Some electron microscopic observations in liquid–crystalline phases in lipid-water systems. *J. Cell Biol.* **12**: 221–229.

81. I. TASAKI, 1968. *Nerve Excitation. A Macromolecular Approach*. Springfield, Ill.: Charles C Thomas, Pub., 201 pp.

82. T. TEORELL, 1953. Transport processes and electrical phenomena in ionic membrane. *Prog. Biophys.* **3**: 305–369.

83. T. TEORELL, 1962. Excitability phenomena in artificial membrane. *Biophys. J. Suppl.* **2**: 27–52.

84. T. E. THOMPSON, 1964. "The properties of bimolecular phospholipid membranes," pp. 83–96, in: M. Locke (ed.), *Cellular Membranes in Development*. New York: Academic Press.

85. T. E. THOMPSON, 1967. Experimental bilayer membrane. *Protoplasma* **63**: 194–196.

86. T. E. THOMPSON and C. HUANG, 1966. The water permeability of lipid bilayer membranes. *Ann. N.Y. Acad. Sci.* **137**: 740–744.

87. H. T. TIEN, 1967. Black lipid membranes: thickness determination and molecular organization by optical methods. *J. Theoret. Biol.* **16**: 97–110.

88. H. T. TIEN and E. A. DAWIDOWICZ, 1966. Black lipid films in aqueous media: a new type of interfacial phenomena. Experimental techniques and thickness measurements. *J. Colloid Interface Sci.* **22**: 438–453.

89. H. T. TIEN and A. L. DIANA, 1967. Some physical properties of bimolecular lipid membranes produced from new lipid solutions. *Nature* **215**: 1199–1200.

90. H. T. TIEN and A. L. DIANA, 1968. Bimolecular lipid membranes: a review and a summary of some recent studies. *Chem. Phys. Lipid* **2**: 55–101.

91. J. M. TOBIAS, 1952. Some optically detectable consequences of activity in nerve. *Cold Spring Harbor Symp. Quant. Biol.* **17**: 15–25.

92. J. M. TOBIAS, 1958. Experimentally altered structure related to function in the lobster giant axon with an extrapolation to molecular mechanisms in excitation. *J. Cell. Comp. Physiol.* **52**: 89–107.

93. J. M. TOBIAS, 1960. Further studies on the nature of the excitable system in nerve: I. Voltage induced axoplasm movement in squid axons. II. Penetration of surviving, excitable axons by proteases. III. Effects of proteases and of phospholipases on lobster giant axon resistance and capacity. *J. Gen. Physiol. Suppl.* **43**: 57–71.

94. J. M. TOBIAS, 1964. A chemically specified molecular mechanism underlying excitation in nerve: a hypothesis. *Nature* **203**: 13–17.

95. J. M. TOBIAS, D. P. AGIN, and P. PAWLOWSKI, 1962. Phospholipid-cholesterol membrane model: control of resistance by ions or current flow. *J. Gen. Physiol.* **45**: 989–1001.

96. J. M. TOBIAS, D. P. AGIN, and R. PAWLOWSKI, 1962. The excitable system in the cell surface. *Circulation* **26**: 1145–1150.

97. J. M. TOBIAS and P. G. NELSON, 1959. "Structure and function in nerve," pp. 248–265, in: R. E. Zirkle (ed.), *A Symposium on Molecular Biology*. Chicago: University of Chicago Press.

98. L. M. TSOFINA, E. A. LIBERMAN, and A. V. BABAKOV, 1966. Production of bimolecular protein-lipid membranes in aqueous solutions. *Nature.* **212**: 681–683.

99. H. ÜSSING, 1952. Some aspects of the application of tracers in permeability studies. *Adv. Enzymol.* **13**: 21–65.

100. V. S. VAIDHYANATHAN, 1965. Statistical mechanical theory of electrolyte diffusion inside a charged membrane. *J. Theoret. Biol.* **8**: 357–366.

101. V. S. VAIDHYANATHAN, 1966. Some comments on steady state transport of ions across a charged biological membrane. *J. Theoret. Biol.* **10**: 159–176.

102. V. S. VAIDHYANATHAN, 1966. Theory of mechano-electric membrane transducers. *J. Theoret. Biol.* **13**: 18–31.

103. V. S. VAIDHYANATHAN and H. M. PHILLIPS, 1966. Mobile ion interactions in the membrane phase. *J. Theoret. Biol.* **13**: 32–47.

104. H. J. VAN DEN BERG, 1965. A new technique for obtaining thin lipid films separating two aqueous media. *J. Mol. Biol.* **12**: 290–291.

105. J. L. WALKER, JR. and G. EISENMAN, 1966. A test of the theory of the steady-state properties of a liquid ion exchange membrane. *Ann. N.Y. Acad. Sci.* **137**: 777–791.

106. J. L. WALKER, JR. and G. EISENMAN, 1966. A test of the theory of the steady-state properties of an ion exchange membrane with mobile sites and dissociated counterions. *Biophys. J.* **6**: 513–533.

107. G. WEISSMANN and G. SESSA, 1967. The action of polyene antibiotics on phospholipid-cholesterol structures. *J. Biol. Chem.* **242**: 616–625.

108. W. WILBRANDT, 1967. Carrier mechanisms. *Protoplasma* **63**: 299–302.

109. W. WILBRANDT and T. ROSENBERG, 1961. The concept of carrier transport and its corollaries in pharmacology. *Pharmacol. Rev.* **13**: 109–183.

INDEX

Racker, E., 37
Rader, R. L., 39
Rathkamp, R., 187
Ray, P. M., 84
Receptor membranes, 186-191
Red blood cell ghosts, 104-106
Red blood cells, cation movements in, 106-108
Redox potential, membrane potential as a, 155-160
Reflection coefficient, 70-72
Refractoriness, 136-137
Reinert, J. C., 45
Resistance, 94, 140-143, 153, 171, 180, 184, 194, 198-203, 229-230, 236
Restricted diffusion, 88
Reticular theory, 127
Robertson, J. D., 9, 15-17, 21, 24-25, 32, 34, 48, 53, 226
Roseman, S., 118
Rosenberg, S. A., 9
Rotunno, C. A., 90, 100-101
Rudin, D. O., 225, 226, 228, 234-238

Sandblom, J. P. 231
Sato, M., 189
Schatzmann, H. J., 108
Schmidt, W. J., 17-18, 26
Schmitt, F. O., 14, 18, 26, 50, 51, 52, 204
Shapiro, H., 14
Shashoua, V. E., 226, 238, 239
Shaw, T. I., 49, 163, 201
Sherrington, C. S., 129, 130
Singer, S. J., 30, 41-44
Sjöstrand, F. S., 24-25, 32, 34
Skou, J. C., 116-118
Smith, Homer, 13
Snell, F. M., 98
Sodium (*see:* Ion involvements)
Sodium pump (*see:* Ion pumps)
Sodium theory, 154-155, 170
Sodium transport across toad bladder, 102
Stabilization (*see:* Excitation)
Standish, M. M., 228
Staverman, A. J., 70
Steim, J. M., 45
Stein, W. D., 48, 50, 70, 79, 118
Steinbach, H. B., 110
Stoeckenius, W., 30
Strophanthin, 108-110, 112-113, 118

Sucrose, 66, 161
Surface tension measurements, 14
Synapses, inhibitory, 182-186
  excitatory, 174-180

Tasaki, I., 163, 165, 208-210, 231
Teorell, T., 70, 163, 225
Tetraethylammonium (TEA), 49, 196-197, 199, 201, 204
Tetrodotoxin (TTX), 49, 191, 201, 204
Thermodynamics, nonequilibrium, 68-70
Time constant, 147
Toad bladder, sodium transport across, 102
Tobias, Julian, 233-234
Tomasi, V., 10
Tosteson, D. C., 107, 112, 120
Tourtellotte, M. T., 39
Transport, (*see:* Permeation of biological membranes, Ion pumps)

Unit membrane theory of Robertson, 15-17
  opposition to, 31-33
Üssing, H. H., 80-81, 89-91, 93, 95, 96

Valinomycin, 226, 237
Valves, 82-83
Van Deenen, L. L. M., 10, 12, 38
Van Den Berg, H. J., 229
Van Golde, L. M. G., 10
Vendenheuvel, F. A., 27, 38
Volta, Alessandro, 191

Waldeyer, Wilhem von, 129
Walker, J. L., Jr., 231
Wallach, D. F. H., 33, 34, 42-43
Walsh, James Joseph, 191
Water, 2, 18, 40, 47-49, 71-86, 89, 100, 102, 156-157, 161, 228-229, 233
Watkins, J. C., 33, 228
Waugh, D. F., 14
Weinstein, D. B., 10
Whittam, R., 108
Wright, E. M., 119

X-ray diffraction studies, 20-21, 25-29, 32, 35, 44-46, 227

Yamamoto, T., 32

Zahler, P. H., 42-43